125
Structure and Bonding

Series Editor: D. M. P. Mingos

Structure and Bonding
Series Editor: D. M. P. Mingos

Recently Published and Forthcoming Volumes

High Energy Density Materials
Volume Editor: Klapötke, T. H.
Vol. 125, 2007

Ferro- and Antiferroelectricity
Volume Editors: Dalal, N. S.,
Bussmann-Holder, A.
Vol. 124, 2007

Photofunctional Transition Metal Complexes
Volume Editor: V. W. W. Yam
Vol. 123, 2007

Single-Molecule Magnets and Related Phenomena
Volume Editor: Winpenny, R.
Vol. 122, 2006

Non-Covalent Multi-Porphyrin Assemblies
Synthesis and Properties
Volume Editor: Alessio, E.
Vol. 121, 2006

Recent Developments in Mercury Sience
Volume Editor: Atwood, David A.
Vol. 120, 2006

Layered Double Hydroxides
Volume Editors: Duan, X., Evans, D. G.
Vol. 119, 2005

Semiconductor Nanocrystals and Silicate Nanoparticles
Volume Editors: Peng, X., Mingos, D. M. P.
Vol. 118, 2005

Magnetic Functions Beyond the Spin-Hamiltonian
Volume Editor: Mingos, D. M. P.
Vol. 117, 2005

Intermolecular Forces and Clusters II
Volume Editor: Wales, D. J.
Vol. 116, 2005

Intermolecular Forces and Clusters I
Volume Editor: Wales, D. J.
Vol. 115, 2005

Superconductivity in Complex Systems
Volume Editors: Müller, K. A.,
Bussmann-Holder, A.
Vol. 114, 2005

Principles and Applications of Density Functional Theory in Inorganic Chemistry II
Volume Editors:
Kaltsoyannis, N., McGrady, J. E.
Vol. 113, 2004

Principles and Applications of Density Functional Theory in Inorganic Chemistry I
Volume Editors:
Kaltsoyannis, N., McGrady, J. E.
Vol. 112, 2004

Supramolecular Assembly via Hydrogen Bonds II
Volume Editor: Mingos, D. M. P.
Vol. 111, 2004

Applications of Evolutionary Computation in Chemistry
Volume Editors: Johnston, R. L.
Vol. 110, 2004

Fullerene-Based Materials
Structures and Properties
Volume Editor: Prassides, K.
Vol. 109, 2004

Supramolecular Assembly via Hydrogen Bonds I
Volume Editor: Mingos, D. M. P.
Vol. 108, 2004

High Energy Density Materials

Volume Editor: T. M. Klapötke

With contributions by
A. J. Bellamy · E. F. C. Byrd · R. D. Chapman · H. Gao
T. M. Klapötke · W. D. Mattson · D. T. Meshri · B. M. Rice
J. M. Shreeve · R. P. Singh · S. Zeman

Springer

The series *Structure and Bonding* publishes critical reviews on topics of research concerned with chemical structure and bonding. The scope of the series spans the entire Periodic Table. It focuses attention on new and developing areas of modern structural and theoretical chemistry such as nanostructures, molecular electronics, designed molecular solids, surfaces, metal clusters and supramolecular structures. Physical and spectroscopic techniques used to determine, examine and model structures fall within the purview of Structure and Bonding to the extent that the focus is on the scientific results obtained and not on specialist information concerning the techniques themselves. Issues associated with the development of bonding models and generalizations that illuminate the reactivity pathways and rates of chemical processes are also relevant.
As a rule, contributions are specially commissioned. The editors and publishers will, however, always be pleased to receive suggestions and supplementary information. Papers are accepted for *Structure and Bonding* in English.
In references *Structure and Bonding* is abbeviated *Struct Bond* and is cited as a journal.

Springer WWW home page: springer.com
Visit the Struct Bond content at springerlink.com

Library of Congress Control Number: 2007926253

ISSN 0081-5993
ISBN 978-3-540-72201-4 Springer Berlin Heidelberg New York
DOI 10.1007/978-3-540-72202-1

Springer is a part of Springer Science+Business Media

springer.com

Cover design: *Design & Production* GmbH, Heidelberg
Typesetting and Production: LE-TEX Jelonek, Schmidt & Vöckler GbR, Leipzig

Printed on acid-free paper 02/3180 YL – 5 4 3 2 1 0

Structure and Bonding
Also Available Electronically

For all customers who have a standing order to Structure and Bonding, we offer the electronic version via SpringerLink free of charge. Please contact your librarian who can receive a password or free access to the full articles by registering at:

springerlink.com

If you do not have a subscription, you can still view the tables of contents of the volumes and the abstract of each article by going to the SpringerLink Homepage, clicking on "Browse by Online Libraries", then "Chemical Sciences", and finally choose Structure and Bonding.

You will find information about the

- Editorial Board
- Aims and Scope
- Instructions for Authors
- Sample Contribution

at springer.com using the search function.

Preface

This book summarizes some recent developments in the area of high-energy high-density (HEDM) materials. Rather than being comprehensive in scope, emphasis is given to structural and bonding features of highly energetic materials with possible applications as high explosives (secondary explosives) or propellants. In this book we do not focus on primary explosives (e.g. lead azide replacements) since by definition the explosive performance (detonation velocity and detonation pressure) of such materials – although very sensitive – are much less energetic than secondary (high) explosives.

Modern HEDMs derive most of their energy (i) from oxidation of the carbon backbone, as in traditional energetic materials, (ii) from ring or cage strain, or (iii) from their very high positive heat of formation. Examples of the first class are traditional explosives, such as TNT, RDX and HMX. Modern nitro-compounds, such as CL-20 or the recently reported hepta- and octanitrocubanes, belong to the second group of explosives and possess very high densities and enhance the energies utilizing substantial cage strain. Members of the third class of compounds are high-nitrogen compounds (up to 85% nitrogen content), such as aminotetrazole and nitrotetrazole derivatives, which show the desired remarkable insensitivity to electrostatic discharge, friction and impact, while having very high positive heats of formation and therefore very high explosive powers.

The synthesis of energetic, non-nuclear materials for military application has been a long-term goal in various academic and military research groups worldwide. Some of the current challenges that face HEDM scientists are:

- Demand for environmentally compatible and toxicologically acceptable explosives and propellants. Examples are replacements for TNT, RDX and HMX since nitro-explosives per se, as well as their environmental transformation products, are toxic.
- Demand of all the services for more destructive energy delivered to the target.
- Minimization of unwanted effects: low collateral damage munitions, i.e. munitions that cause little damage aside from damage to the intended target, have become more important as sites of military interest are increasingly co-located with civilian population centers.

- Continuing interest in insensitivity: insensitive munitions for reduced vulnerability of munitions stores and increased survivability of munitions that are subjected to very stressful conditions under their normal usage.
- Continuing interest in low observable plumes (no signature of a missile, smokeless combustion) and erosion-reduced gun propellants.

The six chapters in this book all focus on various aspects of the structure and bonding of modern HEDMs. While the first four chapters (Bellamy, Shreeve, Klapötke, Chapman) discuss synthetic aspects of energetic materials, chapter five (Rice) gives emphasis to computational aspects of nitrogen-rich HEDMs, while the last chapter (Zeman) is devoted to the sensitivities of high energy compounds. In the synthetic part of this book, the first chapter (Bellamy) describes the most recent aspects of the chemistry of Fox-7, a remarkably insensitive high-nitro explosive. The following two chapters concentrate on modern high-nitrogen compounds: nitrogen-rich heterocycles (Shreeve) and new nitrogen-rich high explosives (Klapötke). The last preparative chapter (Chapman) summarizes the work on the synthesis and structures of cyclic nitramines, new (difluoramino)alkanes and non-nitramines. In the theoretical chapter (Rice), emphasis is given to a variety of computational procedures to predict the properties of energetic nitrogen-rich HEDMs. Finally, the last chapter (Zeman) presents a survey of studies into the sensitivity of HEDMs, focusing mainly on impact and shock sensitivities.

It is no coincidence that three out of six chapters are directed towards the synthesis and computation of high-nitrogen compounds. Nitrogen is unique amongst all other elements of the periodic table insofar that the bond energy per two-electron bond increases from a single over a double to a triple bond. For carbon the situation is the opposite and one might expect acetylene to polymerize in an exothermic reaction whereas dinitrogen is more stable than any other polynitrogen species. Therefore, high-nitrogen HEDMs with N–N bond orders of less than two may provide some solutions to the above-mentioned challenges: very high explosive energy, low toxicity reaction products (N_2), no observable plumes or signatures and reduced gun erosion. It is still too early in the development of such materials to assess whether they will live up to their potential, but they nonetheless represent a very exciting and challenging new area of chemistry.

Thomas M. Klapötke

Contents

Topics in Heterocyclic Chemistry
Series Editor: R. R. Gupta

Recently Published and Forthcoming Volumes

Bioactive Heterocycles I
Volume Editor: S. Eguchi
Volume 6, 2006

Marine Natural Products
Volume Editor: H. Kiyota
Volume 5, 2006

QSAR and Molecular Modeling Studies in Heterocyclic Drugs II
Volume Editor: S. P. Gupta
Volume 4, 2006

QSAR and Molecular Modeling Studies in Heterocyclic Drugs I
Volume Editor: S. P. Gupta
Volume 3, 2006

Heterocyclic Antitumor Antibiotics
Volume Editor: M. Lee
Volume 2, 2006

Microwave-Assisted Synthesis of Heterocycles
Volume Editors: E. Van der Eycken, C. O. Kappe
Volume 1, 2006

Struct Bond (2007) 125: 1–33
DOI 10.1007/430_2006_054

Published online: 25 January 2007

FOX-7 (1,1-Diamino-2,2-dinitroethene)

Anthony J. Bellamy

Cranfield University, Defence Academy of the United Kingdom, Shrivenham,
Swindon SN6 8LA, UK
a.j.bellamy@cranfield.ac.uk

Abstract In this paper, the discovery and synthesis of the explosive 1,1-diamino-2,2-dinitroethene (FOX-7) are described, together with an account of its structural, spectroscopic, and explosive properties. The chemical reactivity of FOX-7 towards nucleophilic substitution (transamination), electrophilic substitution, and acid-base properties is explored, as is its thermal behavior (phase transformations and thermal decomposition). The molecular structure and physical properties of FOX-7 are compared with those of its three isomers (as yet unsynthesized), as derived by theoretical calculations. Finally, the physical properties of FOX-7 are compared to those of various energetic molecules that are structurally related to FOX-7.

Keywords 1,1-Diamino-2,2-dinitroethene · FOX-7 · Synthesis · Properties · Reactions · Isomers · Related molecules

1 Introduction

For more than half a century, the main explosive component of artillery shells and other explosive devices has been RDX (1,3,5-trinitro-1,3,5-

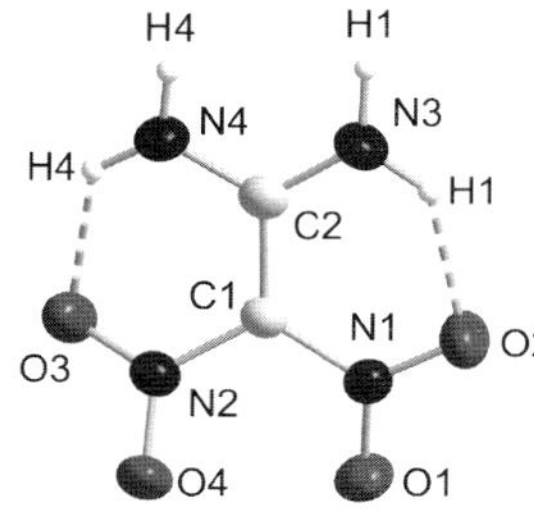

Fig. 1 FOX-7

and C – NO_2 bonds (1.42 and 1.39 Å) and the essentially planar structure of the molecule as a whole (average C – C torsion angle 4.8°). Two strong hydrogen bonds exist between NH and NO on both sides of the molecule. There are four molecules per unit cell whose dimensions are: a 6.941 Å, b 6.569 Å, c 11.315 Å, β 90.55°, V 515.9 $Å^3$, Dx 1.907 Mg m^{-3}.

The supramolecular arrangement of molecules within the crystal is in the form of wave-shaped infinite layers (Fig. 2 [10]), akin to that observed for graphite and triaminotrinitrobenzene (TATB), with intra-layer N – H...O – N hydrogen bonding and inter-layer van der Waals interactions. The distance between layers is ~ 3 Å cf. the layer separation in graphite is ~ 3.5 Å, while that in TATB is ~ 6 Å. The extensive network of hydrogen bonding is probably the reason why 1,1-diamino-2,2-dinitroethene is only slightly soluble in many common solvents (acetone, acetic acid, ethyl acetate, nitromethane, acetonitrile – all < 0.5 g/100ml at 20 °C), the greatest solubility being exhibited in *N*,*N*-dimethylformamide, *N*-methylpyrrolidin-2-one and dimethylsulphox-

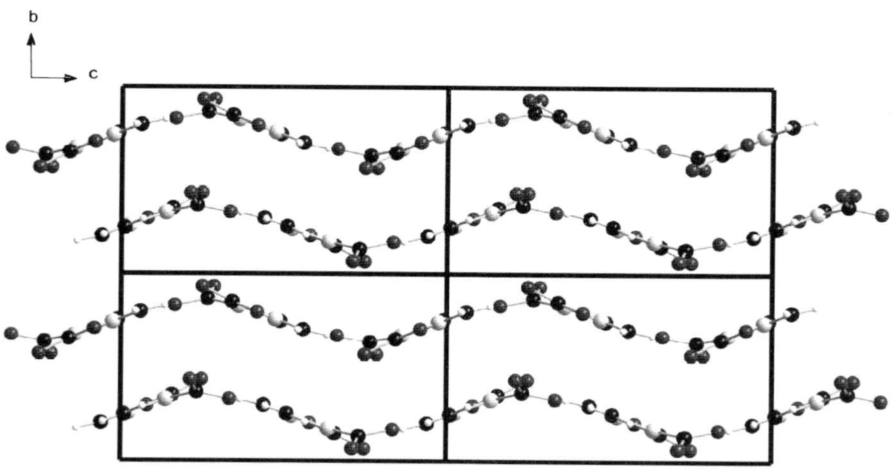

Fig. 2 Crystal structure of FOX-7

ide (21, 32 and ~ 45 g/100ml at 20 °C respectively). It is essentially insoluble in water.

An X-ray diffraction (XRD) study, combined with a range of thermal analysis techniques [differential scanning calorimtery (DSC), thermogravimetry (TG) and thermomechanical analysis (TMA)] [11, 12], has been applied to powdered 1,1-diamino-2,2-dinitroethene [recrystallized from *N*-methylpyrrolidinone (NMP)-water]. This has shown that several other crystalline phases exist which are formed on heating the normal α-polymorph. The first phase transition ($\alpha \rightarrow \beta$) is completely reversible (as observed by DSC) and occurs at 113 °C. A second transition ($\beta \rightarrow \gamma$) occurs at 173 °C, but this is only partially reversible (no reverse transition is observed by DSC until the region for the $\beta \rightarrow \alpha$ transition). Above 200 °C the material begins to decompose (Sect. 4 below). This study also reported the crystallographic parameters for both the α- and the β-polymorphs. The data for the former agreed closely with those reported earlier [8, 9]. The β-polymorph is of the orthorhombic type and there are four molecules per unit cell whose dimensions are: a 6.978 Å, b 6.638 Å, c 11.660 Å, β 90°, V 540.1 Å^3. The supramolecular arrangement of molecules within the crystal of the β-polymorph is also in the form of wave-shaped infinite layers. The two polymorphs are thus very similar in internal structure. The $\alpha \rightarrow \beta$ transition is accompanied by a volume increase ($\equiv$ density decrease) of 1.9%. This might be of significance during the storage of munitions containing FOX-7, although the temperature required to effect this change is rather high.

Thermogravimetry (TG) indicated a 2% mass loss at around 160 °C on the first heating cycle, with no mass loss on repeated temperature cycling. This was attributed to a possible loss of solvent. NMP is known [13] to become occluded when FOX-7 is recrystallized from NMP-water.

A more detailed crystallographic study of the $\alpha \rightarrow \beta$ transition has recently been published [14]. This gives full crystallographic data for a single crystal (recrystallized from *N*,*N*-dimethylformamide or dimethylsulphoxide) of FOX-7 at 5 temperatures in the range – 73 to + 120 °C, the α (monoclinic) $\rightarrow \beta$ (orthorhombic) transition being observed between 100 and 120 °C. Powdered FOX-7 was also studied at 130 and 150 °C. Below the transition temperature the thermal expansion of the unit cell is mainly along the *b* axis i.e. between the layers, but during the transition it is the *c* axis which changes most. The volume change at this point is 1.8 cm^{-3} mol^{-1} (~ 2%). After the transition it is again the *b* axis which is most responsible for the volume change in the unit cell. The crystallographic parameters were unchanged after the $\alpha \rightarrow \beta \rightarrow \alpha$ transitions; after cooling to 20 °C the crystal was still clear and transparent demonstrating that the volume change at the transition had not caused sufficient strain to fracture the crystal. Significantly for the results reported by other workers (see below), no discontinuous changes were observed in the *a*, *b* and *c* cell parameters at temperatures below 100 °C.

In the layered structure (a–c plane) of both polymorphs, the molecular units are held head-to-tail by hydrogen bonds. For the α-polymorph, there are two N – H...O – N hydrogen bonds to each nitro group oxygen (8 in total, of which 6 are intermolecular and 2 intramolecular). For the β-polymorph, one oxygen of each nitro group is involved in 2 hydrogen bonds while the other oxygen is involved in 3 hydrogen bonds (10 in total, of which 8 are intermolecular and 2 intramolecular). Within each FOX-7 molecule both C atoms and both amino N atoms lie within a plane. The two nitro N atoms are slightly out of the plane, while the four oxygen atoms deviate more significantly from the plane.

The $\alpha \rightarrow \beta$ transition is clearly observed by DSC as an endothermic peak at 116.3 °C (onset 114.2 °C, $\Delta H = 18.0 \pm 1.5\,\mathrm{J\,g^{-1}}$). The general picture of phase changes for FOX-7 therefore appears to be:

$$\alpha \underset{113-116\,^{\circ}\mathrm{C}}{\longrightarrow} \beta \underset{173\,^{\circ}\mathrm{C}}{\longrightarrow} \gamma\,.$$

Three other studies, two of them being chronologically earlier than the above work [15–18], report somewhat different observations. Here the temperatures given for the first two transitions are markedly lower, and one further transition is observed. The first phase transition (designated $\alpha \rightarrow \beta$) is here reported to occur at 90 °C and was only detected by X-ray diffraction; it is not visible by DSC. Again a small decrease in density ($1.885 \rightarrow 1.803 \equiv$ increase in volume) is observed. The second transition (designated $\beta \rightarrow \gamma$) occurs at 115 °C (DSC endotherm at 117 °C), and a further transition (designated $\gamma \rightarrow \delta$) occurs at 155 °C (DSC endotherm around 155 °C). The reverse transitions ($\gamma \rightarrow \beta$ and $\beta \rightarrow \alpha$) occur on cooling, but while the $\beta \rightarrow \alpha$ transition is instantaneous, the $\gamma \rightarrow \beta$ (and hence α) transition may take days or even weeks. The second of these studies [16] also demonstrated how particle size can affect the temperature and the rate at which phase transitions occur. For example it was reported that the first transition temperature can vary between $\sim$ 78 °C (for very fine particles) and $\sim$ 95 °C (for coarse particles), but the rate ($\sim$ 3 h for the complete transition) is unaffected by particle size, while the reverse is true for the second transition temperature (116 °C) i.e. the transition temperature is not affected by particle size, but the rate is (10 h and 14 h for fine and coarse particles respectively). The average enthalpy changes involved in the second and third transitions (those visible by DSC) were reported as 21.6 and 17.1 $\mathrm{J\,g^{-1}}$ respectively [17, 18].

Although the FOX-7 samples used in the first two of these three studies [15, 16] were stated to have been recrystallized from dimethylsulphoxide, the behavior described is similar to that of unrecrystallized FOX-7. A broad DSC endotherm at 90–110 °C is the normal observation for unprocessed FOX-7 (Latypov, personal communication), and this is replaced by an endotherm at 113 °C on recrystallization. The TG analysis reproduced in one of the studies [16] shows no mass loss i.e. loss of solvent, in the region of 160 °C, which would have been expected if the material had been recrystallized, again suggesting that unrecrystallized material might have been used. The origin, but not the history of the sample used in the more recent study [17, 18] is reported. It may or may not have been recrystallized at source.

In the absence of more definite information on the pre-treatment of the samples used in the earlier work [15, 16], one is tempted to conclude that the differences in the results reported stem from the differences between the crystallographic nature of unrecrystallized and recrystallized material. Unrecrystallized material will have separated from a rather different medium (acidic aqueous medium) to that used in a recrystallized sample, and at a different rate. Its internal structure may therefore be somewhat different. At the present time (2006) some confusion exists regarding which phase change, if any, occurs below 100 °C, but the transition observed at around 115 °C in all studies [11, 12, 14–18] appears to be the $\alpha \rightarrow \beta$ change (but designated $\beta \rightarrow \gamma$ in [15–18]).

4
Thermal Decomposition of FOX-7

The thermal behavior of FOX-7 at temperatures below 200 °C has already been described above. As the temperature is increased above 200 °C, decomposition, without melting, begins to occur. The appearance of the DSC curve in this region is rather variable and appears to depend upon the history of the sample. Even samples from the same batch may give different DSC curves. For example it has been reported [16] that a sample (1.5 mg) of fine crystals (10 K min^{-1}) exhibited two exotherms, one at $\sim$ 235 °C and another at $\sim$ 275 °C, while another sample from the same batch consisting of a single large crystal (1.5 mg) exhibited a single exotherm at $\sim$ 240 °C. Decomposition is normally complete by 300 °C.

The results from several other groups [17–21] support the general picture of two exothermic DSC peaks (Fig. 3, 10 K min^{-1}, recrystallized sample), with the peak positions and relative size being dependent upon the history of the sample. TG and differential thermal analysis (DTA) results indicate that the mass loss and energy evolution associated with the two DSC exotherms are $\sim$ 38% and 909 J g^{-1} for the low temperature peak, and $\sim$ 45% and 518 J g^{-1} for the high temperature peak [21]. However from what has already been reported above, the figures would probably differ for other FOX-7 samples. In order to examine the processes causing the two DSC peaks, Tiemanis et al. [21] heated a 1 g sample of FOX-7 for 3 days at 185 °C, after which the mass loss was 37%. This sample was then examined thermoanalytically and spectroscopically. By DTA it exhibited a single exotherm at approximately the same temperature as the second exotherm of FOX-7, but no lower temperature phase changes were observed. By TG the mass loss associated

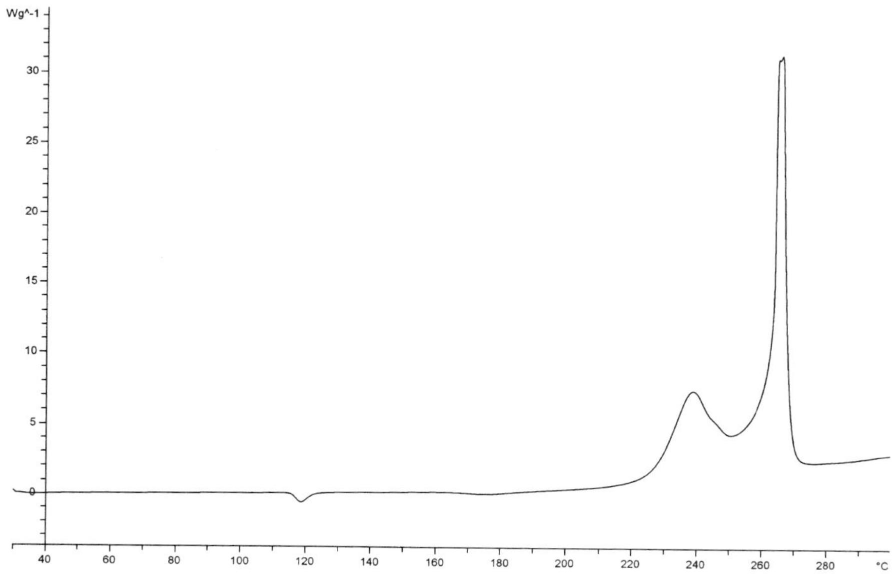

Fig. 3 DSC (10 K min^{-1}) of recrystallized FOX-7

with the exotherm was ∼ 70%. Spectroscopic examination (by ^{1}H, ^{13}C and ^{15}N NMR and FTIR) indicated that the material was essentially FOX-7. This was confirmed by HPLC and powder X-ray diffraction. The authors' suggested explanation for these somewhat surprising results is that the original sample of FOX-7 may have consisted of two phases, part crystalline (detected by XRD) and part amorphous (not detected by XRD), and it is the amorphous phase that decomposes during the first DSC peak leaving the original crystalline part. They point out that the relative amounts of these two phases should vary with the previous processing conditions. Their explanation is supported by the observation that recrystallization of the heat treated material gave FOX-7 which exhibited the normal phase changes on DTA analysis, and the double step mass loss at 200–300 °C on TG analysis.

The actual products formed during decomposition in both decomposition steps have been identified [17, 18] by FTIR as CO_2, HCN, N_2O, NO_2, HOCN, H_2O and NO, while HNO_2 and HCOOH are produced in the second decomposition only. The apparent activation energies for the two steps are reported as 238.3 and 322.4 kJ mol^{-1}, respectively.

5
Spectroscopic Properties of FOX-7

Most of the common spectroscopic data for FOX-7 were reported in the original FOI publication [2]. Nuclear Magnetic Resonance (NMR) spectroscopic data were obtained in dimethyl sulphoxide (DMSO-d_6). The ^1H spectrum (Fig. 4a) shows a single broad peak (at 8.77ppm) for all 4 NH nuclei, while the ^{13}C spectrum (Fig. 4b) exhibits 2 peaks at 128.5 [$C(NO_2)_2$] and 158.8 ppm

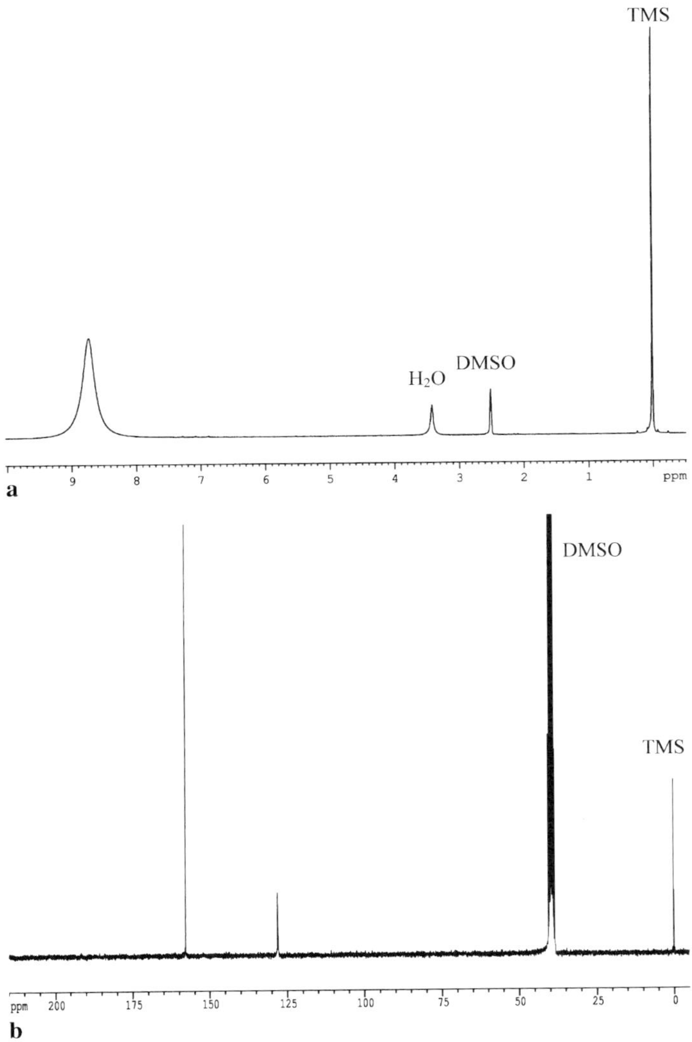

Fig. 4 **a** ^1H NMR spectrum of FOX-7, **b** ^{13}C NMR spectrum of FOX-7

$[C(NH_2)_2]$ [2], the former being significantly less intense (can be as low as 10–15% relative intensity) than the latter. Carbon atoms with nitro groups attached normally give low intensity peaks. The relative peak positions reflect the strong polarization of the electrons of the C = C bond towards the carbon with nitro groups attached, causing greater shielding of that carbon

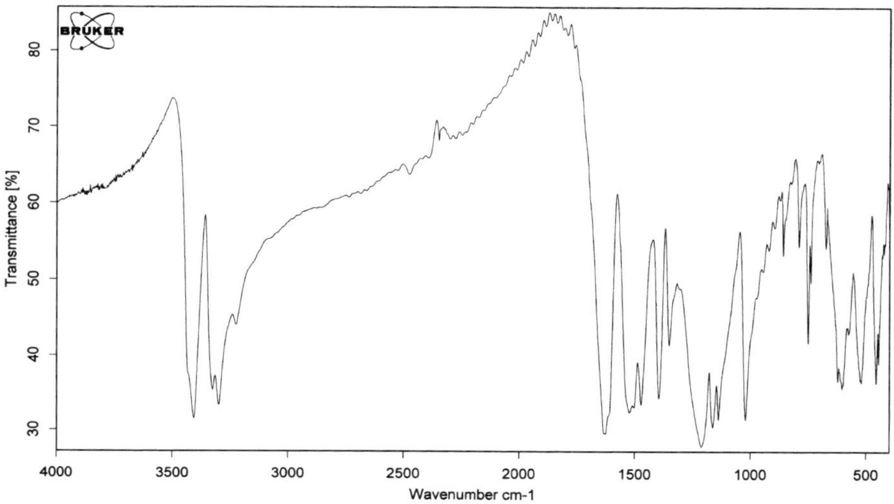

Fig. 5 FTIR spectrum of FOX-7

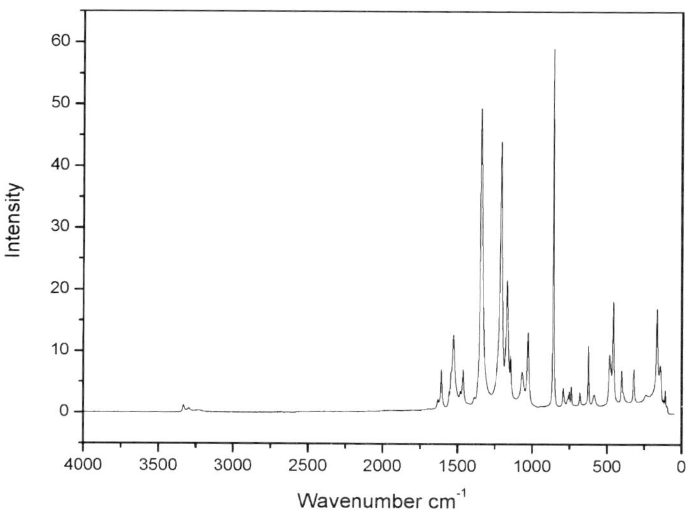

Fig. 6 Raman spectrum of FOX-7

relative to that of the carbon bearing the amino groups. In the ^{15}N spectrum absorptions occur at – 24.0 (NO_2) and – 276.6 ppm (NH_2) [21].

The infrared spectrum of solid FOX-7 (KBr disc, Fig. 5) exhibits [2] peaks at 3417 (NH_2), 3315 (NH_2), 3200 (NH_2), 1638 (NH_2), 1524 (NO_2), 1471, 1400, 1356 (NO_2), 1234, 1176, 1143, 1027, 752, 652, 521, 462 cm^{-1}. The Raman spectrum of solid FOX-7 (neat) is shown in Fig. 6 [10].

FOX-7's ultraviolet spectrum (in water) exhibits three maxima, 210 (ε9500), 278 (ε5000) and 346 nm (ε11 300) (Latypov, unpublished results).

The mass spectrum (70 eV) exhibits a very strong molecular peak at m/z 148 (M^+). A more detailed study [16] reported that the intensity of the parent ion peak indicated that this species was more stable than those encountered with other energetic molecules e.g. RDX, TNT. At both 20 and 70 eV there were peaks at m/z 148, 130 and 18, indicative of loss of H_2O, as also observed with TATB.

6
Explosive Properties of FOX-7

As mentioned in an earlier section, the main international interest in FOX-7 has been as a replacement for the secondary explosive RDX [22]. This section will therefore compare the explosive properties and performance of FOX-7 against that of RDX. Table 1 lists some of the relevant explosive properties for FOX-7 and RDX, and these should help to demonstrate why FOX-7 is an attractive alternative to RDX.

The main reason for requiring a replacement for RDX is that it is a rather vulnerable explosive material with relatively high sensitivity to both impact and friction. By contrast, FOX-7 is much less sensitive to impact and very insensitive to friction. But it is not just sufficient for a replacement explosive to be less sensitive. It must also be capable of performance comparable to that of RDX. In this regard, FOX-7 is a good candidate. While the detonation velocity and detonation pressure are slightly below those of RDX, the maximum density is greater (1.88 versus 1.82 g cm^{-3}), which means that a greater mass ($\sim$ 3%) of explosive material, and potentially more energy, can be contained within a given volume. However, this must be set against the enthalpy of formation of FOX-7 which is negative, while that of RDX is positive. Since both FOX-7 and RDX have the same oxygen balance (– 21.6%), having the atom ratios [$C_nH_{2n}N_{2n}O_{2n}$; $n = 2 \rightarrow$ FOX-7, $n = 3 \rightarrow$ RDX], it appears that FOX-7 contains less energy than RDX. So how large is this difference and in practice is it significant? An attempt to estimate this is made below, but it should be remembered that the actual ΔH_f° values used for FOX-7 and RDX in these calculations may be critical. The values of ΔH_f° that have been reported [25] are surprisingly variable [ΔH_f° (FOX-7) – 53 to – 134 kJ mol^{-1} and ΔH_f° (RDX) + 60 to + 79 kJ mol^{-1}]. The values selected for use below [– 130

Table 1 Some relevant explosive properties of FOX-7 and RDX

Property	FOX-7	RDX
Molecular formula	$C_2H_4N_4O_4$	$C_3H_6N_6O_6$
Molecular mass	148	222
Density ($g\,cm^{-3}$)	1.88	1.82
Enthalpy of formation ($kJ\,mol^{-1}$)	– 130 [23]	+ 67 [23]
Oxygen balance (to CO_2, %)	– 21.6	– 21.6
Decomposition temperature (ARC onset, °C)	219–223 [24]	195–199 [24]
Detonation velocity (experimental, $m\,s^{-1}$)	8870 [22]	8930 [22]
Detonation pressure (calculated, GPa)	33.96 [22]	34.63 [22]
Impact sensitivity		
Dropweight test (BAM, 2 kg, cm)	126 [16]	38 [16]
Rotter Impact (F of I)	110–140 [7]	80 [7]
Friction sensitivity		
(Julius–Petri, > 35 kp)	> 350 [16]	120 [16]
Electrostatic Discharge Test	4.5 J ignition, 0.45 J no ignition [7]	4.5 J ignition, 0.45 J no ignition [7]

FOX-7: 1,1-diamino-2,2-dinitroethene
RDX: 1,3,5-trinitro-1,3,5-triazacyclohexane

and + 67 $kJ\,mol^{-1}$, respectively] were selected because they were both from the same reference [23].

In order to estimate the amount of energy available for release on detonation, we need to apply the Wilson–Kistiakowsky rules [26], which state that (for an explosive with an oxygen balance not below – 40%; FOX-7 and RDX, with oxygen balances of – 21.6%, both fall within this group):

(i) carbon atoms are converted to CO
(ii) any remaining oxygen is used to convert hydrogen atoms to H_2O
(iii) any oxygen remaining after (ii) is satisfied is used to convert CO to CO_2
(iv) all nitrogen atoms are converted to N_2.

Thus, applying these rules to FOX-7 and RDX, the following ratios of detonation products are predicted:

$$\text{FOX-7:} \quad C_2H_4N_4O_4 \rightarrow 2CO + 2H_2O + 2N_2$$
$$\text{RDX:} \quad C_3H_6N_6O_6 \rightarrow 3CO + 3H_2O + 3N_2\,.$$

The energy released in these two processes is then calculated by subtracting the enthalpy of formation of the explosive from the sum of the enthalpies of formation of the detonation products (all in units kJ mol^{-1}):

$$\Delta H^\circ_{det}(\text{FOX-7}) = 2\Delta H_f^\circ(\text{CO}) + 2\Delta H_f^\circ(\text{H}_2\text{O}) - \Delta H_f^\circ(\text{FOX-7})$$
$$= 2 \times (-110) + 2 \times (-242) - (-130)$$
$$= -574 \text{ kJ mol}^{-1} = -3.88 \text{ MJ kg}^{-1} = -7.29 \text{ MJ dm}^{-3}$$

$$\Delta H^\circ_{det}(\text{RDX}) = 3\Delta H_f^\circ(\text{CO}) + 3\Delta H_f^\circ(\text{H}_2\text{O}) - \Delta H_f^\circ(\text{RDX})$$
$$= 3 \times (-110) + 3 \times (-242) - (+67)$$
$$= -1123 \text{ kJ mol}^{-1} = -5.06 \text{ MJ kg}^{-1} = -9.21 \text{ MJ dm}^{-3}.$$

An experimental value for ΔH°_{det} (FOX-7) is -4.86 ± 0.06 MJ kg^{-1} [27], significantly larger than the above calculated value. Selection of a less negative value for ΔH_f° (FOX-7) in the above calculation would have resulted in a calculated ΔH°_{det} (FOX-7) closer to the experimental value. An experimental value for ΔH°_{det} (RDX-wax 94/6), using the same technique, is -5.28 MJ kg^{-1} [28]. The calculated value for ΔH°_{det} (RDX) would have been closer to the experimental value if a more positive value for ΔH_f° (RDX) had been chosen. As already stated, the values of ΔH_f° chosen for the calculations were selected because they were both from the same reference and were therefore assumed to be mutually consistent. The experimental values for ΔH°_{det} suggest that this assumption may be incorrect.

Using the calculated rather than the experimental values for ΔH°_{det}, it is clear that on a mole basis only about half the amount of energy is released by FOX-7 compared to RDX. However, it is volume that limits the mass of energetic material that can be contained within a piece of ordnance, and on the basis of volume (via mass and then density) ~ 21% less energy is released by FOX-7 relative to an equal volume of RDX. [Using the experimental values for ΔH°_{det} instead, FOX-7 would release only 5% less energy than an equal volume of RDX]. Thus, although it looked initially as though FOX-7 might be greatly inferior to RDX regarding energy output, from an energy density viewpoint, the position of FOX-7 looks significantly closer to that of RDX. And, as has already been stated, the detonation velocity and detonation pressure are only slightly below those of RDX. It is also worth noting that the volume of gas generated on detonation (per mole of explosive) is greater for RDX (9 mol versus 6 mol $\equiv$ 202 dm^3 versus 134 dm^3 at STP), but per unit mass they are the same (0.91 dm^3 g^{-1}), and per unit volume of explosive, FOX-7 generates slightly more gas (1.66 dm^3 cm^{-3} versus 1.71 dm^3 cm^{-3}).

Of course these calculations are based on rather idealized behavior. In reality, other factors, such as actual density of the charge (which may be significantly lower than the crystal density, the latter being the maximum value attainable) and non-ideal detonation behavior, have a bearing on the comparison. Some actual performance testing has been reported ([29, 30], Sect. 10) and indicates that FOX-7 is comparable with RDX in its explosive output.

Other reported testing has included the range of tests that are normally conducted in order to ascertain the suitability of an energetic material for service. This includes production of crystals of the desired size and shape, large-scale sensitivity and thermal stability tests, and compatibility with other components in an explosive formulation. The overall picture from these tests is that the future for FOX-7 as a replacement for RDX seems very promising.

7
Chemical Reactions of FOX-7

The location of two electron-donating, amino groups on one carbon, and two electron-withdrawing, nitro groups on the other carbon should make 1,1-diamino-2,2-dinitroethene a good example of a push-pull alkene [31], with strong polarization of the double bond (Scheme 5).

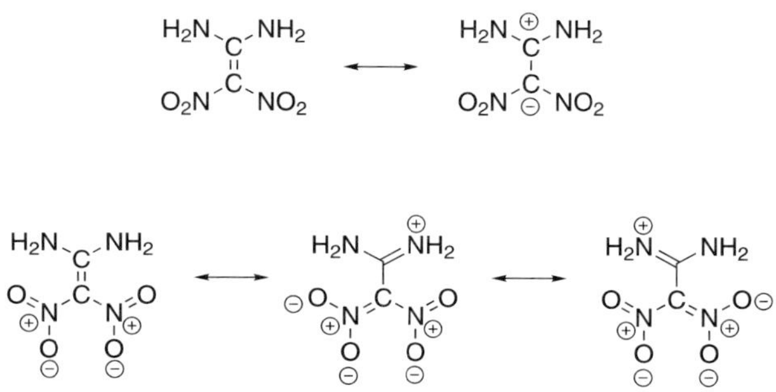

Scheme 5 FOX-7 as a "push-pull" alkene

The main consequences of this polarization are (i) the carbon bearing the amino groups (C-1) should be relatively electron deficient and susceptible to attack by nucleophiles, (ii) the H atoms should be more acidic than those of simple amines, and (iii) the amine N atoms should be poor nucleophiles. All of these aspects have been demonstrated to be present in the chemistry of 1,1-diamino-2,2-dinitroethene.

Simple nitro-enamines have been reported to undergo transamination reactions via an addition-elimination mechanism. For example, 1-dimethylamino-2-nitroethene reacts with amines to release dimethylamine, accompanied by the formation of a new nitro-enamine (Scheme 6 where R^1R^2NH = piperidine, morpholine and pyrrolidine) [32, 33].

1,1-Diamino-2,2-dinitroethene, which is a double nitro-enamine, reacts with primary alkylamines in *N*-methylpyrrolidin-2-one (NMP) at 90 °C to give both single and double substitution products [34–36], with loss of ammonia. 3-Aminopentane reacts quantitatively to give the mono-substituted

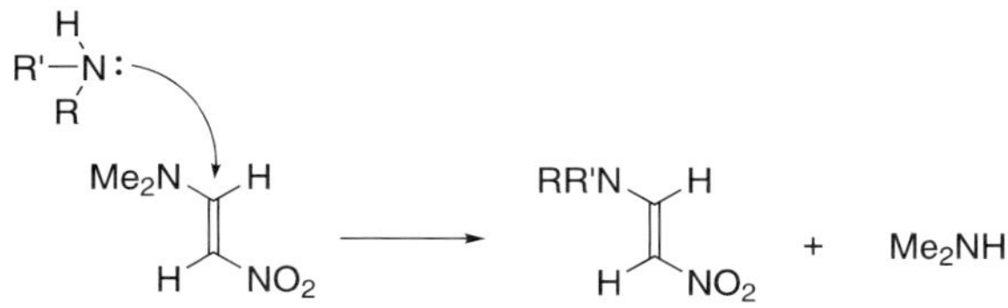

Scheme 6 Transamination of a nitro-enamine

product 1-amino-1-(pent-3-ylamino)-2,2-dinitroethene, while less hindered primary amines, e.g., 1-aminobutane, 2-aminoethanol, 3-aminopropanol, give mixtures of both mono- and bis-substitution (Scheme 7), the ω-aminoalkan-1-ols affording water-soluble products.

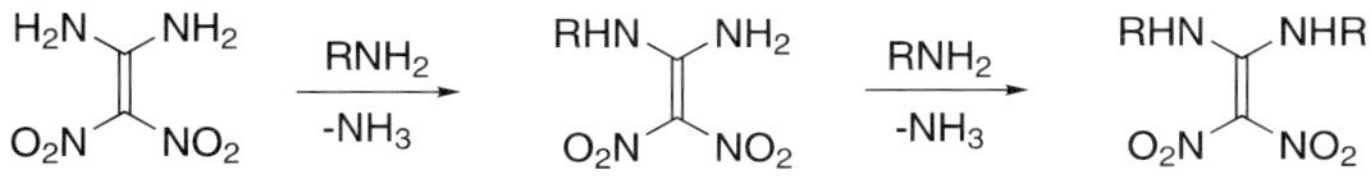

Scheme 7 Transamination of 1,1-diamino-2,2-dinitroethene with a primary alkylamine

The di-amines 1,2-diaminoethane and 1,3-diaminopropane react at both amine functions to give the same cyclic products (Scheme 8) that Baum et al. [1] had obtained by reacting the same di-amines with 1,1-diiodo-2,2-dinitroethene.

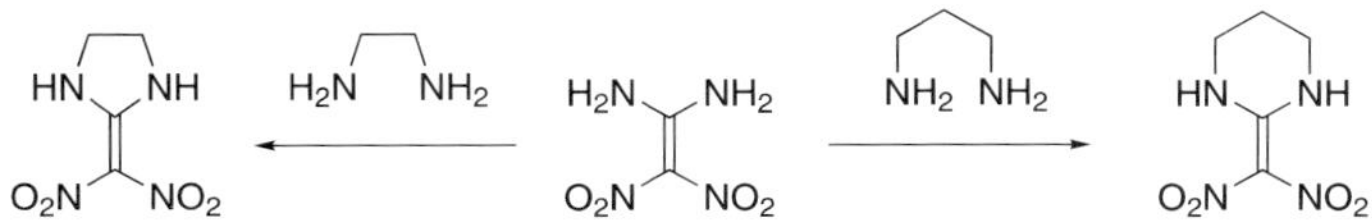

Scheme 8 Transamination of 1,1-diamino-2,2-dinitroethene with α,ω-diamines

In the absence of solvent, 1,2-diaminoethane reacts further to effect displacement of both nitro groups (Scheme 9).

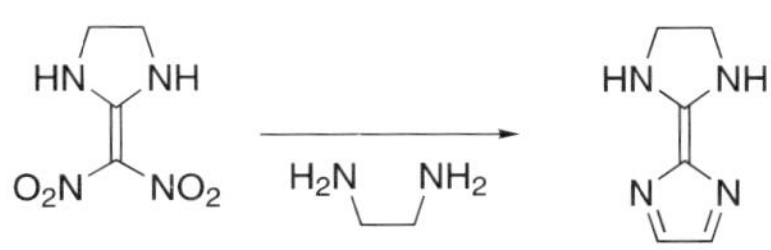

Scheme 9 Further reaction of 2-dinitromethyleneimidazolidine

A similar transamination reaction occurs with hydrazine, but in this case only mono-substitution is achieved. Reaction with excess hydrazine only serves to form the hydrazinium salt of the conjugate base of the mono-substituted product (Scheme 10).

Reaction with guanidine, instead of effecting transamination, causes deprotonation of 1,1-diamino-2,2-dinitroethene to give its guanidinium salt

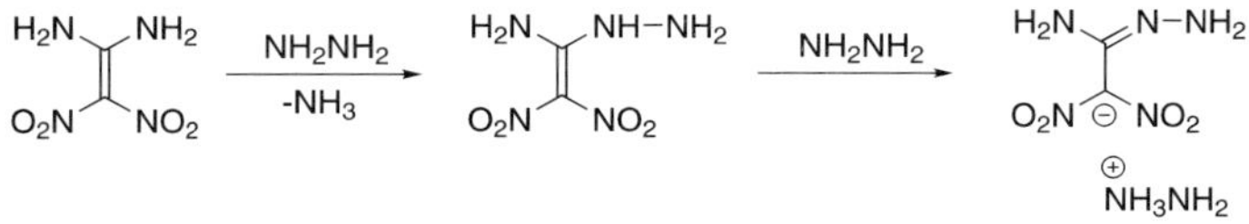

Scheme 10 Transamination of 1,1-diamino-2,2-dinitroethene with hydrazine

(Scheme 11), and thus exhibits another aspect of the reactions of 1,1-diamino-2,2-dinitroethene.

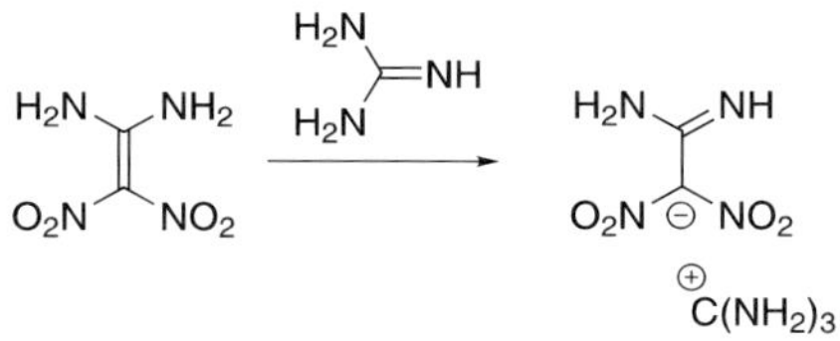

Scheme 11 Reaction of 1,1-diamino-2,2-dinitroethene with guanidine

Indeed, this is a general reaction with common bases like KOH. The potassium salt of 1,1-diamino-2,2-dinitroethene may be isolated as a white, crystalline solid on treatment with cold aqueous KOH, and the free 1,1-diamino-2,2-dinitroethene may be regenerated on acidification. The pKa of 1,1-diamino-2,2-dinitroethene was found [36, 37] to be approximately 10.6, the conjugate base existing in the pH range 11–13 (Scheme 12). Both 1,1-diamino-2,2-dinitroethene and its conjugate base may be detected by UV and ^{13}C NMR spectroscopy.

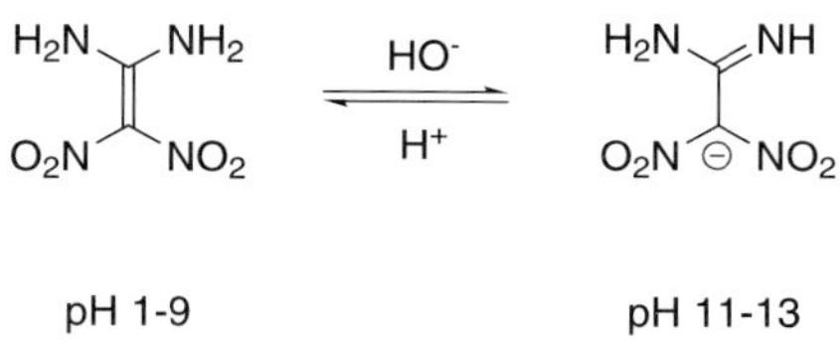

Scheme 12 Acid-base equilibria of 1,1-diamino-2,2-dinitroethene

The corresponding pKa of the mono-hydrazino derivative (see above) is lower (value ∼ 6.7) than that of 1,1-diamino-2,2-dinitroethene and explains why excess hydrazine causes deprotonation of the initial mono-transamination product. At pH < 2 the protonated form appears to be generated (Scheme 13). Again UV and ^{13}C NMR spectroscopy may be used to detect the interconverting species [36, 37].

Heating 1,1-diamino-2,2-dinitroethene in the presence of aqueous KOH (70 °C) causes hydrolytic cleavage of the C = C bond, presumably via nucleo-

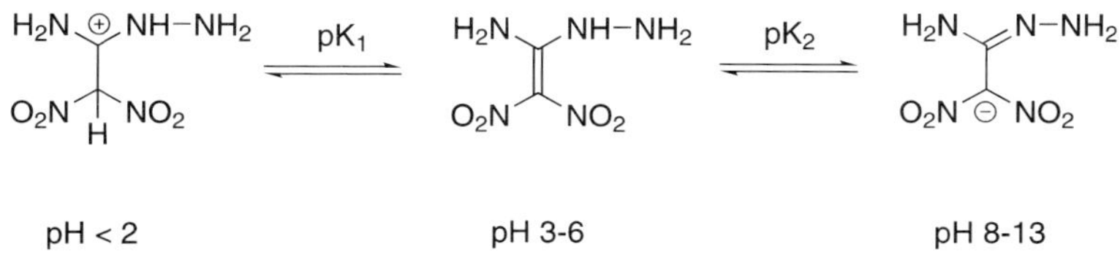

Scheme 13 Acid-base equilibria of 1-amino-1-hydrazino-2,2-dinitroethene

philic attack by HO^- at C-1, and generates the potassium salt of dinitromethane (Scheme 14).

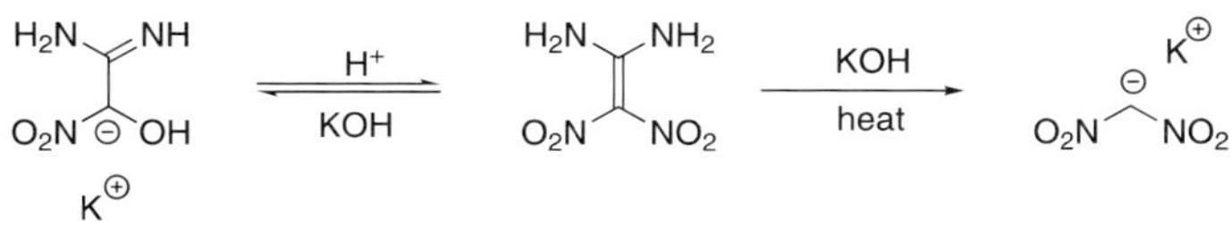

Scheme 14 Base hydrolysis of 1,1-diamino-2,2-dinitroethene

The observation by Baum et al. [1] that the ammonium salt of cyanodinitromethane (**8**) and not 1,1-diamino-2,2-dinitroethene is formed when 1,1-diiodo-2,2-dinitroethene is treated with ammonia (Scheme 1), prompted a study of the behavior of ammonium cyanodinitromethanide (**8**) and 1,1-diamino-2,2-dinitroethene in liquid ammonia, in the hope that ammonium cyanodinitromethanide might isomerize to 1,1-diamino-2,2-dinitroethene (Scheme 15) (Bellamy et al., unpublished results).

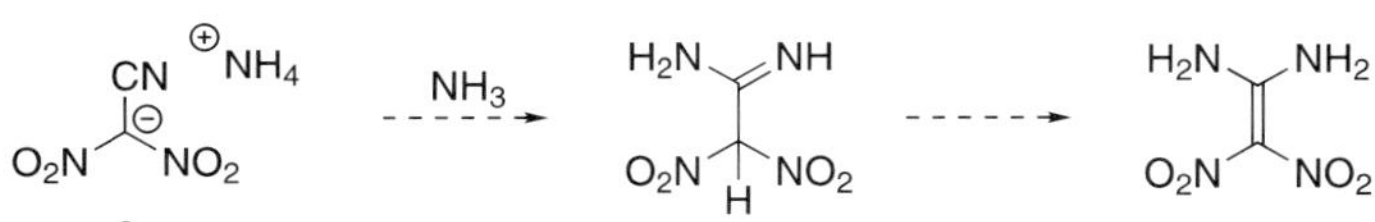

Scheme 15 Hypothetical isomerization of ammonium cyanodinitromethanide (**8**) to 1,1-diamino-2,2-dinitroethene

It was found that 1,1-diamino-2,2-dinitroethene reacts with liquid ammonia, albeit under more extreme conditions (100 °C/110 bar) than used by Baum et al. [1], to give a mixture of ammonium cyanodinitromethanide (**8**) and the guanidinium salt of dinitromethane (**19**), the latter predominating (ratio 1 : 20, $t_{1/2} \sim 6$ h). Furthermore, treatment of ammonium cyanodinitromethanide under the same conditions gave back starting material containing only a small amount of 1,1-diamino-2,2-dinitroethene (0.25%), and no guanidinium dinitromethanide. Thus conditions which would appear to offer the possibility for 1,1-diamino-2,2-dinitroethene to be formed from ammonium cyanodinitromethanide, appear to favor the latter. Moreover, since no guanidinium dinitromethanide was formed when ammonium cyanodinitromethanide was subjected to pressure and heat, it must be formed from 1,1-diamino-2,2-dinitroethene alone. One has to conclude therefore that the desired isomerization to 1,1-diamino-2,2-dinitroethene is both unfavorable and slow. The formation of ammonium cyanodinitromethanide from 1,1-diamino-2,2-dinitroethene presumably results from the reverse of the intended reaction (Scheme 16), while guanidinium dinitromethanide is formed by conjugate addition of ammonia to 1,1-diamino-2,2-dinitroethene (the initial step in a transamination reaction), followed by C – C bond cleavage (Scheme 17), reflecting both the basic and the nucleophilic properties of ammonia, respectively.

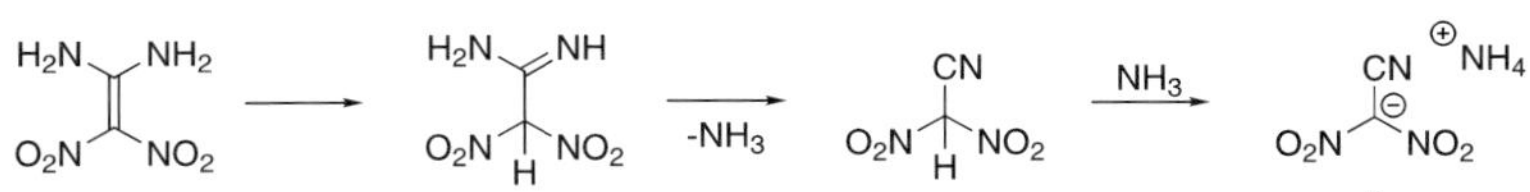

8

Scheme 16 Isomerization of 1,1-diamino-2,2-dinitroethene to ammonium cyanodinitromethanide (**8**)

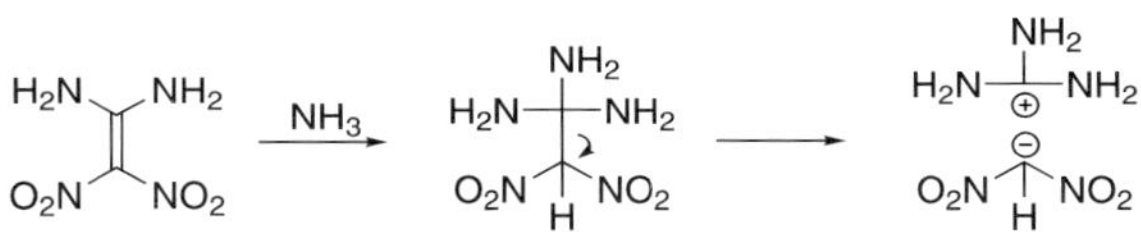

19

Scheme 17 Isomerization of 1,1-diamino-2,2-dinitroethene to guanidinium dinitromethanide (**19**)

An indication of the low reactivity of the amino nitrogen atoms in 1,1-diamino-2,2-dinitroethene was evident from the original FOI publication [2]. It was reported that 1,1-diamino-2,2-dinitroethene would not react with 1,2-dibromoethane, even in hot *N*,*N*-dimethylformamide (DMF), without prior activation to the conjugate base with sodium hydride (2 equivalents). Thereafter reaction at 115 °C gave 2-dinitromethyleneimidazolidine, the same product that was prepared by Baum et al. [1] by reacting 1,1-diiodo-2,2-dinitroethene with 1,2-diaminoethane (Scheme 18). However, 1,1-diamino-2,2-dinitroethene was found to react more readily (without NaH activation) with oxalyl chloride to give 2-dinitromethyleneimidazolidine-4,5-dione, the intermediate isolated from the nitrations of 2-methylimidazole and 2-methylimidazolidine-4,5-dione (Scheme 19).

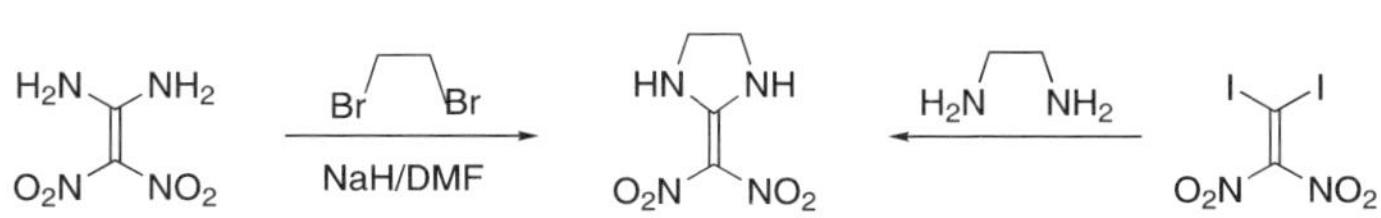

Scheme 18 Alkylation of 1,1-diamino-2,2-dinitroethene

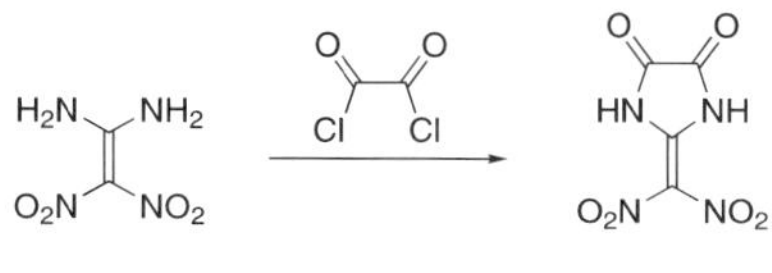

13

Scheme 19 Acylation of 1,1-diamino-2,2-dinitroethene with oxalyl chloride

The chemical reactivity of 1,1-diamino-2,2-dinitroethene has been further explored in a joint investigation by FOI (Sweden) and SNPE (France) [38]. No evidence was found for the participation of the C = C bond of 1,1-diamino-2,2-dinitroethene in [2 + 1] cycloadditions (with Cl_2C:) or [3 + 2] cycloadditions (with benzyl azide, ethyl diazoacetate and benzonitrile oxide). Furthermore, acetylation on the amino groups with acetyl chloride only succeeded if a catalyst was present, the mono-*N*-acetyl derivative being produced (Scheme 20).

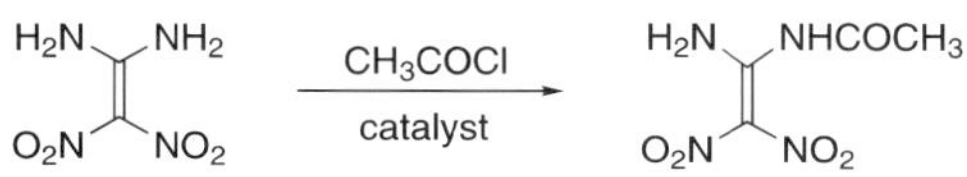

Scheme 20 Acetylation of 1,1-diamino-2,2-dinitroethene

More successful reactions involved treatment with (i) *N*-chlorosuccinimide (CBS), (ii) *N*-bromosuccinimide (NBS) and (iii) nitric acid in trifluoroacetic anhydride/trifluoroacetic acid (HNO_3/TFAA/TFA). All three reagents gave products in which the electrophilic species (Cl, Br and NO_2 respectively) had added at both an amino N-atom and C-2 to produce 1-chloro/bromo/nitro-1,1-dinitro-2-(*N*-chloro/bromo/nitro-amidino)ethane, respectively (Scheme 21). In the case of the nitration, this product is analogous to those obtained on nitration of 1,1-bis(alkylamino)-2,2-dinitroethenes by K. Baum and N.V. Nguyen [39].

Reagents: (i) X = Cl, NCS/MeOH/20°C/3h
(ii) X = Br, NBS/MeOH/20°C/3h
(iii) X = NO_2, HNO_3/TFAA/TFA/-5 to 5°C

Scheme 21 Halogenation and nitration of 1,1-diamino-2,2-dinitroethene

8
Alternative Attempts to Synthesize FOX-7

At the present time (2006) the preferred starting material for the manufacture of FOX-7 is 4,6-dihydroxy-2-methylpyrimidine, despite the fact that for every 10 kg of FOX-7 there is > 5 kg of unwanted dinitromethane produced as a by-product and two equivalents of nitric acid are wasted. It would obviously be preferable if these disadvantages could be avoided and consequently the nitration of a number of other, closely related heterocyclic systems has been studied as potential alternative precursors [40–42]. These were chosen because they were unable or unlikely to nitrate at position 5. Perhaps the most obvious candidate to investigate was 6-methyl-1,3,5-triazine-2,4(1*H*,3*H*)-dione (**20**), since this molecule has nitrogen instead of carbon between the carbonyl groups.

When 6-methyl-1,3,5-triazine-2,4(1*H*,3*H*)-dione (**20**) was nitrated under the conditions normally used for the 2-methylpyrimidine derivative (mixed acid, 25 °C), there was no evidence for nitration even after several days. When the temperature was raised to 80 °C, nitration did occur, but not to give the anticipated product. After 20 h at 80 °C, a quantitative yield of 2,4,6-trihydroxy-1,3,5-triazine (**21**, cyanuric acid) was obtained, accompanied by tetranitromethane (Scheme 22). Thus it appears that nitration has occurred initially as anticipated, but then the nitration has progressed further with cleavage of the original $C-CH_3$ bond and full nitration of the methyl carbon. This cleavage is analogous to that which occurs in the nitration of 2-methylimidazole (**9**, Scheme 2), where parabanic acid (**10**) is a major product. It is likely that cleavage to parabanic acid also occurs to some extent in the nitration of 2-methylimidazolidine-4,5-dione (**15**, Scheme 3) since the yield of the intermediate **13** is only 67%. It is possible that cleavage does not occur in the methylpyrimidine case because the tetra-nitrated derivative **17** (Scheme 4) separates from the nitration medium. When the methyltriazine **20** was nitrated with nitric acid alone at 20 °C, cyanuric acid was still the product isolated.

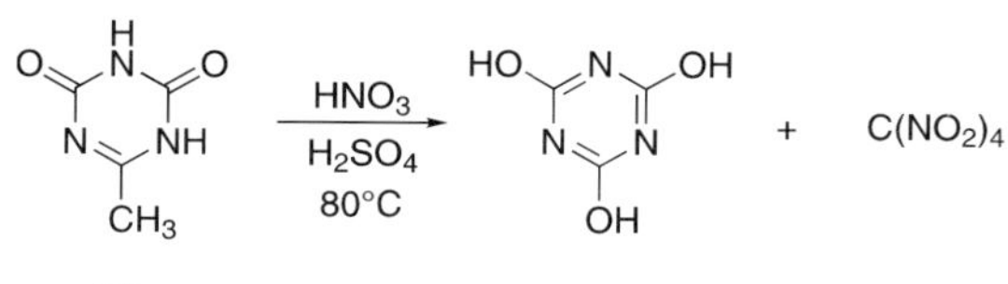

Scheme 22 Nitration of 6-methyl-1,3,5-triazine-2,4(1*H*,3*H*)-dione (**20**)

A second triazine system that was also investigated was 2,4-dimethoxy-6-methyl-1,3,5-triazine [**22**, the synthetic precursor of 6-methyl-1,3,5-triazine-2,4(1*H*,3*H*)-dione]. Mixed acid nitration of this at 20 °C again gave over nitration at the methyl carbon, but in this case the product, 2,4-dimethoxy-6-trinitromethyl-1,3,5-triazine (**23**), had not undergone C – C bond cleavage. Reductive removal of one of the nitro groups (KI/MeOH) did produce the potassium salt of 2,4-dimethoxy-6-dinitromethyl-1,3,5-triazine (**24**) but this could not be hydrolyzed to FOX-7, cyanuric acid again being the final product (Scheme 23).

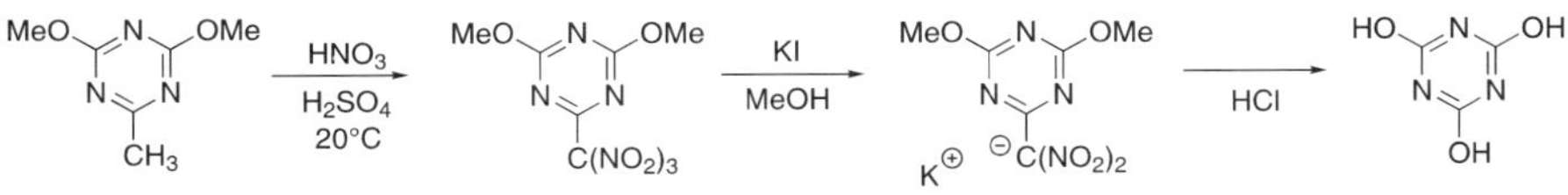

22 23 24 21

Scheme 23 Mixed acid nitration of 4,6-dimethoxy-2-methyl-1,3,5-triazine (**22**)

Finally, when 2,4-dimethoxy-6-methyl-1,3,5-triazine was nitrated with nitric acid alone at 20 °C, the product was 3,4-bis(3′,5′-dimethoxy-1,3,5-triazinyl)-1,2,5-oxadiazole 2-oxide (**27**). This product was presumably formed by the normal dimerisation of a nitrile oxide to a furazan N-oxide, the nitrile oxide in this case being 3,5-dimethoxy-1,3,5-triazinonitrile oxide (**26**). The latter would be formed by tautomerism and elimination of water from the mono-nitration product 3,5-dimethoxy-2-nitromethyl-1,3,5-triazine (**25**) before it could be further nitrated to the product formed under mixed acid conditions (Scheme 24).

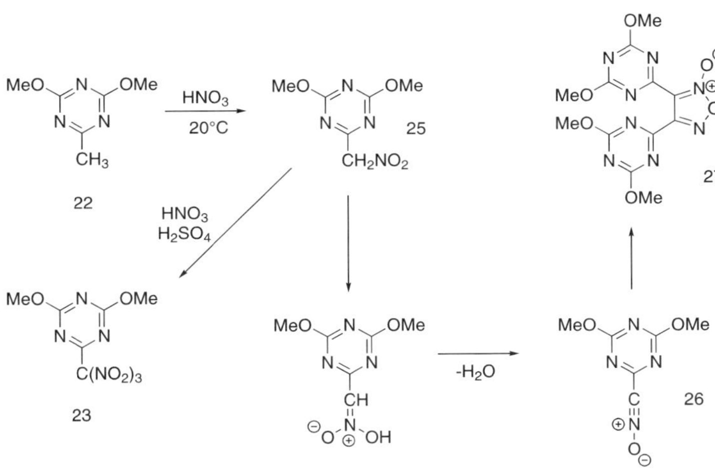

Scheme 24 Alternative modes of nitration of 4,6-dimethoxy-2-methyl-1,3,5-triazine (**22**)

Two other 2-methylpyrimidine systems have been studied, 5,5-dimethyl- and 5,5-diethyl-2-methylpyrimidine-4,6-dione (**28** and **29** respectively) [42]. These systems are more closely similar to the parent system, but here

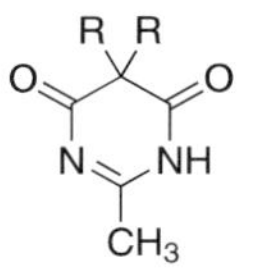

28 R = Me

29 R = Et

nitration at C-5 is blocked by the alkyl groups. Unfortunately, neither system gave FOX-7 on nitration. Interestingly, nitration of 4,6-dihydroxy-2,5-dimethylpyrimidine (**30**), a system which is structurally even closer to 4,6-dihydroxy-2-methylpyrimidine (**16**) was reported [42] to give trinitromethane, plus presumably 2-methylbarbituric acid, instead of the expected 2-dinitromethylene-5-methyl-5-nitrodihydropyrimidine-4,6(1*H*,5*H*)-dione (**31**) (Scheme 25).

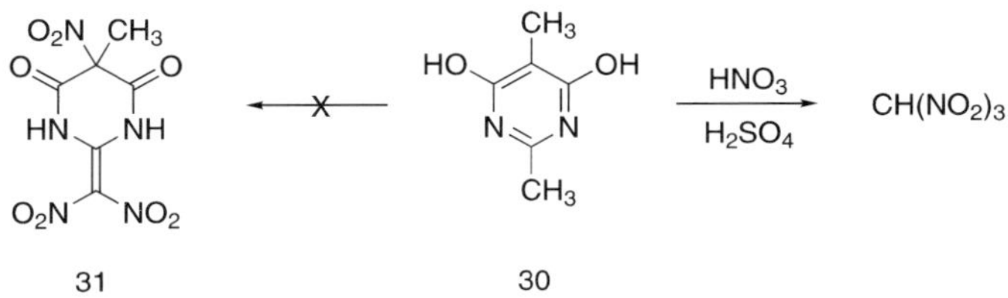

Scheme 25 Nitration of 4,6-dihydroxy-2,5-dimethylpyrimidine (**30**)

Thus, 4,6-dihydroxy-2-methylpyrimidine (**16**) still remains the preferred precursor of FOX-7, and its success probably hinges on the insolubility of its tetranitro derivative, this preventing further nitration at the methyl carbon.

9
Isomers of FOX-7

While FOX-7 (**2**) shows great promise as a secondary explosive, it is not ideal and it is interesting to compare its properties with other, structurally related molecules. How does it compare for example with its isomers, *trans*- (**32**) and *cis*-(**33**) 1,2-diamino-1,2-dinitroethene and 1,3-dinitro-1,3-diazacyclobutane (**34**)? Would they be more or less energetic than FOX-7? Would they have more desirable physical properties, e.g., higher density, greater thermal sta-

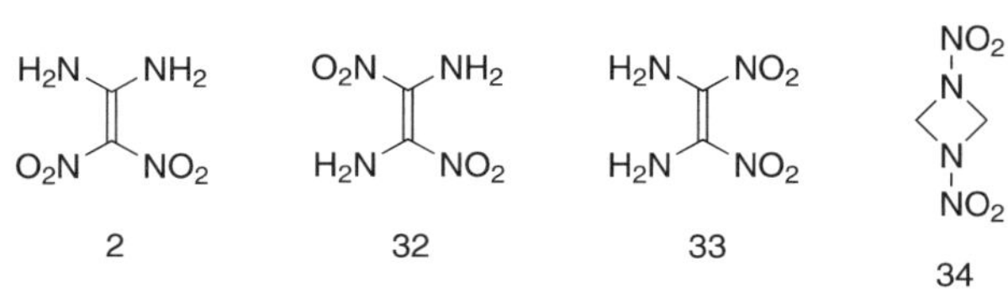

bility, lower sensitivity to impact etc? Unfortunately, despite various attempts, none of these has, as yet, been synthesized. Therefore the best that can be done is to compare their theoretically calculated properties.

Several groups have attempted to address this problem and although this approach may have limitations, it can be used as a pointer for future synthetic targets. Politzer et al. [43] were the first to compute the relative energies of the three isomeric diaminodinitroethenes. In fact, their seminal paper was accepted for publication three months before the first successful report [2] of the synthesis of FOX-7 was submitted. They concluded that the relative stabilities and molecular geometries of all three molecules are largely dictated by the available "push-pull" electronic delocalizations and intramolecular hydrogen bonds. Their computed structures are shown in Fig. 7 [44].

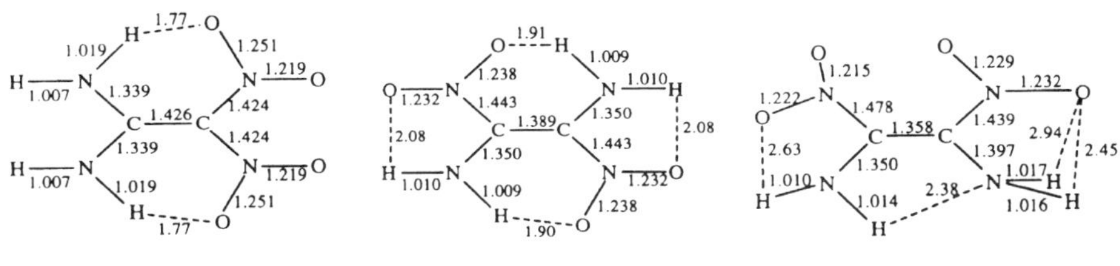

FOX-7 *trans*-isomer *cis*-isomer

Fig. 7 Computed structures of the isomeric diaminodinitroethenes

In the gas phase, the heats of formation are – 4, + 4 and + 63 kJ mol^{-1} for FOX-7, *trans*- and *cis*-diaminodinitroethene, respectively, which, by means of the calculated heats of sublimation, translates to – 113, – 80 and – 29 kJ mol^{-1} respectively for the solids. FOX-7 is thus calculated to be the solid of lowest energy. Conversely, since all three molecules are isomeric, FOX-7 would liberate the least amount of energy on decomposition (detonation) to CO, H_2O and N_2. The heats of such a reaction are calculated to be – 592, – 630 and – 680 kJ mol^{-1}, respectively, roughly half that emitted by RDX (experimental value – 1134 kJ mol^{-1}). However, on a kJ g^{-1} basis, which is the more important parameter for energetic performance, *cis*-diaminodinitroethene is much closer to RDX (– 4.59 vs – 5.11 kJ g^{-1}).

Regarding the molecular geometries, all of the bonds attached to each carbon are essentially coplanar in all three diaminodinitroethenes. The bonds to the oxygen-bearing nitrogen atoms are also coplanar, but those to the hydrogen-bearing nitrogen atoms are somewhat pyramidal. For FOX-7 and the *trans*-isomer, the molecules are almost planar, but the *cis*-isomer has one nitro group and one amino group which are rotated out of the plane in order to facilitate weak hydrogen bonding and to avoid steric interactions between neighboring *syn* nitro groups. FOX-7 has two relatively strong (short, both 1.77 Å) intramolecular hydrogen bonds, the *trans*-isomer has four weaker (longer, two each at 1.90 and 2.08 Å) ones, while the *cis*-isomer also has four hydrogen bonds but these are very weak (2.38–2.94 Å). The "push-pull" effect exerted by the electron donating amino groups and the electron withdrawing nitro groups, is, as expected, strongest in FOX-7 and least in the *cis*-isomer, where non-planarity of one nitro group and one amino group reduces the orbital overlap.

Two further theoretical studies of the isomers of diaminodinitroethene [45, 46] gave similar results to those found by the Politzer group. The study by Sorescu et al. [45] was particularly detailed and gave good agreement with the parameters calculated by Politzer et al. for molecules in vacuo. In the case of FOX-7, the study was also extended to the crystalline state and these results were in accord with the previously reported crystallographic data [8, 9]. This agreement adds confidence to the results obtained for the other two isomers. However, calculation of the lattice parameters for FOX-7 at 25 K intervals over the temperature range 25–450 K failed to detect the experimentally observed phase change ($\alpha \rightarrow \beta$) in the region of 373–393 K.

A comparison of the various bond lengths, bond angles and crystallographic data for FOX-7 as calculated by the studies of Politzer et al. [43] and Sorescu et al. [45], and those found experimentally [8, 9, 14] is given in Table 2. The largest error appears to be in the estimation of the length, and hence strength, of the intramolecular hydrogen bonds. The more recent study is somewhat better in this respect, but still underestimates the length. However there is some divergence on the crystallographically determined length also.

One further theoretical study [30] on FOX-7 alone, in addition to using CHEETAH to calculate the detonation velocity and detonation pressure for FOX-7, also reported the experimentally determined performance of both FOX-7 [velocity of detonation 8870 m s^{-1}, detonation pressure 34.0 GPa, 91% performance compared to RDX] and RDX [8930 m s^{-1} and 35.64 GPa, respectively].

Returning to the third, un-synthesized isomer of FOX-7, 1,3-dinitro-1,3-diazacyclobutane (**34**), the four-membered ring analogue of RDX, its molecular geometry and heat of formation have been calculated by the Politzer group [47]. The bond lengths were computed to be: C – C 1.466 Å, N – N 1.355 Å, N – O 1.191 Å; the bond angles were: N – C – N 87.3°, C – N – C 92.7°, C – N – N 119.9°, N – N – O 116.4°, O – N – O 127.2°; and the dihedral angles were: N – C – N – N 127.1°, C – N – N – O 34.6° and 147.7°. The heat of formation at 25 °C was calculated to be + 218 kJ mol^{-1}, 26 kJ mol^{-1} greater than that for RDX. 1,3-Dinitro-1,3-diazacyclobutane therefore contains considerably more energy than FOX-7 (– 113 kJ mol^{-1}) and its other isomers (*trans* – 80 kJ mol^{-1}, *cis* – 29 kJ mol^{-1}). These data enable one to estimate the heat of reaction to form CO, H_2O and N_2 (detonation) as about – 922 kJ mol^{-1} [≡ – 6.23 kJ g^{-1} vs – 5.28 kJ g^{-1} (experimental) for RDX]. Thus, from an energy release viewpoint 1,3-dinitro-1,3-diazacyclobutane would appear to be a worthwhile target molecule for synthesis. However it is most probable that

Table 2 Measured and computed bond lengths, bond angles and crystallographic data of FOX-7

	Bemm et al. [8] 173 K	Evers et al. [14] 200 K	Gilardi [9] 294 K	Politzer et al. [43] 0 K	Sorescu et al. [45]
Interatomic distances (Å)					
C2 – N3	1.399	1.389	1.400	1.424	1.410
C2 – N4	1.426	1.417	1.418	1.424	1.423
C1 – C2	1.456	1.460	1.446	1.426	1.465
C1 – N1	1.319	1.309	1.310	1.339	1.331
C1 – N2	1.325	1.320	1.324	1.339	1.336
N3 – O2	1.249	1.243	1.247	1.251	1.271
N3 – O1	1.252	1.254	1.239	1.219	1.262
N4 – O4	1.242	1.237	1.237	1.251	1.262
N4 – O3	1.242	1.238	1.231	1.219	1.264
O2...H2	1.97	1.87	1.96	1.77	1.88
O4...H3	2.03	2.04	1.98	1.77	1.92
Bond angles (°)					
N3 – C2 – N4	116.3	116.5	115.9	–	116.7
N3 – C2 – C1	123.9	124.1	123.7	–	123.5
N4 – C2 – C1	119.8	119.4	120.3	–	119.9
N1 – C1 – N2	118.4	118.2	118.2	–	118.2
N1 – C1 – C2	120.8	120.7	121.3	–	121.0
N2 – C1 – C2	120.7	121.1	120.4	–	120.7
O1 – N3 – O2	120.9	120.3	121.0	–	120.7
O2 – N3 – C2	118.9	119.1	118.6	–	119.0
O1 – N3 – C2	120.1	120.5	120.4	–	120.7
O3 – N4 – O4	121.0	121.7	120.7	–	120.6
O4 – N4 – C2	118.6	118.4	118.3	–	118.8
O3 – N4 – C2	120.4	119.9	121.0	–	120.6
Crystallographic data					
a (Å)	6.941	6.921	6.940	–	6.978
b (Å)	6.569	6.552	6.637	–	6.776
c (Å)	11.315	11.274	11.341	–	11.336
β (°)	90.55	90.06	90.61	–	90.80
Z	4	4	4	–	4
V (Å^3)	515.9	511.2	522.3	–	536.0

it would be less thermally and chemically stable than either its other isomers or RDX. Nevertheless one must never allow tentative theoretical arguments to persuade one not to attempt to synthesize a promising molecule.

Other properties of 1,3-dinitro-1,3-diazacyclobutane (**34**) have been predicted in a series of publications by Zeman et al. [48–52]. By using data collected for a wide range of energetic nitramines and other polynitro compounds, and by seeking trends in various properties as a function of different structural parameters, it was possible to predict the same properties for the un-synthesized 1,3-dinitro-1,3-diazacyclobutane (**34**). For example, the DSC melting point was predicted to have an onset of 161 °C and a peak at 163 °C, with a heat of fusion of 26.32 kJ mol^{-1} [48]. Other predicted properties included heat of sublimation (94.2 kJ mol^{-1}), Arrhenius parameters for thermal decomposition (Ea 153.10 kJ mol^{-1}, log A 14.1), impact sensitivity, detonation characteristics (detonation velocity 8.46 km s^{-1}), and ^{15}N NMR chemical shifts (nitro – 27.83 ppm, amino – 203.62 ppm).

10
Compounds Structurally Related to FOX-7

It was stated in Sect. 1 that 1,1-diamino-2,2-dinitroethene is structurally similar to several other known energetic molecules viz. 1,3,5-triamino-2,4,6-trinitrobenzene (TATB, **3**), 2,6-diamino-3,5-dinitropyridine (DADNP, **4**), 2,4,6-triamino-3,5-dinitropyridine (TADNP, **5**), 2,6-diamino-3,5-dinitropyrazine (**6**) and 2,5-diamino-3,6-dinitropyrazine (**7**). All of these molecules possess a similar juxtaposition of amino and nitro groups, and it is the interplay, predominantly through hydrogen bonding, of these two groups that is believed to be responsible for decreased sensitivity towards external stimuli e.g. impact and friction, high melting points and high thermal stability. TATB (**3**) in particular is known to be extremely insensitive to impact and has a high decomposition temperature. One could think of these related structures as FOX-7 with extra carbon and/or nitrogen atoms inserted between the carbon atoms bearing the amino and nitro groups. In some cases, there are extra nitro and/or amino groups present. That these molecules are structurally related to FOX-7 is reflected in their crystal structures, most, if not all, existing as sheets of molecules with strong intermolecular hydrogen bonding. In addition, there is strong intramolecular hydrogen bonding between vicinal N – H and N – O groups.

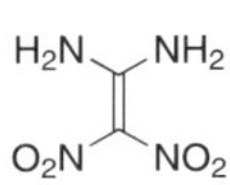

2

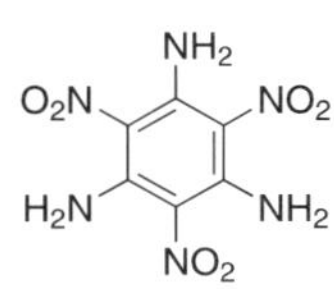

3

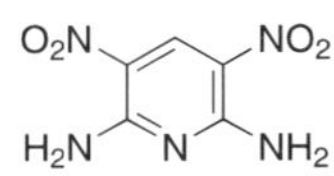

4

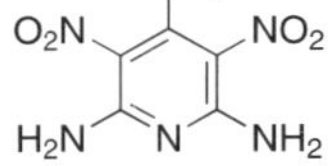

5

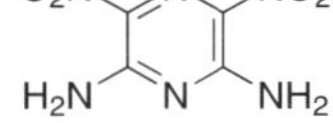

6

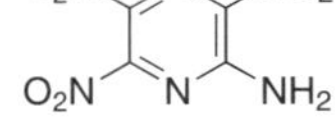

7

Since one of the objectives in designing a new energetic molecule is to incorporate as much oxygen as possible into the molecule, so that a maximum number of the carbon atoms are converted to CO_2, rather than CO or even C, during detonation, it is the *N*-oxide derivatives of the above listed N-heterocycles that are greater interest as explosives. Thus it is 2,6-diamino-3,5-dinitropyridine 1-oxide (DADNP-O, **35**) rather than 2,6-diamino-3,5-dinitropyridine (DADNP, **4**) [53, 54], 2,4,6-triamino-3,5-dinitropyridine 1-oxide (TADNP-O, **36**) rather than 2,4,6-triamino-3,5-dinitropyridine (TADNP, **5**) [54], 2,6-diamino-3,5-dinitropyrazine 1-oxide (**37**) [55–57] rather than 2,6-diamino-3,5-dinitropyrazine (**6**), and 2,5-diamino-3,6-dinitropyrazine 1,4-dioxide (**38**) rather than 2,5-diamino-3,6-dinitropyrazine (**7**) [58] that are of greater interest as explosives. In all of these cases the un-oxidized form is an intermediate in the synthesis of the *N*-oxide. However, it is the un-oxidized forms that are closer structurally to 1,1-diamino-2,2-dinitroethene.

Apart from FOX-7, only one of the energetic molecules in this list, 1,3,5-triamino-2,4,6-trinitrobenzene (TATB, **3**), is already established as an important, widely used explosive. Two of the others, 2,6-diamino- (**35**) and 2,4,6-triamino-3,5-dinitropyridine N-oxide (**36**), and have been around for 10–15 years and have yet to become widely used. This is in part because they are not much better than TATB in terms of explosive properties and thermal stability. One molecule which does promise to be significantly better than TATB is 2,6-diamino-3,5-dinitropyrazine 1-oxide (**37**). This has only relatively recently appeared on the scene and has yet to be fully investigated.

So how does FOX-7 compare, in terms of its explosive performance, physical properties and ease of manufacture, with these related molecules (see Table 3)? Only 2,6-diamino-3,5-dinitropyrazine 1-oxide (**37**) has potential explosive performance comparable with FOX-7, and it also has a higher density than FOX-7, but its production involves, at present, four synthetic steps. The other molecules show inferior explosive performance, partly due to their poor

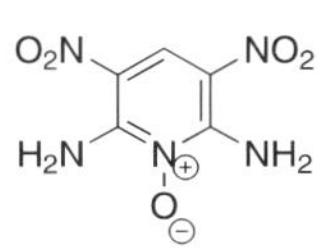

35

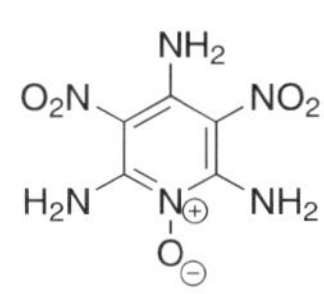

36

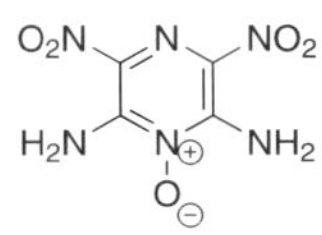

37

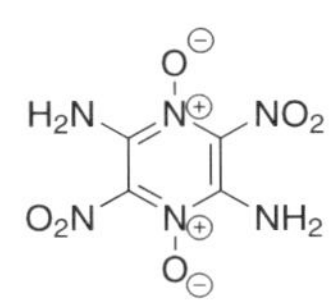

38

Table 3 Comparison of the physical properties of FOX-7 and structurally related energetic molecules

Compound	Melting or decomposition temperature (°C)	Density ($g\,cm^{-3}$) obs (calc)	Detonation velocity ($m\,s^{-1}$) calc* (obs)	Detonation pressure (kbar) calc* (obs)
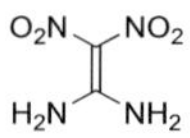	235	1.88	8740 (8870)	360 (340)
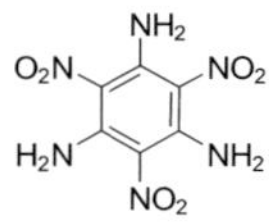	350	1.93	7870 (7620 at 96% max density)	278 (285)
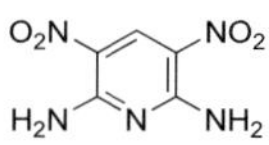	345	1.75	7480 (6800 at 97% max density)	242
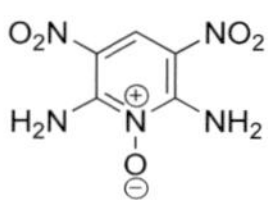	> 340	1.84	7840 (7030 at 94% max density)	276
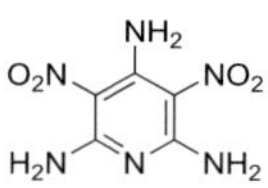	353	1.82	7670	260
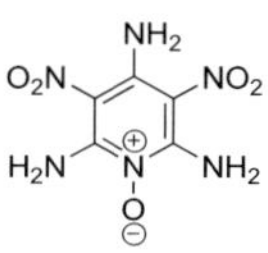	308	1.88	8010	291
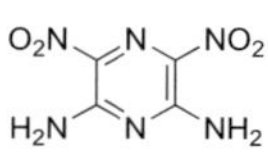	357	–	8465	334
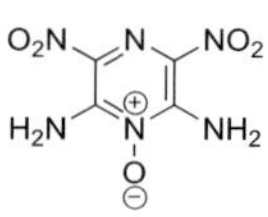	343	1.91	8730	358
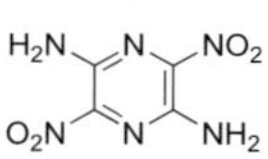	288	(1.88)	8465	334
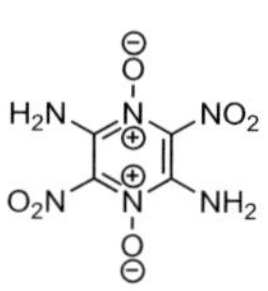	–	(1.91)	8930	397

* Calculated according to Rothstein and Petersen's method [59–61]

oxygen balance, a consequence of which is that the oxygen which is available in the molecule, is insufficient to oxidize the carbon atoms beyond the CO stage. A considerable amount of the potential energy (if sufficient oxygen were present to oxidize all carbon atoms to CO_2) is thus not released. Perhaps the only deficiency which FOX-7 has is its lower thermal stability. All of the N-oxides in this list have decomposition temperatures well in excess of 300 °C and are potential high temperature explosives. However none can be manufactured, in high yield, in a single step from a readily available precursor and for a general purpose explosive that is a great advantage.

Acknowledgements I wish to thank Professor Thomas Klapötke (Ludwig-Maximilian University, Munich, Germany), Dr. Nikolaj Latypov and Dr. Patrick Goede (FOI, Sweden), and Dr. Alessandro Contini (Cranfield University, UK) for their help.

References

1. Baum K, Bigelow SS, Nguyen NV, Archibald TA, Gilardi R, Flippen-Anderson JL (1992) J Org Chem 57:235
2. Latypov NV, Bergman J, Langlet A, Wellmar U, Bemm U (1998) Tetrahedron 54:11525
3. Latypov NV, Langlet A, Wellmar U (1999) WO Patent 99/03818
4. Latypov NV, Langlet A, Wellmar U, Goede P, Bergman J (2001) Insensitive Munitions and Energetic Materials Technology Symposium (NDIA) Bordeaux, France, p 620
5. Latypov NV, Johansson M, Holmgren E, Sizova EV, Sizov VV, Bellamy AJ (2007) Org Proc Res Dev 11:56
6. Astrat'ev AA, Dashko DV, Marshin AY, Stepanov AI, Urazgil'deev NA (2001) Russian J Org Chem 37:729
7. Lockert IJ (2001) DSTO-TR-1238
8. Bemm U, Östmark H (1997) Acta Cryst C54:1997
9. Gilardi R (1999) Cambridge Crystallographic Data Centre, CCDC 127539
10. Klapötke TM (2006) Ludwig-Maximilian University, Munich, Germany
11. Kempa PB, Herrmann M, Molina Metzger FJ, Thome V, Kjellström A, Latypov N (2004) International Annual Conference of ICT, Karlsruhe, Germany, 71:1–15
12. Kempa PB, Herrmann M (2005) Part Part Syst Char 22:418
13. Bellamy AJ, Brzoska E (2003) J Ener Mater 21:43
14. Evers J, Klapötke TM, Mayer P, Oehlinger G, Welch J (2006) Inorg Chem 45:4996
15. Bemm U, Eriksson L (2001) Insensitive Munitions and Energetic Materials Technology Symposium (NDIA), Bordeaux, France, p 775
16. Östmark H, Bergmann H, Bemm U, Goede P, Holmgren E, Johannson M, Langlet A, Latypov NV, Pettersson A, Pettersson M-L, Wingborg N (2001) International Annual Conference of ICT, Karlsruhe, Germany, 26:1–21
17. Burnham AK, Weese RK, Wang R, Kwok QSM, Jones DEG (2005) International Annual Conference of ICT, Karlsruhe, Germany, 150:1–12
18. Burnham AK, Weese RK, Wang R, Kwok QSM, Jones DEG (2005) Proc North Am Therm Anal SOC CONF 33:51
19. Sinditskii VP, Levshenkov AI, Egorshev VY, Serushkin VV (2003) Proceedings of the 30th International Pyrotechnics Seminar, St Malo, France, p 299

20. Garmasheva NV, Chemagina IV, Filin VP, Kazakova MB, Loboiko BG (2004) New Trends in Research of Energetic Materials. 7th International Seminar. Pardubice, Czech Republic, p 116
21. Ticmanis U, Kaiser M, Pantel G, Fuhr I, Teipel U (2004) International Annual Conference of ICT, Karlsruhe, Germany, 70:1–13
22. Janzon B, Bergman H, Eldstäter C, Lamnevik, Östmark H (2002) 20th International Symposium on Ballistics. Orlando, FL, p 686
23. Matyushin YN, Afanas'ev GT, Lebedev VP, Manov MN, Pepekin VI (2003) International Annual Conference of ICT, Karlsruhe, Germany, 119:1–13
24. Bergman H, Östmark H, Pettersson A, Pettersson M-L, Bemm U, Hihkiö M (1999) NDIA Conference, Tampa, FL
25. Fraunhofer Institut for Chemische Technologie (2004) ICT Database of Thermochemical Values, Version 7.0
26. Akhavan J (2004) The Chemistry of Explosives, 2nd edn. Royal Soc Chem, p 78
27. Trzcinski WA, Cudzilo S, Chylek Z, Szymanczyk L (2006) International Annual Conference of ICT, Karlsruhe, Germany, 57:1–10
28. Cudzilo S, Trebinski R, Trzcinski WA, Wolanski P (1998) International Annual Conference of ICT, Karlsruhe, Germany, 150:1–8
29. Wild R, Teipel U (2004) International Annual Conference of ICT, Karlsruhe, Germany, 69:1–9
30. Cai H-Q, Shu Y-J, Yu W-F, Zeng G-Y, Cheng B-B (2004) Ener Mater 12:124
31. Rajappa S (1981) Tetrahedron 37:1453, (section 4.3)
32. Marchetti L, Passalacqua V (1967) Ann Chim 57:1266
33. Fetell AI, Feuer H (1978) J Org Chem 43:1238
34. Bellamy AJ, Latypov NV, Goede P (2002) J Chem Res (S)257
35. Bellamy AJ, Latypov NV, Goede P (2002) J Chem Research (M)0641–0661
36. Bellamy AJ, Goede P, Sandberg C, Latypov NV (2002) International Annual Conference of ICT, Karlsruhe, Germany, 3:1–9
37. Sandberg C, Latypov N, Goede P, Tryman R, Bellamy AJ (2002) New Trends in Research of Energetic Materials. 5th International Seminar. Pardubice, Czech Republic, p 292
38. Hervé G, Jacob G, Latypov N (2005) Tetrahedron 61:6743
39. Baum K, Nguyen NV (1992) J Org Chem 57:3026
40. Bellamy AJ, Latypov NV, Goede P (2003) J Chem Res (S)529
41. Bellamy AJ, Latypov NV, Goede P (2003) J Chem Res (M)0943–0958
42. Bellamy AJ, Latypov NV, Goede P (2004) New Trends in Research of Energetic Materials. 7th International Seminar. Pardubice, Czech Republic, p 75
43. Politzer P, Concha MC, Grice ME, Murray JS, Lane P, Habibollazadeh D (1998) J Mol Struc 452:75
44. Politzer P, Concha MC, Grice ME, Murray JS, Lane P, Habibollazadeh D (1998) Computational investigation of the structures and relative stabilities of amino/nitro derivatives of ethylene. J Mol Struc 452:77, reprinted with permission from Elsevier
45. Sorescu DC, Boatz JA, Thompson DL (2001) J Phys Chem A 105:5010
46. Ji G-F, Xiao H-M, Dong H-S, Gong X-D, Li J-S, Wang Z-Y (2001) Acta Chim Sin 59:39
47. Grice ME, Habibollahzadeh D, Politzer P (1994) J Chem Phys 100:4706
48. Zeman S (1997) Themochim Acta 302:11
49. Zeman S (1999) J Ener Mater 17:305
50. Zeman S (1999) Themochim Acta 333:121
51. Zeman S, Krupka M (2003) Propell Explos Pyrotech 28:301

52. Zeman S (2003) Propell Explos Pyrotech 28:308
53. Licht H, Ritter H (1993) International Annual Conference of ICT 6:1–8
54. Hollins RA, Merwin LH, Nissan RA, Wilson WS, Gilardi R (1996) J Het Chem 33:895
55. Pagoria PF, Mitchell AR, Schmidt RD, Simpson RL, Garcia F, Forbes J, Cutting J, Lee R, Swansiger R, Hoffman DM (1998) Insensitive Munitions & Energetic Materials Technology Symposium, San Diego, CA I-3-1–5
56. Pagoria P, Lee GS, Mitchell AR, Schmidt RD (2001) Insensitive Munitions & Energetic Materials Technology Symposium, Bordeaux, France 2:655
57. Kerth J, Kuglstatter W (2001) International Annual Conference of ICT, Karlsruhe, Germany, 166:1–11
58. Philbin SP, Millar RW, Coombes RG (2000) Propell Explos Pyrotech 25:302
59. Rothstein LR, Petersen R (1979) Propell Explos 4:56
60. Rothstein LR, Petersen R (1979) Propell Explos 4:86
61. Rothstein LR, Petersen R (1981) Propell Explos 6:91

Struct Bond (2007) 125: 35–83
DOI 10.1007/430_2006_055

Published online: 25 January 2007

Nitrogen-Rich Heterocycles

Rajendra P. Singh[1] · Haixiang Gao[2] · Dayal T. Meshri[1] · Jean'ne M. Shreeve[2] (✉)

[1]Advance Research Chemicals, Inc., 1110 W. Keystone Avenue, Catoosa, OK 74015, USA

[2]Department of Chemistry, University of Idaho, Moscow, ID 83844-2343, USA
jshreeve@uidaho.edu

Abstract Many advantages accrue from nitrogen-rich heterocyclic compounds compared to traditional molecular energetic compounds. Utilization of heterocyclic nitrogen-containing cations and anions in energetic salts gives rise to lower vapor pressures, higher heats of formation and higher densities. Additionally, smaller amounts of hydrogen and carbon contribute to a better oxygen balance than normally is found with their carbocyclic analogues. Nitrogen-rich compounds are promising high energetic materials that may be more acceptable than their alternatives for both industrial and military uses since a higher percentage of their decomposition products will be dinitrogen.

Keywords Nitrogen rich · Energetic · Heterocycles · Heat of formation · Density · Thermochemistry

1 Introduction

Energetic materials store relatively large amounts of available energy that is compact and readily deliverable. Explosives are employed when very rapid rates of energy application and high pressures are essential, for example to produce high intensity shock waves in air, water, rock, and water such as in mining, road construction, and other civil engineering purposes; in shaped charges and many specialty devices requiring high rates of energy transmission; and for initiation of detonation phenomena. However, wherever a readily controllable source of energy is required for periods of time ranging from milliseconds in guns to seconds in rockets, propellants which evolve gases are employed as a working fluid for propelling projectiles and rockets, driving turbines, operating pumps, and starting engines. For industrial purposes, more than 90% of the energetic compositions used are based on ammonium nitrate [1, 2]. The single-component explosives most commonly employed for military compositions are RDX, HMX, TNT, nitrocellulose, and nitroglycerin [1–16].

Research programs worldwide endeavor to discover high-energy density materials (HEDM) with high performance and/or enhanced insensitivity with respect to thermal shock, friction, and electrostatic discharge. These modern HEDMs derive most of their energy either from: (i) their very high positive heats of formation; or (ii) oxidation of the carbon backbone, as with traditional energetic materials [17, 18]. The first group of energetic materials is exemplified by modern nitro-compounds, such as: CL-20 (2,4,6,8,10,12-(hexanitro-hexaaza) cyclododecane), TNAZ (1,3,3-trinitroazetidine), FOX-7 (1,1-diamino-2,2-dinitroethene), hepta- and octanitrocubanes (having high densities and energies utilizing substantial cage strain), 3,3′-azobis (6-amino-1,2,4,5-tetrazine), and tetrazole azide. TNT (trinitrotoluene), RDX (1,3,5-trinitro-1,3,5-triazacyclohexane), and HMX (1,3,5,7-tetranitro-1,3,5,7 tetraazacyclooctane) are found in the second group [19–23].

High-nitrogen compounds form a unique class of energetic materials whose energy is derived from their very high heats of formation directly attributable to the large number of inherently energetic N – N and C – N bonds rather than from the overall heats of combustion [24]. Heterocyclic-based molecular compounds have most often been utilized in energetic roles due to higher heats of formation, density, and oxygen balance than those of their carbocyclic analogues. In the cases where heterocyclic rings (containing amino, nitro, or azide substituents) are paired with nitrate, perchlorate, dinitramide, or picrate anions, highly energetic salts are formed. With the exception of perchlorate, these salts are more environmentally acceptable since a higher percentage of their decomposition products is dinitrogen. Another area of expanding interest is based on high energy salts in which both the cation and the anion are high-nitrogen species. All of these energetic salts possess advantages over non-ionic molecules due to lower vapor pressures and higher densities. While the use of high nitrogen compounds is likely to be limited because of cost, the main advantages of such compounds would be for low flame temperature (low gun barrel corrosion), tactical missiles, and low signature among others.

Over the past several years a relatively large number of publications describing energetic nitrogen-rich salts have appeared. This review which covers this chemistry from January 1999 through late 2006 is timely and arises because of our continuing involvement in this subject. By covering the current advances in this rapidly growing field, the review should be of great interest to the scientific community.

2 Triazolium-Based Heterocycles

Triazoles are five-membered aromatic heterocycles that contain three nitrogen atoms located at the 1,2,3 or 1,2,4 positions in the ring. A large number of ionic compounds that contain a triazole derivative are described as energetic materials.

2.1 1,2,3- and 1,2,4-Triazolium Heterocycles

1,2,3-triazole and 1,2,4-triazole have positive heats of formation of 272 kJ/mol and 109 kJ/mol, respectively [25, 26]. The salts comprised of the protonated heterocyclic cations paired with perchlorate, nitrate, or dinitramide anions were initially synthesized [27–29]. Reactions of 1,2,4-triazole or its 1,2,3-isomer with concentrated nitric, perchloric, or dinitramidic acid gave excellent yields of 1,2,4- or 1,2,3-triazolium nitrate (**1a**) or (**2a**), perchlorate (**1b**) or (**2b**), and dinitramide (**1c**) or (**2c**), respectively (Scheme 1). All the salts were characterized by Raman, IR, and NMR spectroscopy, density determination, elemental analysis, DSC (differential scanning calorimetry), TGA (thermogravimetric analysis), and initial safety testing. The melting points of 1,2,3-triazolium salts are lower than their 1,2,4-triazolium analogues (Table 1). Some of these salts (**1b–c**, **2a–c**) have melting points $< 100\,^{\circ}C$ and thus fall into the ionic liquid class.

These salts are energetic and are stable at moderate temperatures, with the two perchlorate salts the most thermally stable, followed by the nitrates, and the dinitramides the least. A single crystal X-ray diffraction study of **1b**

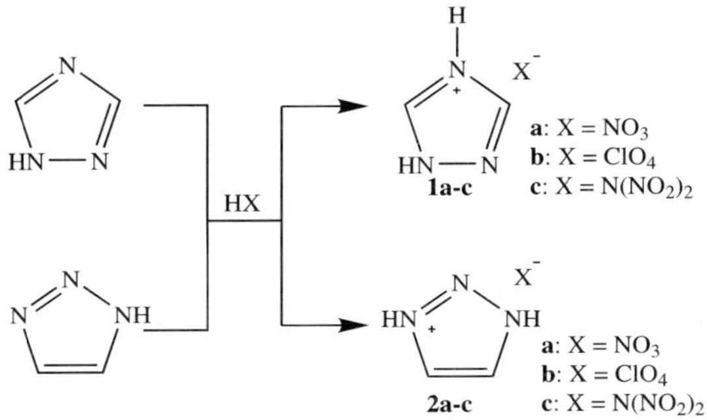

Scheme 1

Table 1 Yields, thermal characteristics, and densities of 1,2,4- and 1,2,3-triazolium salts

Compound	Yield (%)	Anion	Tm °C	Td °C	Density (calc.) g/cm^3	ΔH_f° kJ/mol
1a	99	NO_3^-	137	182	1.64 (1.55)	– 42.7
1b	93	ClO_4^-	89	285	1.96 (1.75)	– 1.3
1c	93	$N(NO_2)_2^-$	75	120	1.66 (1.64)	149.9
2a	99	NO_3^-	68	125	1.57 (1.55)	42.9
2b	85	ClO_4^-	73	200	1.79 (1.77)	90.4
2c	99	$N(NO_2)_2^-$	61	80	1.66 (1.64)	228.2

Tm, melting point, Td, thermal degradation temperature, ΔH_f° standard heat of formation (calculated by Gaussian 03)

showed that significant hydrogen bonding exists between the perchlorate anion and the protonated 1,2,4-triazolium ring, which likely contributes to its relatively high density. Interestingly, the 1,2,3-triazolium salts are in general less dense, which suggests less efficient packing in the crystal.

2.2
Triazolium Heterocycles Containing Amino Substituents

Incorporation of amino groups into a heterocyclic triazole ring is one of the simplest routes to enhance thermal stability [8]. The N-amino group behaves as an electron-withdrawing group in these high-nitrogen heterocycles that, when paired with perchlorate, nitrate, or dinitramide anions, form energetic salts. 4-Amino-1,2,4-triazolium nitrate (**3a**), perchlorate (**3b**), and dinitramide (**3c**) are formed by the reaction of 4-amino-1,2,4-triazole with the concentrated acids, HNO_3, $HClO_4$, and $HN(NO_2)_2$, respectively, under reaction conditions analogous to those utilized for (**1a–c**) and (**2a–c**) [27–29].

1-Amino-1,2,4-triazole with concentrated HNO_3 and $HClO_4$ led to the formation of the corresponding salts (**4a**) and (**4b**) in excellent yields (Scheme 2) [30]. Iodomethane with 1-amino-1,2,4-triazole gave a quaternary salt (**5**) that underwent metathesis with $AgNO_3$ and $AgClO_4$ to give high yields of the corresponding nitrate (**6a**) and perchlorate (**6b**) salts, respectively. Similarly when 1-methyl-4-amino-1,2,4-triazolium iodide (**7**) was metathesized with $AgNO_3$ and $AgClO_4$, the salts, 1-methyl-4-amino-1,2,4-triazolium nitrate (**8a**) and 1-methyl-4-amino-1,2,4-triazolium perchlorate (**8b**), were formed [31]. The syntheses of 1,5-diamino-1,2,4-triazolium nitrate (**9a**) and 1,5-diamino-1,2,4-triazolium perchlorate (**9b**) were also reported (Scheme 2) [30]. Characterization by elemental analysis, and vibrational, multinuclear NMR, and mass spectra confirmed these structures. Single crystal X-ray structural analysis of 1-methyl-4-amino-1,2,4-triazolium per-

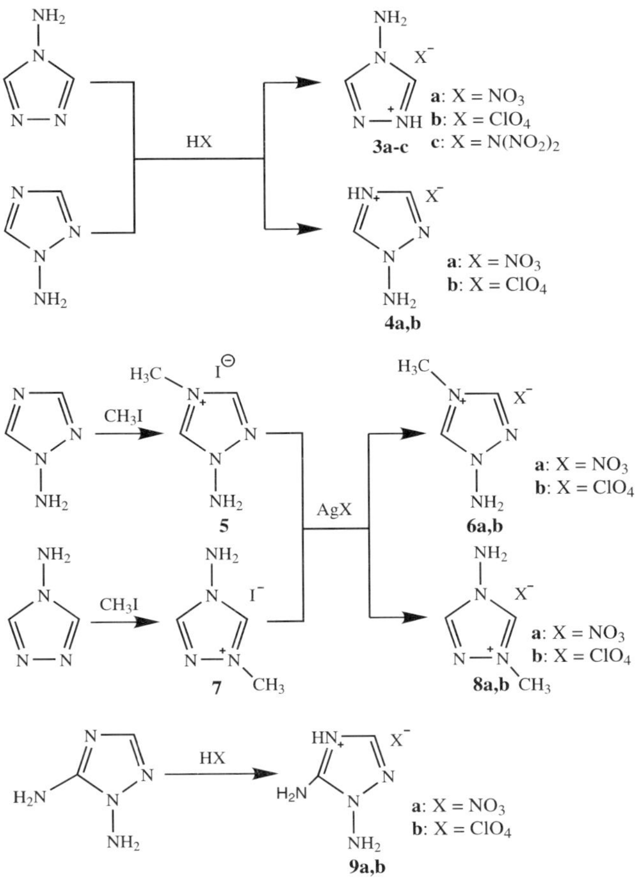

Scheme 2

chlorate (**8b**) showed the existence of significant hydrogen bonding between the perchlorate anion and the amino group, and that methylation had indeed occurred at N-1. Physical and thermal characteristics of these salts are tabulated in Table 2 [31].

In Table 2 certain interesting facts regarding thermal characteristics of the compounds are shown. For example, the position of the methyl group on the ring does not impact the melting points of 1-amino-4-methyl-1,2,4-triazolium nitrate (**6a**) (T_g – 62 °C) and 1-methyl-4-amino-1,2,4-triazolium nitrate (**8a**) (T_g – 60 °C). The analogous perchlorate salts (**6b**) and (**8b**) melt at 91 °C and 86 °C, respectively. However, the nitrate salts, 1-amino-1,2,4-triazolium nitrate (**3a**) (m.p. 69 °C) and 4-amino-1,2,4-triazolium nitrate (**4a**) (m.p. 121 °C), are considerably higher melting than their methyl-substituted analogs. Apparently the opportunity for hydrogen bonding, as well as modified packing effects and reduced lattice energies, is markedly reduced in the methylated salts. Thermal degradation temperatures for both perchlorate and nitrate salts are increased when a methyl group is present in the ring. Introduction of a second amino functionality causes the melting point (**4b** vs. **9b**) to increase and the enthalpy of formation to become more positive, but appears to have essentially little impact on other properties. The standard molar enthalpies of formation for perchlorate salts are invariably higher than those of nitrates (Table 2).

The synthesis and characterization of a series of nitrate salts, **10a–d**, based on 1-alkyl-4-amino-1,2,4-triazolium cations (alkyl = methyldecyl, isopropyl,

Table 2 Density and thermal characteristics of amino-1,2,4-triazolium salts

Compound	Anion	Tm [a] °C	Td [a] °C	Density (calc.) g/cm^3	ΔH_f° [c] kJ/mol
3a	NO_3^-	69	180	1.60 (1.58)	77.1
3b	ClO_4^-	84	210	1.81 (1.78)	117.2
3c	$N(NO_2)_2^-$	20	146	– (1.66)	273.6
4a	NO_3^-	121	149	1.69 (1.75)	72.8
4b	ClO_4^-	91	235	1.80 (1.92)	126.6
6a	NO_3^-	– 62	217	1.51 (1.60)	45.7
6b	ClO_4^-	91	235	1.66 (1.77)	91.3
8a	NO_3^-	– 60	221	1.55 (1.60)	57.6
8b	ClO_4^-	86	259	1.80 (1.92)	106.9
9a	NO_3^-	159	149	1.65 (1.76)	196.2
9b	ClO_4^-	138	–	1.83 (1.93)	240.9

[a] melting point
[b] thermal degradation temperature
[c] standard heat of formation (calculated by Gaussian 03)

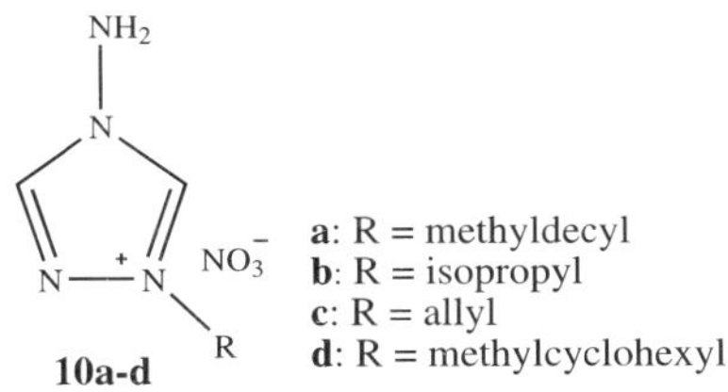

10a-d

Fig. 1 1-Alkyl-4-amino-1,2,4-triazolium nitrates

allyl, and methylcyclopropyl) have been reported (Fig. 1) [32–34]. Single crystal X-ray diffraction studies of 1-isopropyl-4-amino-1,2,4-triazolium nitrate (**10b**) (triclinic, P-1 symmetry) and 1-methylcyclohexyl-4-amino-1,2,4-triazolium nitrate (**10d**) (monoclinic, $P2_1/n$ symmetry) indicated that in both cations, the amino group is twisted and does not saddle the triazole ring in a symmetric fashion. This most likely arises from the extensive hydrogen bonding within the crystal structure which, however, does not affect the bond distances in the structures in a significant manner [32–34]. 1-Amino-3-substituted-1,2,3-triazolium salts have been examined [34].

2.3 Triazolium Heterocycles Containing Azido Substituents

Substitution of the hydrogen atoms of azoles by various nitrogen-containing energetic functional groups occurs in a straightforward manner. When endothermic moieties, such as the azido group, are incorporated into azoles, the heats of formation of the azoles are increased [31, 35, 36]. For example, the standard enthalpy of formation for 3-azido-1,2,4-triazole ($\Delta H_f =$ + 458 kJ/mol) is approximately four times larger than that of 1-H-1,2,4-triazole [36]. These azido-substituted azoles, when paired with nitrate, perchlorate, or azolate, e.g., 4,5-dinitro-imidazolate or 5-nitrotetrazolate anions, formed energetic salts [31, 37].

Quaternization of various nitrogen-containing energetic salts based on azido-1,2,4-triazoles (Scheme 3) [31] occurred readily in good yields with concentrated nitric or perchloric acid in methanol to give 3-azido-1,2,4-triazolium nitrate (**11a**), 3-azido-1,2,4-triazolium perchlorate (**11b**), 5-methyl-3-azido-1,2,4-triazolium nitrate (**11c**), and 3,5-diazido-1,2,4-triazolium nitrate (**11d**). Under similar reaction conditions, 1-methyl-3-azido-1,2,4-triazole reacted with concentrated nitric and perchloric acids to give 1-methyl-3-azido-1,2,4-triazolium nitrate (**12a**), and 1-methyl-3-azido-1,2,4-triazolium perchlorate (**12b**), respectively. A quaternary salt (**13**) formed when iodomethane was reacted with 1-methyl-3-azido-1,2,4-triazole. Subsequent metathetical reactions with $AgNO_3$ and $AgClO_4$ led to 1,4-dimethyl-3-azido-1,2,4-triazolium nitrate (**14a**) and 1,4-dimethyl-3-azido-1,2,4-triazolium perchlorate (**14b**), respectively.

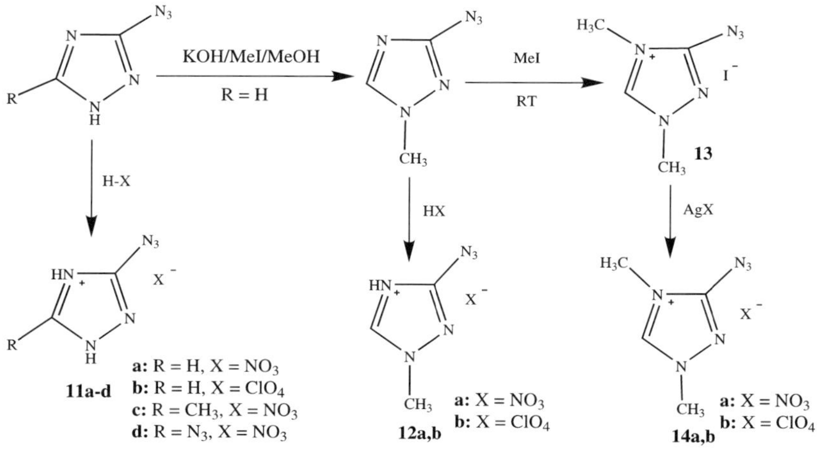

Scheme 3

These azido triazolium salts were characterized by elemental analysis, IR, ^{1}H and ^{13}C NMR, and mass spectra, and DSC studies. Examination of the crystal structures of **11a** and **14a** illustrated the influence of significant hydrogen bonding between the nitrate anion and the protonated 1,2,4-triazole ring for **11a**. This accounts for its higher density, 1.76 g/cm^3 and m.p. 147 °C, compared with that of **14a** ($d = 1.58$ g/cm^3, m.p. 98 °C), in which the presence of the methyl group on the triazole ring reduced the opportunity for hydrogen bonding (Table 3).

The chemistry of energetic salts and ionic liquids containing azidoethyl-substituted triazolium rings was examined [37]. Reaction of 1-(2-azidoethyl)-1,2,4-triazole with nitric acid, perchloric acid, 4,5-dinitroimidazole, and 5-nitrotetrazole gave 1-(2-azidoethyl)-1,2,4-triazolium nitrate (**15a**), 1-(2-azidoethyl)-1,2,4-triazolium perchlorate (**15b**), 1-(2-azidoethyl)-1,2,4-triazolium 4,5-dinitroimidazolate (**15c**), and 1-(2-azidoethyl)-1,2,4-triazolium 5-nitrotetrazolate (**15d**) salts in > 97% yield (Scheme 4). 1-(2-Azidoethyl)-1,2,4-triazole was also reacted with iodomethane to form 1-(2-azidoethyl)-4-methyl-1,2,4-triazolium iodide (**16**), which upon metathesis with the silver salts of nitric and perchloric acids formed 1-(2-azidoethyl)-4-methyl-1,2-4-triazolium nitrate (**17a**), and 1-(2-azidoethyl)-4-methyl-1,2-4-triazolium perchlorate (**17b**), respectively, in very good yields. The melting points of all these salts were found to be lower than 100 °C, and most of them are liquids at room temperature. The standard heats of formation for the perchlorate salts are more positive than those of the nitrate analogues. Most of these salts showed good thermal stability, moderate densities, and positive heats of formation (Table 3).

Reactions of 1-(2-azidoethyl)-3-azido-1,2,4-triazole with concentrated nitric acid and 5-nitrotetrazole gave 1-(2-azidoethyl)-3-azido-1,2,4-triazolium

Table 3 Density and thermal data of azido and azidoethyl-1,2,4-triazolium salts

Compound	Anion	Tm [a] °C	Td [b] °C	Density (calc.) g/cm^3	ΔH_f° [c] kJ/mol
11a	NO_3^-	147	174	1.76 (1.80)	326.3
11b	ClO_4^-	123	154	–	–
11c	NO_3^-	118	136	1.61 (1.68)	301.4
11d	NO_3^-	97	136	–	712.5
12a	NO_3^-	66	139	1.63 (1.66)	301.4
12b	ClO_4^-	55	187	1.66 (1.80)	353.2
14a	NO_3^-	98	129	– (1.58)	279.2
14b	ClO_4^-	68	147	–	
15a	NO_3^-	99	170	1.60 (–)	287.5
15b	ClO_4^-	– 56	150	1.61 (–)	338.7
15c	4,5-dinitro imidazolate	85	140	1.62 (–)	538.6
15d	5-nitro tetrazolate	– 42	164	1.51 (–)	744.4
17a	NO_3^-	– 57	119	1.49 (–)	274.4
17b	ClO_4^-	– 52	192	1.60 (–)	315.5
18a	NO_3^-	– 54	142	1.58 (–)	651.5
18b	5-nitro tetrazolate	– 46	141	1.52 (–)	1098.9
19a	NO_3^-	70	153	1.57 (–)	405.2
19b	ClO_4^-	– 46	218	1.63 (–)	450.6
20a	NO_3^-	– 56	143	1.45 (–)	283.3
20b	ClO_4^-	63	152	1.59 (–)	321.8

[a] melting point
[b] thermal degradation temperature
[c] standard heat of formation (calculated by Gaussian 03)

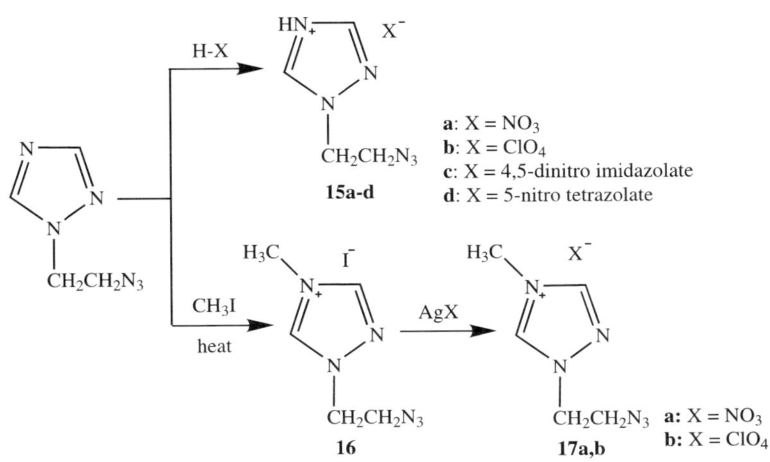

Scheme 4

nitrate (**18a**) and 1-(2-azidoethyl)-3-azido-1,2,4-triazolium 5-nitro tetrazolate (**18b**). Metathesis of 1-(2-azidoethyl)-4-amino-1,2,4-triazolium bromide and 1-methyl-4-(2-azidoethyl)-1,2,4-triazolium bromide with $AgNO_3$ and $AgClO_4$ gave good yields of 1-(2-azidoethyl)-4-amino-1,2,4-triazolium nitrate (**19a**), 1-(2-azidoethyl)-4-amino-1,2,4-triazolium perchlorate (**19b**), 1-methyl-4-(2-azidoethyl)-1,2,4-triazolium nitrate (**20a**), and 1-ethyl-(2-azidoethyl)-1,2,4-triazolium perchlorate (**20b**) (Scheme 5) [37].

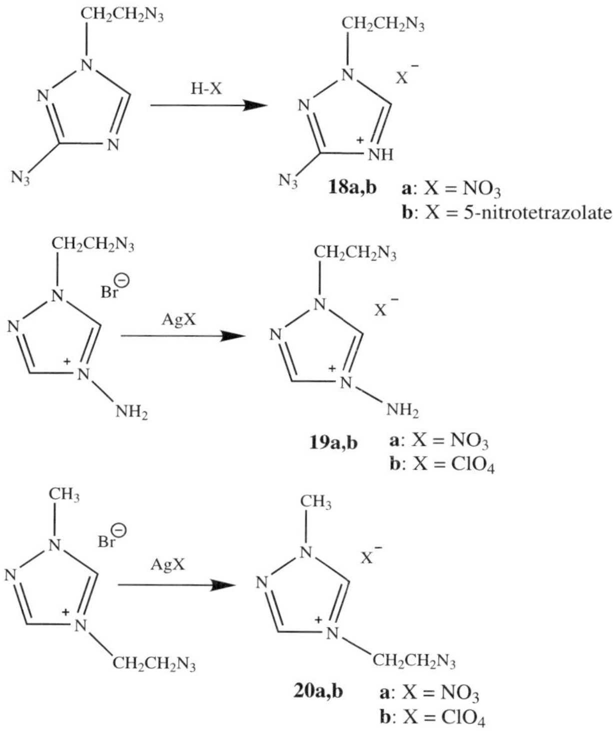

Scheme 5

Density and thermal characteristics of azido- and azidoethyl-containing triazolium derivatives are summarized in Table 3. The data show that the position of a substituent group on the triazolium ring played an important role, e.g., T_g (– 57 °C) for 1-(2-azidoethyl)-4-methyl-1,2,4-triazolium nitrate (**17a**) and 1-methyl-4-(2-azidoethyl)-1,2,4-triazolium nitrate (**20a**) (T_g – 56 °C) are essentially the same, but their decomposition temperatures (**17a,** T_d, 119 °C; **20a,** T_d143 °C) differ markedly [37]. For the analogous perchlorate salts, the melting points and thermal degradation temperatures are T_g, – 52 °C and T_d 192 °C for **17b**, and T_m, 63 °C and T_d 152 °C for **20b**, respectively.

Density and enthalpy of formation are important characteristics of energetic salts and are governed by their molecular structures. As is shown in Table 3, in general, the densities for the salts having perchlorate as anions are higher than those of the analogous nitrates. Comparing the standard molar enthalpy (ΔH_f°) of the salts, when perchlorate is used as an anion, the positive heats of formation are higher than those of the corresponding nitrates. Considering **15a** to **15d**, the impact of the anion on heat of formation of 1-(2-azidoethyl) 1,2,4-triazolium salts decreases in the order 5-nitro-tetrazolate > 4,5-dinitro-imidazolate > perchlorate > nitrate.

2.4
Triazolium Heterocycles Containing Fluoroalkyl Substituents

Substituents at carbon atoms of 1,2,4-triazoles was efficiently achieved by designing triazoles with the required substituents in place [38, 39]. Through this process various fluoroalkyl-substituted 1,2,4-triazolium salts were obtained [40], first via reaction of 1,3,5-substituted triazoles (**21a–c**) with iodomethane to form quaternary salts (**22a–c**), which when metathesized with various metal salts resulted in ionic liquids and energetic salts (**23a–h**) (Scheme 6). The yields of products are usually excellent.

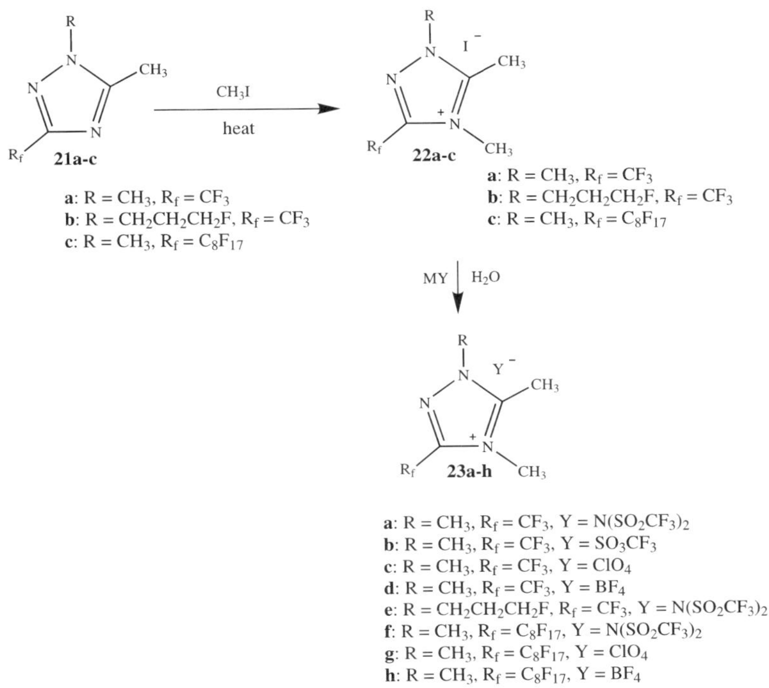

Scheme 6

Characterization of these salts was achieved by elemental analyses, IR, multinuclear NMR, and GC mass spectra. The bis(trifluoromethanesulfonyl)-amide salts (**23a,e,f**) are insoluble in water, which helped in purification. The solubilities of these salts in organic solvents increase with increasing dielectric constant of the solvent. Their melting points decrease in the order $SO_3CF_3^- > BF_4^- > ClO_4^- > N(SO_2CF_3)_2^-$ [40], which is attributable to the nature of the diffuse charge on the latter anion. Salts with larger cations have lower melting points due to less efficient packing in the solid phase.

Reactions of 1-alkyl-1,2,4-triazoles with polyfluoroalkyl halides resulted in quaternization of the triazole ring at N-4 to yield **24a–n** (Scheme 7). Metathe-

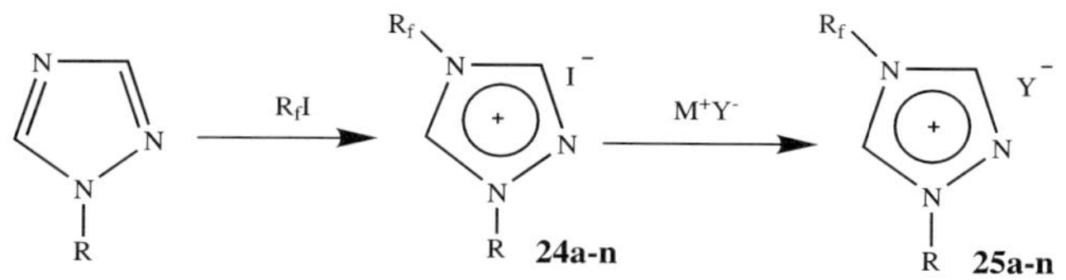

Comp.	R	R_f	Y	*Tm* (°C)[a]	*T*(°C)[b]
25a	CH_3	$(CH_2)_2CF_3$	$N(SO_2CF_3)_2$	-58 (Tg)	376
25b	$(CH_2)_3CH_3$	$(CH_2)_2CF_3$	$N(SO_2CF_3)_2$	-67 (Tg)	395
25c	$(CH_2)_6CH_3$	$(CH_2)_2CF_3$	$N(SO_2CF_3)_2$	-67 (Tg)	402
25d	$(CH_2)_9CH_3$	$(CH_2)_2CF_3$	$N(SO_2CF_3)_2$	-68 (Tg)	400
25e	$(CH_2)_3CH_3$	$(CH_2)_2(CF_2)_3CF_3$	$N(SO_2CF_3)_2$	69	394
25f	$(CH_2)_6CH_3$	$(CH_2)_2(CF_2)_3CF_3$	$N(SO_2CF_3)_2$	55	396
25g	$(CH_2)_9CH_3$	$(CH_2)_2(CF_2)_3CF_3$	$N(SO_2CF_3)_2$	53	396
25h	CH_3	$(CH_2)_2(CF_2)_5CF_3$	$N(SO_2CF_3)_2$	62	381
25i	$(CH_2)_3CH_3$	$(CH_2)_2(CF_2)_5CF_3$	$N(SO_2CF_3)_2$	53	396
25j	$(CH_2)_6CH_3$	$(CH_2)_2(CF_2)_5CF_3$	$N(SO_2CF_3)_2$	22	405
25k	$(CH_2)_3CH_3$	$(CH_2)_2(CF_2)_3CF_3$	PF_6	296	357
25l	$(CH_2)_6CH_3$	$(CH_2)_2(CF_2)_3CF_3$	PF_6	298	354
25m	$(CH_2)_3CH_3$	$(CH_2)_2(CF_2)_5CF_3$	PF_6	289	361
25n	$(CH_2)_3CH_3$	$(CH_2)_2(CF_2)_3CF_3$	SO_3CF_3	173	391

[a] Phase transition temperature. [b] Thermal degradation.

Scheme 7

sis of **24a–n** with various metal salts gave excellent yields of salts which include the bis(trifluoromethanesulfonyl)amide salts of 1-methyl-4-(3,3,3-trifluoropropyl)-1,2,4-triazolium (**25a**), 1-butyl-4-(3,3,3-trifluoropropyl)-1,2,4-triazolium (**25b**), 1-heptyl-4-(3,3,3-trifluoropropyl)-1,2,4-triazolium (**25c**), 1-decyl-4-(3,3,3-trifluoropropyl)-1,2,4-triazolium (**25d**), 1-butyl-4-(1*H*,1*H*,2*H*,2*H*-perfluorohexyl)-1,2,4-triazolium (**25e**), 1-heptyl-4-(1*H*,1*H*,2*H*,2*H*-perfluorohexyl)-1,2,4-triazolium (**25f**), 1-decyl-4-(1*H*,1*H*,2*H*,2*H*-perfluorohexyl)-1,2,4-triazolium (**25g**), 1-methyl-4-(1*H*,1*H*,2*H*,2H-perfluorooctyl)-1,2,4-triazolium (**25h**), 1-butyl-4-(1*H*,1*H*,2*H*,2*H*-perfluorooctyl)-1,2,4-triazolium (**25i**), 1-heptyl-4-(1*H*,1*H*,2*H*,2*H*-perfluorooctyl)-1,2,4-triazolium (**25j**); the hexafluorophosphate salts of 1-butyl-4-(1*H*,1*H*,2*H*,2*H*-perfluorohexyl)-1,2,4-triazolium (**25k**), 1-heptyl-4-(1*H*,1*H*,2*H*,2*H*-perfluorohexyl)-1,2,4-triazolium (**25l**), 1-butyl-4-(1*H*,1*H*,2*H*,2*H*- perfluorooctyl)-1,2,4-triazolium (**25m**); and the trifluoromethane-sulfonate salt of 1-butyl-4-(1*H*,1*H*,2*H*,2*H*-perfluorohexyl)-1,2,4-triazolium (**25n**) (Scheme 7) [41,42]. Salts containing longer alkyl and polyfluoroalkyl substituents have lower melting points because of the less efficient packing in the solid. The densities were found to be higher with elongation of the fluoroalkyl substituent and with higher fluorine concentration. Cations with bulkier alkyl groups resulted in salts of lower density.

A high-yield, efficient procedure to synthesize functionalized alkyl/fluoroalkyl-containing triazolium quaternary salts and triazolium ionic liquids led to triazolium cations that contained covalently bound anionic sites, such as, fluorocarboxy, sulfonate, fluorohomoallylic, and fluoroalkanol groups [43, 44]. These were converted by metathetical reactions with fluorine-containing anions to low-melting salts (**26a,b, 27, 28a,b, 29c,d, 30**) (Scheme 8) [43]. These functionalized triazolium salts exhibit many characteristics associated with excellent ionic liquids with respect to stability in air and water, and when heated. All are liquids at 25 °C.

A synthetic route to 1-alkyl-1,2,4-triazolium 4-nitroimides was developed based on alkylation of the metal salts of 4-nitramino-1,2,4-triazole with halo- and dihaloalkanes (Scheme 9) [44]. These salts are 1-methyl-1,2,4-triazolum 4-nitroimide (**31a**), 1-ethyl-1,2,4-triazolum 4-nitroimide (**31b**), 1-(2-oxapropyl)-1,2,4-triazolium 4-nitroimide (**32a**), 1-ethoxycarbonylmethyl-1,2,4-triazolium 4-nitroimide (**32b**), 1-methoxymethyl-1,2,4-triazolium 4-nitroimide (**33a**), 1-(4-fluoro-4,4-dinitro-2-oxabutyl)-1,2,4-triazolium 4-nitroimide (**33b**), 1-(2,3-dihydroxypropyl)-1,2,4-triazolium 4-nitroimide (**34**), bis(nitroimido-1,2,4-triazolium-1-yl)ethane (**35**), 1,3-bis(4-nitroimido-1,2,4-triazolium-1-yl)-2-oxapropane (**36**), and 1,3-bis(4-nitroimido-1,2,4-triazolium-1-yl)-3-oxapentane (**37**).

These salts are solid and were characterized by IR, NMR and elemental analysis. In the IR spectra the characteristic absorption bands due to an N-nitroimido group bonded to a heterocycle are observed at 1280–1300 and 1390–1415 cm^{-1}. The nitramino group that would have shown absorption bands at 1550–1620 cm^{-1} is absent. In the ^{1}H NMR spectra, the signals

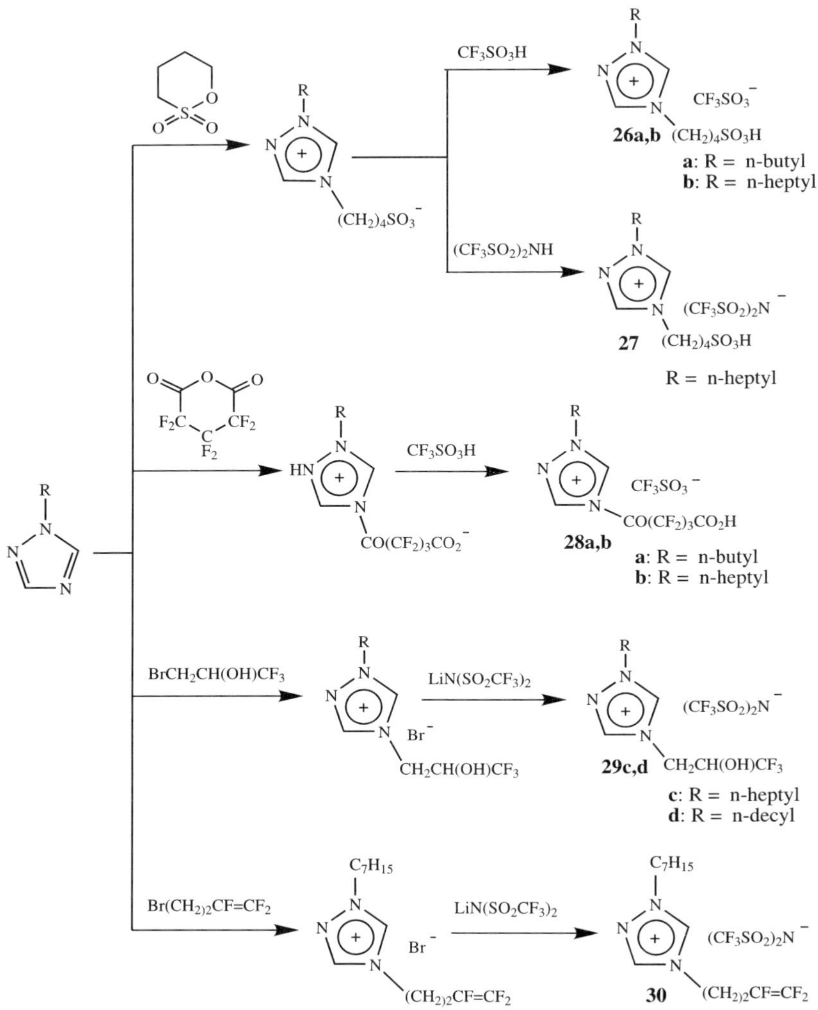

Scheme 8

for the protons of the triazole ring are nonequivalent and are shifted downfield relative to the signals for the protons in the starting material which is evidence that supports the imide structure. This conclusion is supported by the ^{13}C NMR data. However, no thermal or physical properties were reported.

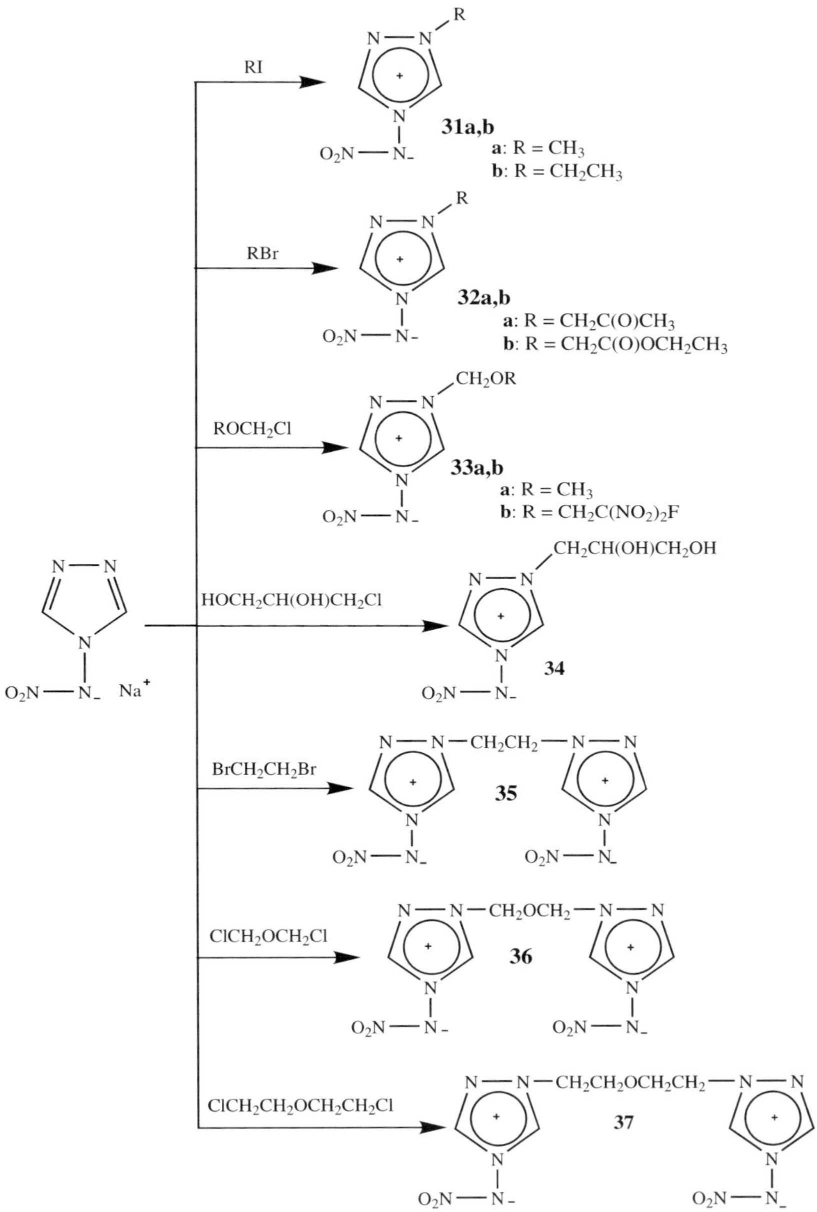
RI
31a,b
a: R = CH3
b: R = CH2CH3
RBr
32a,b
a: R = CH2C(O)CH3
b: R = CH2C(O)OCH2CH3
ROCH2Cl
33a,b
a: R = CH3
b: R = CH2C(NO2)2F
HOCH2CH(OH)CH2Cl
34
BrCH2CH2Br
35
ClCH2OCH2Cl
36
ClCH2CH2OCH2CH2Cl
37

Scheme 9

2.5
1,2,4-Triazoliun Azolate Heterocycles

Lithium dicyano-1,2,3-triazolate was reported as a useful electrolyte (**38**) [45]. 1-Butyl-3-methylimidazolium 3,5-dinitro-1,2,4-triazolate (**39**) (Fig. 2) (m.p. 32 °C, T_d239 °C) resulted from the metathesis of 1-butyl-3-methylimidazolium halide with potassium 3,5-dinitro-1,2,4-triazolate [46–49]. However, no energetic characterization was carried out for this material. Some novel ionic liquids made up of azolium cations and anions were also reported. These salts are: 1-ethyl-3-methylimidazolium 1,2,4-triazolate (**40**) (T_g – 76 °C, T_d207 °C, η 60.2 *cP* at 25 °C) and tetrazolate (**41**) (T_g – 89 °C, η 42.5 *cP* at 25 °C) (Fig. 2). Both **40** and **41** were prepared by the coupling reactions of 1-ethyl-3-methylimidazolium hydroxide with triazole or tetrazole, respectively [50].

On the basis of the fact that both 4,5-dinitro-imidazole and 5-nitro-tetrazole are strong NH acids ($pK_a = 0.8$ for 5-nitro-tetrazole) [51], energetic salts with various 1,2,4 triazolium derivatives as cations and 4,5-dinitro-imidazolate and 5-nitrotetrazolate anions were obtained [52]. 4,5-Dinitro-imidazole and 5-nitro-tetrazole were reacted with various N-4-substituted derivatives of 1,2,4-triazoles to produce energetic ionic salts including: 1,2,4-triazolium 4,5-dinitro-imidazolate (**42a**), 1-methyl-1,2,4-triazolium 4,5-dinitro-imidazolate (**42b**), 3-azido-1,2,4-triazolium 4,5-dinitro-imidazolate (**42c**), 1-methyl-3-azido-1,2,4-triazolium 4,5-dinitro-imidazolate (**42d**), 1,2,4-triazolium 5-nitro-tetrazolate (**43a**), 1-methyl-1,2,4-triazolium 5-nitrotetrazolate (**43b**), 3-azido-1,2,4-triazolium 5-nitro-tetrazolate (**43c**), and 1-methyl-3-azido-1,2,4-triazolium 5-nitro-tetrazolate (**43d**) (Scheme 10), as well as 4-amino-1,2,4-triazolium 4,5-dinitro-imidazolate (**42e**) and 4-amino-1,2,4-triazolium 5-nitro-tetrazolate (**43e**).

Densities and thermochemical characteristics of the azolium azolates are given in Table 4. When 5-nitro-tetrazolate was used as the anion, the melting points of the compounds were lower than those of the corresponding

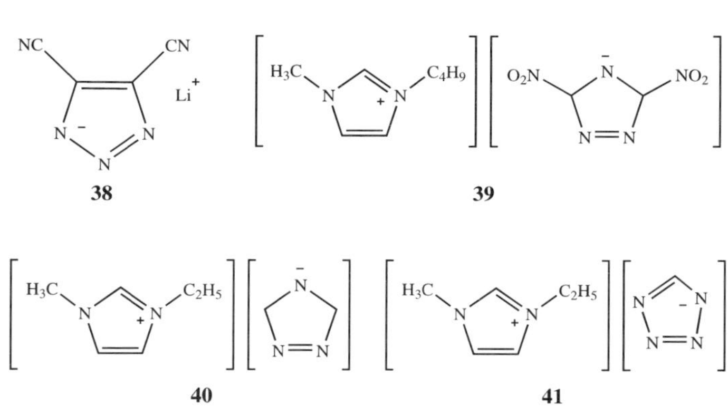

Fig. 2 Ionic salts containing azole ions

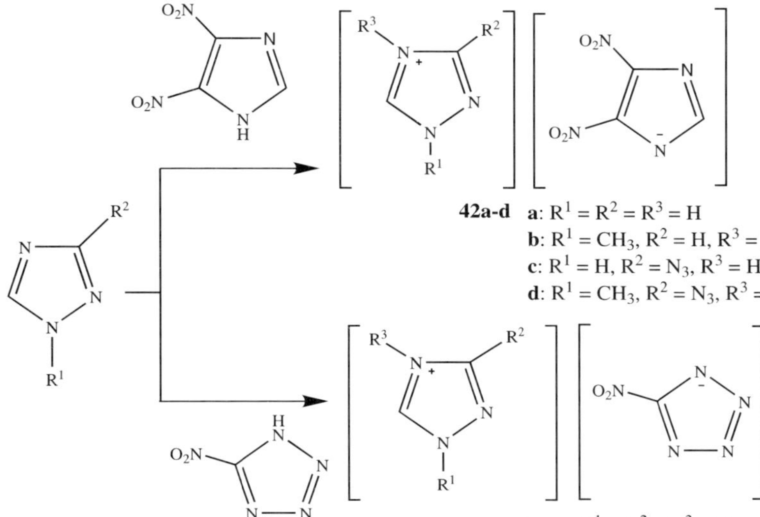

42a-d a: $R^1 = R^2 = R^3 = H$
b: $R^1 = CH_3$, $R^2 = H$, $R^3 = H$
c: $R^1 = H$, $R^2 = N_3$, $R^3 = H$
d: $R^1 = CH_3$, $R^2 = N_3$, $R^3 = H$

43a-d a: $R^1 = R^2 = R^3 = H$
b: $R^1 = CH_3$, $R^2 = H$, $R^3 = H$
c: $R^1 = H$, $R^2 = N_3$, $R^3 = H$
d: $R^1 = CH_3$, $R^2 = N_3$, $R^3 = H$

Scheme 10

Table 4 Density and thermal characteristics of triazolium azolate salts

Compound	Anion	Tm [a] °C	Td [b] °C	α [c] %	Density (calc.) g/cm^3	ΔH_f° [c] kJ/mol
42a	4,5-dinitroimidazolate	156	165	32	1.73	232.1
42b	4,5-dinitroimidazolate	102	150	25	1.66	200.6
42c	4,5-dinitroimidazolate	92	158	33	1.70	593.5
42d	4,5-dinitroimidazolate	80	145	26	1.60	560.5
42e	4,5-dinitroimidazolate	137	149	30	1.65	348.7
43a	5-nitrotetrazolate	137	183	25	1.53	436.2
43b	5-nitrotetrazolate	62	163	18	1.52	402.7
43c	5-nitrotetrazolate	– 35	161	26	1.53	800.9
43d	5-nitrotetrazolate	– 38	141	19	1.45	768.5
43e	5-nitrotetrazolate	102	190	23	1.58	545.2

[a] melting point
[b] thermal degradation temperature
[c] oxygen coefficient
[d] standard heat of formation (calculated by Gaussian 03)

4,5-dinitro-imidazolates, while the thermal decomposition temperatures were higher. The opportunity for hydrogen bonding, in addition to modified packing effects and reduced lattice energies, with the anion is reduced in methyl-substituted compounds, resulting in lower melting points. The densities exceed 1.45 g/cm^3 for all the compounds. Energetic salts that incorporate the nitro group substantially improve the oxygen balance of the corresponding compounds and eventually result in higher heats of combustion and detonation processes. The oxygen coefficients (α) [53] for the new salts are between 0.18 and 0.33 (Table 4) and are within the range reported for known energetic compounds [54].

3
Tetrazolium-Based Heterocycles

Tetrazoles are unsaturated five-membered heterocycles that contain four nitrogen atoms in the ring. The enthalpies of energetic chemical systems are governed by their molecular structure. Moving from imidazole ($\Delta H_f{}^\circ$, + 58.5 kJ/mol) to 1,2,4-triazole ($\Delta H_f{}^\circ$, + 109.0 kJ/mol) to tetrazole ($\Delta H_f{}^\circ$ + 237.2 kJ/mol) [35], there is an increasing positive trend in heats of formation. Since the generation of molecular nitrogen as an end-product of propulsion or explosion is highly desired in order to avoid environmental pollution and health risks, as well as to reduce detectible plume signatures, compounds containing a backbone of directly linked nitrogen atoms (catenated nitrogen) are of great interest.

3.1
Tetrazolium Heterocycles Containing Amino and Azido Substituents

Aminotetrazoles have a high nitrogen content and, despite their large positive enthalpies of formation [55], are thermally stable [56]. They are, therefore, prospective high energy materials. The synthesis of various tetrazolium salts containing amino substituents on the ring via reactions of 1-amino-5-methyl tetrazole, and 2-amino-5-methyl tetrazole with iodomethane resulted in quaternary salts (**44**) and (**46**). Metathesis with $AgNO_3$ or $AgClO_4$ formed 1-amino-4,5-dimethyltetrazolium nitrate (**45a**) or perchlorate (**45b**), and 2-amino-4,5-dimethyltetrazolium nitrate (**47a**) or perchlorate (**47b**) in high yields (Scheme 11) [30, 31]. Elemental analyses, and IR, multinuclear NMR, and mass spectral analyses support the syntheses (Table 5).

A high yield synthesis of 5-amino-tetrazolium nitrate (**48**) is obtained from the reaction of 5-amino-tetrazole with nitric acid (Scheme 12) [57]. Its characterization was accomplished through IR, Raman, and multinuclear (^{1}H, ^{13}C, ^{15}N) NMR spectroscopy, and DSC. Bomb calorimetry, sensitivity measurements, and ab initio calculations were also performed. From the combined

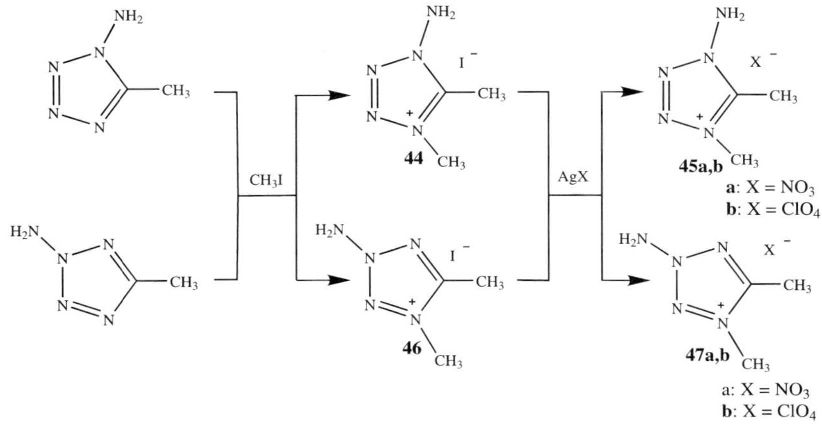

Scheme 11

Table 5 Densities and thermal characteristics of substituted amino, aminomethyl, and polymethyl tetrazolium salts

Compound	Anion	Tm [a] °C	Td [b] °C	Density (calc.) g/cm^3	ΔH_f° [c] kJ/mol
45a	NO_3^-	- 59	170	1.50 (1.55)	141.1
45b	ClO_4^-	51	182	1.81 (1.78)	183.8
47a	NO_3^-	94	173	– (1.55)	132.0
47b	ClO_4^-	140	238	1.65 (1.71)	179.5
50a	NO_3^-	121	181	1.51 [d]	174.6 [e]
50b	$N(NO_2)_2^-$	85	184	1.72 [d]	385.6 [e]
50c	N_3^-	135	137	1.42 [d]	676.6 [e]

[a] melting point
[b] thermal degradation temperature
[c] standard heat of formation (calculated by Gaussian 03)
[d] X-ray (calc)
[e] [59]

experimental and theoretical calculations, **48** is predicted to be a powerful and promising explosive with a good oxygen balance and low sensitivity.

The syntheses of 1,5-diamino-4-methyltetrazolium salts via quaternization, e.g., of 1,5-diamino-1*H*-tetrazole with iodomethane, gave 1,5-diamino-4-methyl-tetrazolium iodide (**49**) in 86% yield [58–63]. Subsequent metathesis of **49** with silver nitrate, silver dinitramide, or silver azide led to 1,5-diamino-4-methyl-tetrazolium nitrate (**50a**), 1,5,-diamino-4-methyltetrazolium dinitramide (**50b**), or 1,5-diamino-4-methyltetrazolium azide (**50c**) in good yields. Most of these salts exhibited good thermal stability. Salt **50b** as

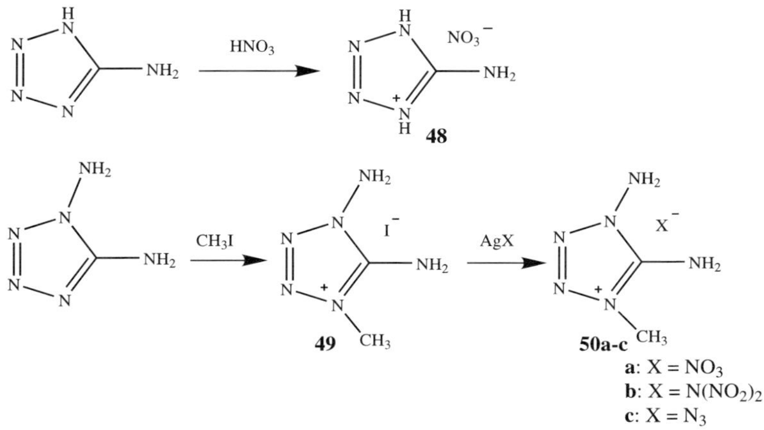

Scheme 12

well as **45a, 45b**, and **47a** melt at < 100 °C which qualifies them as energetic ionic liquids.

These salts were characterized by IR, Raman and NMR spectroscopy, mass spectrometry, elemental analysis, X-ray, and initial safety testing (impact and friction sensitivity). Low impact sensitivities were demonstrated. Densities and thermochemical characteristics of substituted amino, amino-methyl, and polymethyl tetrazolium salts are summarized in Table 5. All of these new salts exhibit thermal stabilities > 170 °C based on *DSC/TGA* studies (except the azide). The densities of 1-amino-4,5-dimethyl tetrazolium perchlorate (**45b**) and 1-methyl-4,5-diamino tetrazolium dinitramide (**50b**) are markedly higher than the others.

Synthesis and characterization of 1-nitrotetrazolato-2-nitro-2-azapropane (**52**) and 1,4-dimethyl-5-aminotetrazolium 5-nitrotetrazolate (**53**) are reported in high yields (Scheme 13). Compounds **52** and **53** are white powders and were characterized by single crystal X-ray analysis. Compound **52** explodes when thrown in a flame. Both **52** and **53** have low impact sensitivity [64, 65].

Nitrogen-rich energetic heterocycles, N-nitroso- (**57a,c**) and N-nitraminotetrazole (**58a–c**) were synthesized from the corresponding aminotetrazoles (**56a–c**) by the direct nitration with acetic anhydride/HNO_3 (Scheme 14). The conversion of **57a,c** with peroxytrifluoroacetic acid yielded the corresponding nitraamines in high yields (**58a**: 82%, **58c**: 80%). Compounds **56a,c**, **57a,c**, **58a–c** were characterized by single crystal X-ray analysis. The calculated heats of formation for **58a** and **58c** were found to be positive (11.7 and 357.0 kJ mol^{-1}, respectively) and the calculated detonation velocity of 5988 ms^{-1} (**58a**) is similar to the values of TNT and nitroglycerin [66].

The chemistry of 5-azidotetrazole, 5-azido-1-phenyltetrazole, and tetrazolylpentazole has been reported. The 5-pentazolyl and 5-azido derivatives

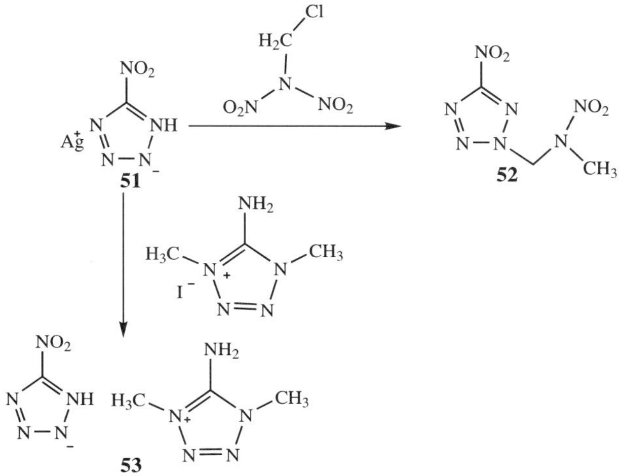

Scheme 13

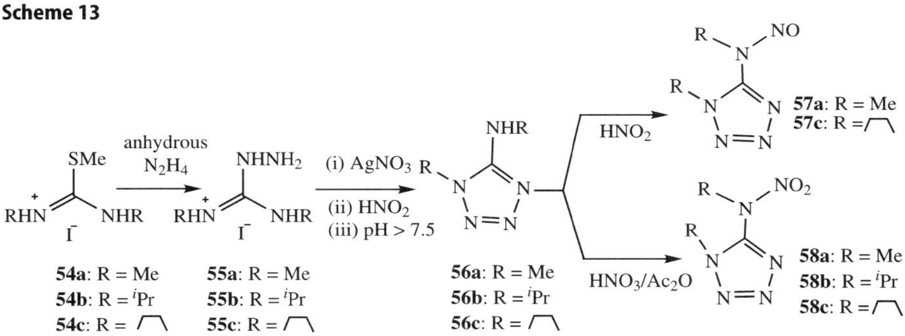

Scheme 14

(**60a,b**) of tetrazole include large amounts of nitrogen at 90.6 and 88.3 weight percent. While the pentazole compound decomposes at low temperature, 5-azido tetrazole is stable at room temperature but extremely sensitive (and should be synthesized in only small amounts). 5-Azido-1-phenyl tetrazole (**60b**) is found to be stable and characterized by single crystal X-ray analysis (Scheme 15) [63, 67].

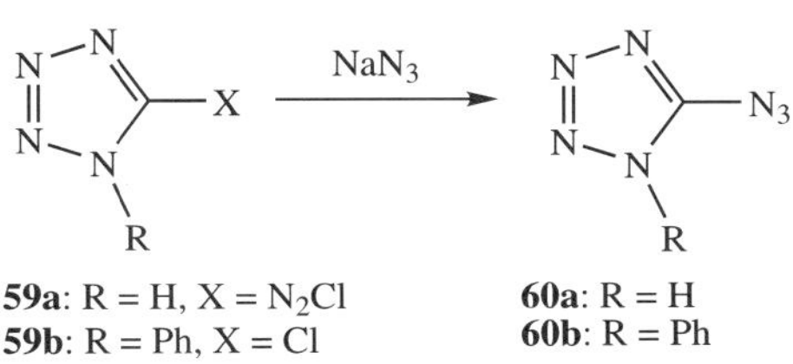

59a: R = H, X = N_2Cl
59b: R = Ph, X = Cl

60a: R = H
60b: R = Ph

Scheme 15

3.2 Tetrazolium Heterocycles Containing 5,5′-Azotetrazolate Anions

In the 1890s, Thiele first prepared sodium 5,5′-azotetrazolate (**61**) from oxidation of 5-amino-tetrazole with potassium permanganate [68–71]. Reaction of strontium and barium chlorides with an aqueous solution of **61** gave strontium 5,5′-azotetrazolate (**62a**) and barium 5,5′-azotetrazolate (**62b**), respectively [72]. Both salts were insoluble in water. Barium 5,5′-azotetrazolate (**62b**), was used with various metal sulfates to produce the corresponding 5,5′-azotetrazolate salts (**63a–d, 64a,b,** and **65a–f**), in water (Scheme 16). These metal salts often contain water of crystallization, and loss of water has been observed during storage. Upon loss of water, the sensitivity

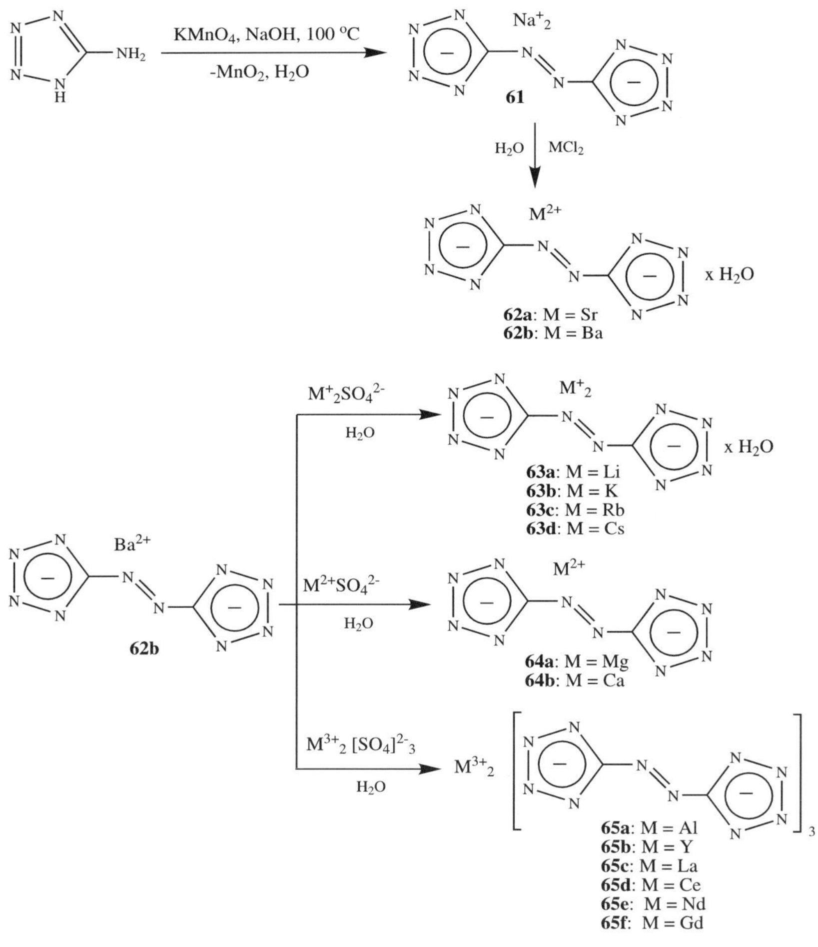

Scheme 16

of the compounds to shock and friction increased drastically. The salts were characterized by IR, Raman, and NMR spectroscopy, and their thermal properties were studied by DSC and TGA. X-ray crystal studies on sodium 5,5′-azotetrazolate pentahydrate (**61**), barium 5,5′-azotetrazolate pentahydrate (**62b**), lithium 5,5′-azotetrazolate hexahydrate (**63a**), rubidium 5,5′-azotetrazolate dihydrate (**63c**), calcium 5,5′-azotetrazolate octahydrate (**64b**), yttrium 5,5′-azotetrazolate docosahydrate (**65b**), and gadolinium 5,5′-azotetrazolate hydrate (**65f**) show that the water molecules were either in the coordination sphere of the cation or bound by hydrogen bonds. The azotetrazolate ion is not connected to the harder cations such as calcium or yttrium, and the salt decomposed in water, due to its acidity, to form tetrazolhydrazine with concomitant evolution of nitrogen. The free acid 5,5′-azotetrazolate, synthesized from the sodium salt with HBF_4, was decomposed within seconds at room temperature but could be retained at – 30 °C.

Salts of 5,5′-azotetrazolate with protonated nitrogen bases (e.g., ammonium guanidinium and triaminoguanidinium) are unique gas-generating agents producing little smoke or residue, which may lead to a variety of applications, including gas generators and explosives [73, 74]. Salts of the 5,5′-azotetrazolate dianion with various methylated ammonium (**66a–e**) and hydrazinium (**67a–d,f**) cations were obtained from the reactions of barium 5,5-azotetrazolate (**62b**) with appropriate ammonium and hydrazinium sulfates [73–78]. However, N,N,N-trimethylhydrazinium 5,5′-azotetrazolate, $[(CH_3)_3N\text{-}NH_2]_2^+\ [N_4C-N=N-CN_4]^{2-}$ (**67e**) was obtained by the metathesis of trimethylhydrazinium iodide and silver 5,5-azotetrazolate (Scheme 17).

Other 5,5′azotetrazolates reported include salts that contain cations such as: bis(hexahydropyridazinium) (**67g**), bis[N-amino-1-azonia-cyclohexane] (**67h**), bis[ethylene-dihydrazinium (+1)] (**67i**), ethylenedihydrazinium (+2) (**67j**), bis[N-amino-*N′*,*N′*-dimethyl-guanidinium] (**67k**), diguanidinium (**67l**), bis(triaminoguanidinium) (**67m**), and bis(azidoformamidinium) (**67n**) (Fig. 3) [76, 79]. These salts were characterized by IR, Raman, and NMR spectroscopy, and elemental analyses; X-ray crystal structures of **66a,c,e**, **67a,e,j**, and **67l** were determined.

The salts were insensitive to shock, friction, or electric discharge (∼ 20 kV). Detonation was not observed in either the drop hammer test (5 kg, 50 cm) or when the salts were ground forcefully in a mortar. None of these salts melts, but rather decomposes at specific temperatures with rapid gas evolution. Above the decomposition point, explosion occurred upon rapid heating to give nitrogen gas as the main product, especially if the compound was compressed before heating. The decomposition temperature decreased for salts with NH^+ functions as the number of methyl groups increased. The hydrazine salts also produced large amounts of hydrogen, which decreased with the increasing number of methyl groups. The ammonium compounds (**66a–c**) produced only small amounts of hydrogen, whereas none was detected for **66d,e** [76]. The formation of methane was observed for all salts.

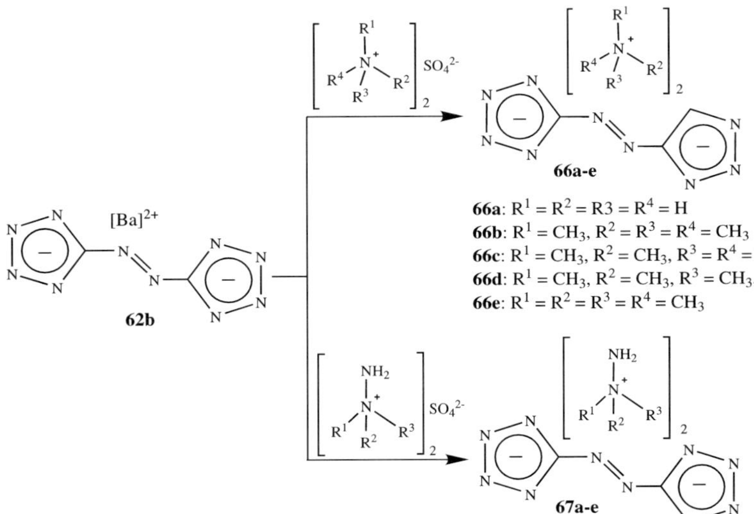

66a: $R^1 = R^2 = R3 = R^4 = H$
66b: $R^1 = CH_3$, $R^2 = R^3 = R^4 = CH_3$
66c: $R^1 = CH_3$, $R^2 = CH_3$, $R^3 = R^4 = H$
66d: $R^1 = CH_3$, $R^2 = CH_3$, $R^3 = CH_3$, $R^4 = H$
66e: $R^1 = R^2 = R^3 = R^4 = CH_3$

67a: $R^1 = R^2 = R^3 = H$
67b: $R^1 = CH_3$, $R^2 = R^3 = H$
67c: $R^1 = CH_3$, $R^2 = CH_3 = R^3 = H$
67d: $R^1 = C_2H_5$, $R^2 = R^3 = H$
67e: $R^1 = C(CH_3)_3$, $R^2 = R^3 = H$
67f: $R^1 = HOCH_2CH_2$, $R^2 = R^3 = H$

Scheme 17

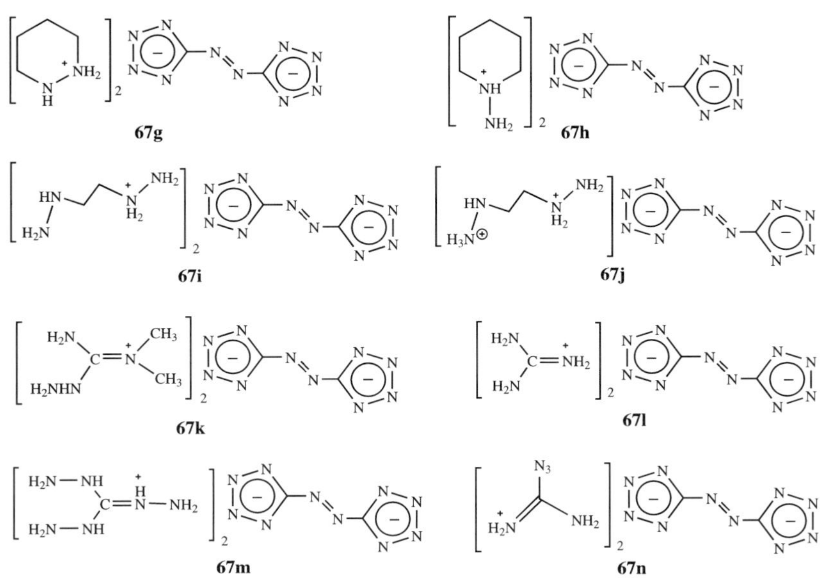

Fig. 3 Salts of 5,5′-azotetrazolate dianions with hydrazinium derivative cations

Heats of formation of hydrazinium [75], ammonium 5,5′-azotetrazolates [79], and guanidinium [24, 80], were measured as $\Delta H_f^\circ = +858$, $+443$, and $+410\ \mathrm{kJ\,mol^{-1}}$, respectively. In order to compare these properties with azotetrazolates that contain heterocyclic cations, imidazolium, triazolium, and tetrazolium derivatives were synthesized and characterized (Scheme 18) [81]. Surprisingly, while most azotetrazolates exhibit melting points in excess of 160 °C, [bis(1-butyl-3-methyl-imidazolium)] 5,5′-azotetrazolate (**68a**) was a liquid at 25 °C with a melting point of 3 °C, similar to its 3,5-dinitrotetrazolate analog [46].

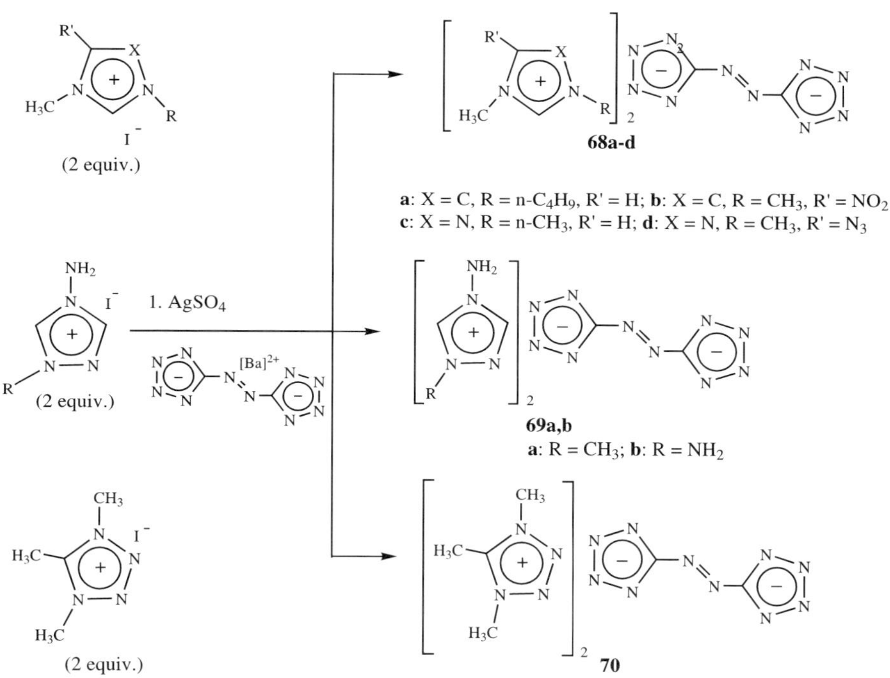

Scheme 18

Although the quaternized salts of 1,4 dimethyl-3-azido triazolium (**68d**), bis(1,4-diamino-1,2,4-triazolium) (**69b**), and 1,4,5-trimethyl tetrazolium (**70**) 5,5′-azotetrazolate have lower nitrogen content than the bis(triaminoguanidinium) derivative, their heats of formation are much higher at + 1852.4, + 1620.0, and + 1334.8 kJ/mol, respectively, with densities of ~ 1.46–1.59 g cm^{-3} (Table 6). The structure of **69a**, determined by single crystal X-ray analysis, shows that the unit cell was packed as a layered structure with hydrogen bonds and an interlayer distance of 3.04 Å [82–87]. In contrast to the metal azotetrazolates [72], none of these salts (except **70**) was solvated. Compounds (**68b–70**) decomposed violently upon melting. Microwave of **69a**

Table 6 Densities and thermal characteristics of azotetrazolate derivatives

Compound	Tm [a] °C	Density (calc.) g/cm^3	ΔH_f° [b] kJ/mol
67l	262	1.54	486.5
67m	209 [c]	1.60	1171.2
67n	142 [c]	1.62	–
68a	3	1.26	896.9
68b	145	1.54	1094.2
68c	182	1.42	1136.2
68d	155	1.55	1852.4
69a	180	1.57	1129.5
69b	–	1.59	1620.0
70	189	1.46	1334.8

[a] melting point
[b] standard heat of formation (calculated by Gaussian 03)
[c] thermal decomposition temperature

at 200 °C caused a violent decomposition leading to the formation of a carbon black powder. With the exception of **69b**, which spontaneously evolved nitrogen gas, most of the salts were stable at room temperature for at least two months.

3.3 Bistetrazolate Heterocycles

Although there are several reports of 5,5′-bistetrazolate [82–87] and iminobis(5-tetrazolate) [88–90] salts, most are described in the patent literature. Some processes have been developed for preparing various 5,5′-tetrazolate salts, which were very useful as slightly toxic and easy to handle gas-generating agents for air bags and as high molecular weight foaming agents. In methanol at reflux, bistriazole or iminobis(5-tetrazole) readily quaternized 4-amino-1,2,4-triazole to prepare [bis(4-amino-triazolium)] 5,5′-bistetrazolate (**71a**) or [bis(4-amino-triazolium)] iminobis(5-tetrazolate) (**71b**) (Scheme 19) [81].

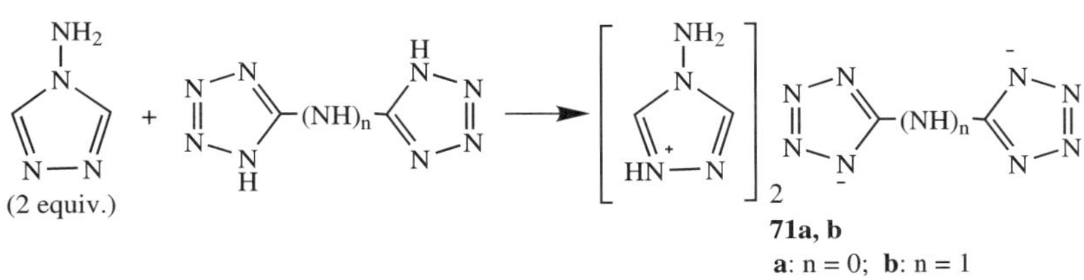

Scheme 19

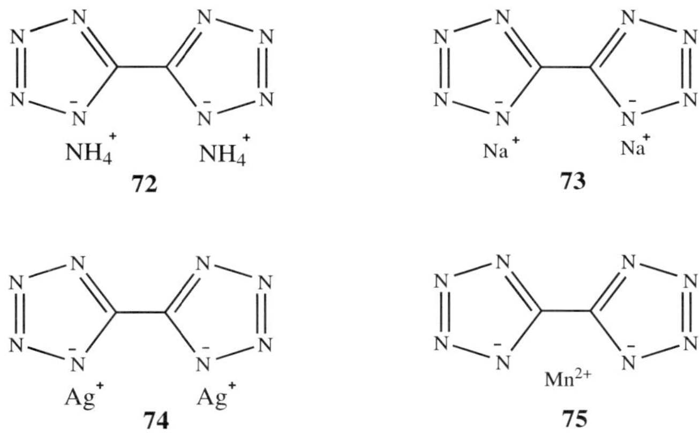

Fig. 4 5,5′-bistetrazolate salts

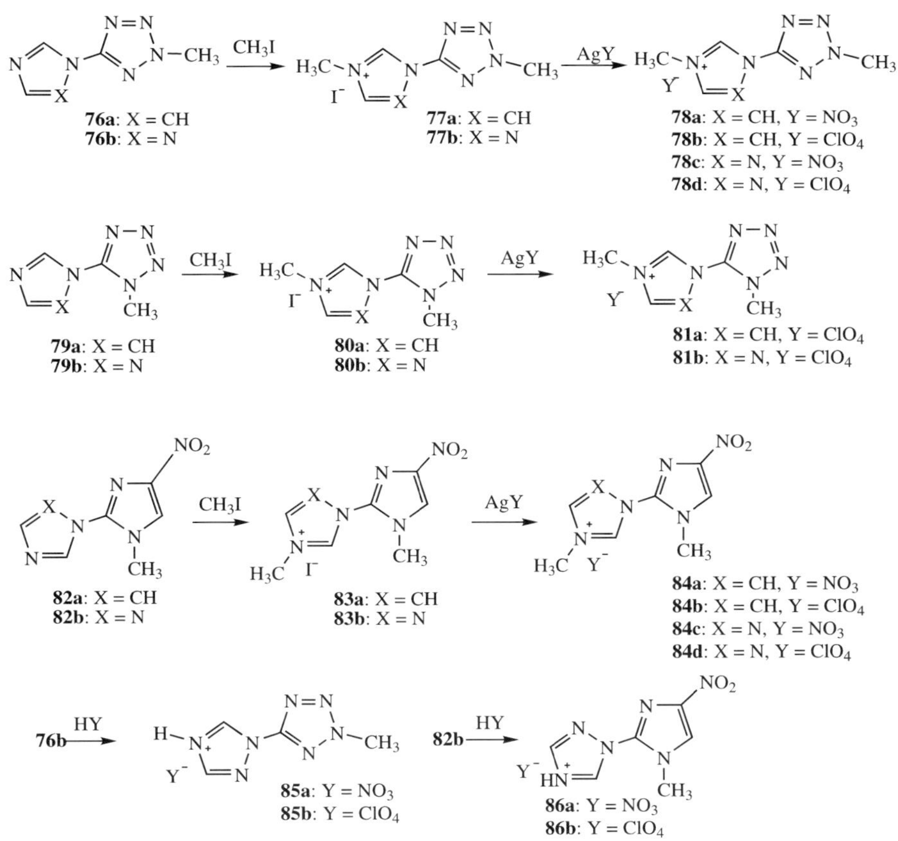

Scheme 20

The diammonium (**72**), disodium (**73**), disilver (**74**), and manganese(II) (**75**) 5,5′ bistetrazolates are also known (Fig. 4) [82–87]. Chemistry of copper complexes with bis(tetrazolyl)amine has been also studied and some of them are of interest as additives in pyrotechnics and ammonium perchlorate-based propellants [86].

The chemistry of various nitrogen-rich energetic bicyclic azolium salts has been described. The thermally stable bicyclic azolium salts (**78a–d, 81a–b, 84a–d, 85a,b, 86a,b**) with densities ranging between 1.52–1.67 g/cm^3, were prepared by quaternization with nitric acid or perchloric acid or

Table 7 The structures and properties of bicyclic azolium salts

Compound	Y	No	Tm [a] °C	Td [b] °C	*d* g/cm^3	$\Delta_f H_m$ [c] kJ/mol
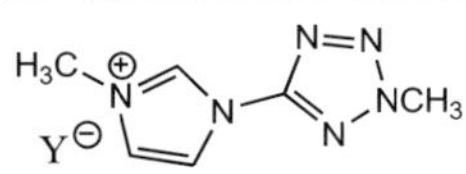	NO_3	**78a**	79	175	1.519(1.531)	–
	ClO_4	**78b**	110	283	1.631(1.653)	–
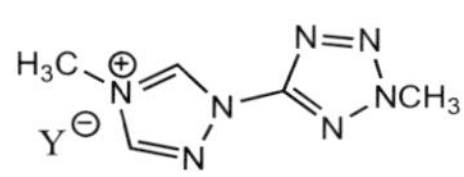	NO_3	**78c**	151	151	1.515(1.574)	310.8
	ClO_4	**78d**	141	293	1.639(1.695)	357.7
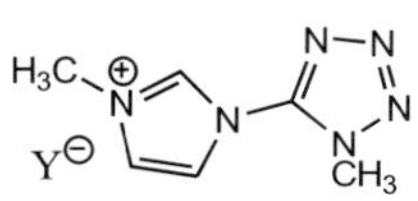	ClO_4	**81a**	159	295	1.648(1.653)	–
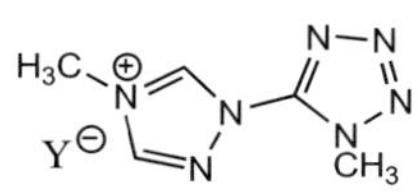	ClO_4	**81b**	152	240	1.645(1.695)	412.3
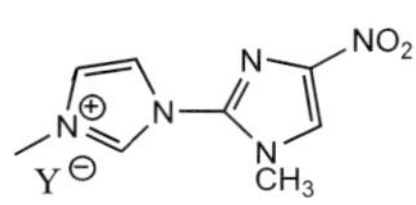	NO_3	**84a**	125	190	1.581(1.574)	–
	ClO_4	**84b**	169	289	1.641(1.678)	–
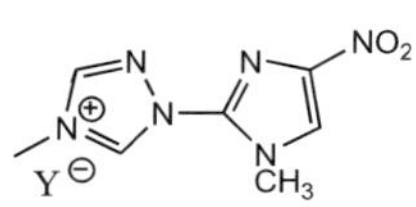	NO_3	**84c**	–	145	1.566(1.612)	–
	ClO_4	**84d**	227	282	1.652(1.716)	–
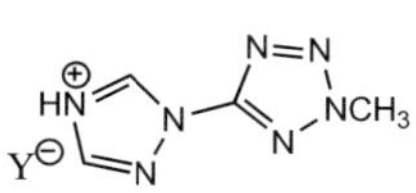	NO_3	**85a**	154	166	1.565(1.644)	341.8
	ClO_4	**85b**	128	229	1.648(1.773)	385.5
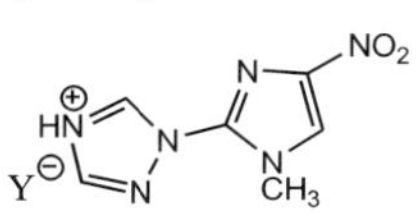	NO_3	**86a**	153	165	1.617(1.676)	209.9
	ClO_4	**86b**	–	175	1.674(1.784)	345.6

[a] melting point
[b] thermal decomposition temperature
[c] standard heat of formation (calculated by Gaussian 03)

iodomethane followed by metathesis reactions with silver nitrate and silver perchlorate (Scheme 20, Table 7). The structures of **81b** and **86b** were confirmed by single-crystal X-ray analysis [91].

4
Urotropinium-Based Heterocycles

The chemistry of high-energy density materials (HEDM) containing only C, H, N, and O atoms is of great interest. Urotropine, a nitrogen-rich cage molecule, when paired with energetic anions, formed energetic salts. The initial preparation of urotropinium salts was carried out in the 1950s [92], and subsequently the syntheses of a variety of them were reported [93]. Urotropinium nitrate (**87**) [colorless crystals, m.p. 157–161 °C (dec.), density (from X-ray) 1.47 g/cm^3] was prepared by the reaction of urotropine with nitric acid. N-methylurotropinium azide (**88a**) [white solid, m.p. 165–170 °C (dec.), density 1.4 g/cm^3], N-methylurotropinium dinitramide (**88b**) [white solid, m.p. 121–124 °C (dec.), density 1.46 g/cm^3], and N-methylurotropinium azotetrazolate (**88c**) [yellow crystals, m.p. 181–184 °C (dec.), density 1.46 g/cm^3] were prepared from either the corresponding iodide or sulfate (Scheme 21). Because of the high sensitivity and explosive nature of anhydrous silver azide, an alternative route for **88a** using sodium azide with N,N-dimethylurotropinium diiodide was utilized [94]. All the salts

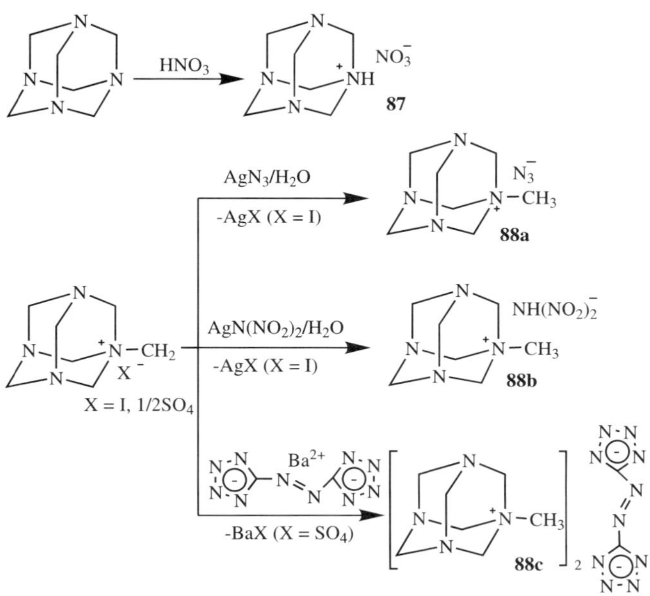

Scheme 21

were handled easily, were insensitive to air or light, and were soluble in polar organic solvents. They were characterized by analytical and spectroscopic (IR, Raman, ^{1}H, ^{13}C, ^{14}N NMR) methods, and X-ray diffraction techniques.

The X-ray crystal structure of **87** (monoclinic, space group $P2_1/C$) consisted of a urotropinium cation linked to a planar nitrate group via hydrogen bridges. The structures of **88a–c** showed separation of anions and cations without contact. The azide ion in **88a** (monoclinic, space group $P2_1/m$) is linear; the dinitramide ion in **88b** (monoclinic, space group $P2_1$) is asymmetric; and the azotetrazolate anion in **88c** (monoclinic, space group C_2/m) is planar [93].

The syntheses of urotropinium and N-methylurotropinium salts were extended to include several other energetic organic and inorganic anions, such as 3,5-dinitro-pyrazolate, 4,5-dinitro-imidazolate, 3,5-dinitro-1,2,4-triazolate, 5-nitro-tetrazolate, perchlorate, nitrate, and azide [94]. In methanol solution, urotropine was found to react readily with 3,5-dinitro-pyrazole, 4,5-dinitro-imidazole, 3,5-dinitro-1,2,4-triazole, and 5-nitro-tetrazole to form urotropinium 3,5-dinitro-pyrazolate (**89a**), 4,5-dinitro-imidazolate (**89b**), 3,5-dinitro-1,2,4-triazolate (**89c**), and 5-nitro-tetrazolate (**89d**). Reaction of silver salts of 3,5-dinitro-1,2,4-triazole, nitric, perchloric, and hydrofluoric acid with N-methylurotropinium iodide led to N-methylurotropinium 3,5-dinitro-1,2,-triazolate (**90a**), nitrate (**90b**), perchlorate (**90c**), and fluoride (**90d**)

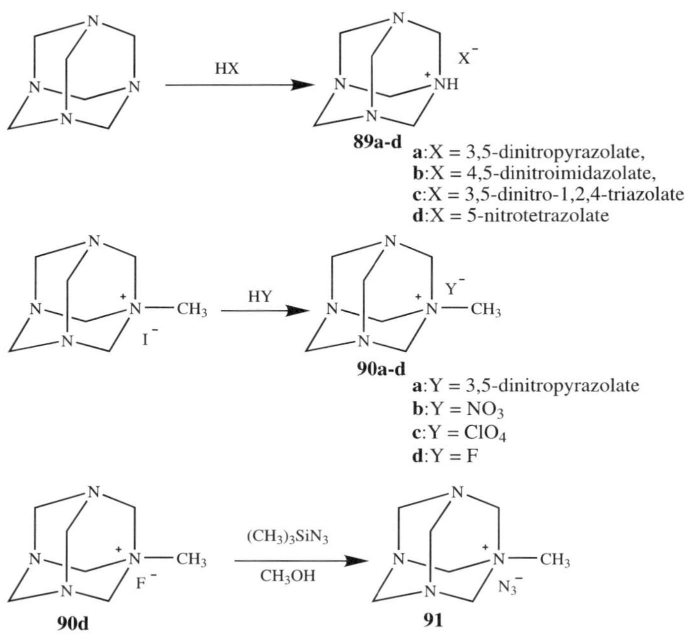

Scheme 22

Table 8 Phase transition, decomposition temperature, density, nitrogen content (N), oxygen coefficient (α), and thermochemical results for urotropinium salts at 298.2 K

Compound	Tm [a] °C	Td [b] °C	Density (calc) g/cm^3	N [c] %	α %	ΔH_f° [e] kJ/mol
89a	180	184	1.56	37.6	16	299.0
89b	183	183	1.41	37.6	16	254.6
89c	177	182	1.72	42.1	18	268.7
89d	166	169	1.48	49.4	10	435.3
90a	170	202	1.45	40.2	16	250.7
90b	194	196	1.42	32.2	14	- 30.6
90c	200	205	1.47	22.0	–	14.5
91	165	176	1.35	49.7	0	468.8

[a] melting point;
[b] thermal decomposition temperature;
[c] nitrogen content;
[d] oxygen coefficient;
[e] standard heat of formation (calculated by Gaussian 03)

salts. N-methylurotropinium azide (**91**) is made by the reaction of **90d** with $(CH_3)_3SiN_3$ (Scheme 22, Table 8) [94]. The structure of **90b** was confirmed by single crystal X-ray analysis.

5 Tetrazine-Based Heterocycles

There has been considerable interest in the study of the reactivity and properties of various tetrazine derivatives. The 1,2,4,5-tetrazine ring system is electroactive and has a high electron affinity. Tetrazines possess high positive heats of formation and large crystal densities—properties important in energetic materials applications. Additionally, they seem to be insensitive to destructive stimuli such as friction, impact, and electrostatic discharge.

The synthesis and properties of various ionic 1,2,4,5-tetrazine explosives and energetic materials including 3,6-diguanidino-1,2,4,5-tetrazine diperchlorate (**92a**), 3,6-diguanidino-1,2,4,5-tetrazine dinitrate (**92b**), 3,6-diguanidino-1,2,4,5-tetrazine-1,4-di-N-oxide diperchlorate (**93a**), 3,6-diguanidino-1,2,4,5-tetrazine-1,4-di-N-oxide dinitrate (**93b**), disilver 3,6-bis-nitroguanyl tetrazine (**94**), bistriamino-guanidinium 3,6-bis-nitro-guanyl tetrazine (**95**), 3,6-dihydrazino-1,2,4,5-tetrazine diperchlorate (**96a**), 3,6-dihydrazino-1,2,4,5-tetrazine dinitrate (**96b**), 3,6-dihydrazino-1,2,4,5-tetrazine bis-(dinitramide) (**96c**), and 3,6-dihydrazino-1,2,4,5-tetrazine bis-dinitroimidazolate (**97**) were reported (Fig. 5) [95–98].

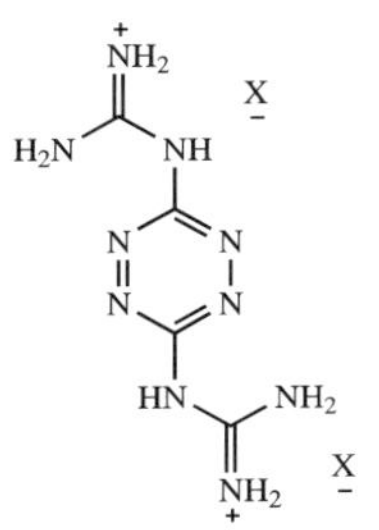

92a,b
a: X = ClO_4
b: X = NO_3

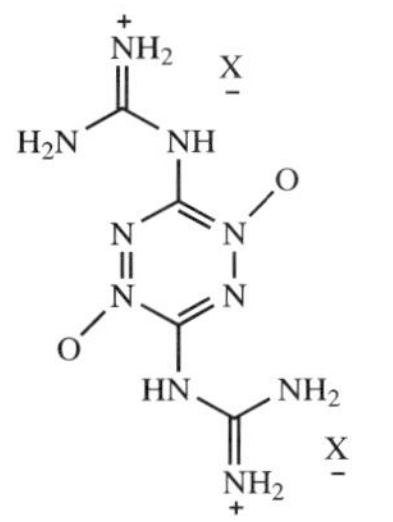

93a,b
a: X = ClO_4
b: X = NO_3

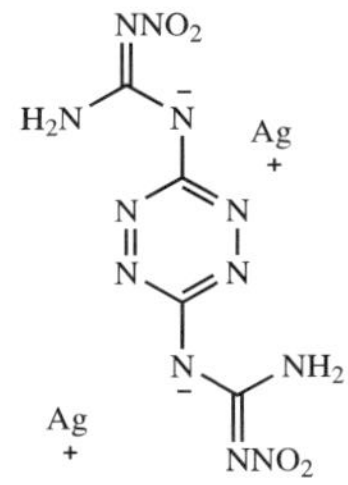

94

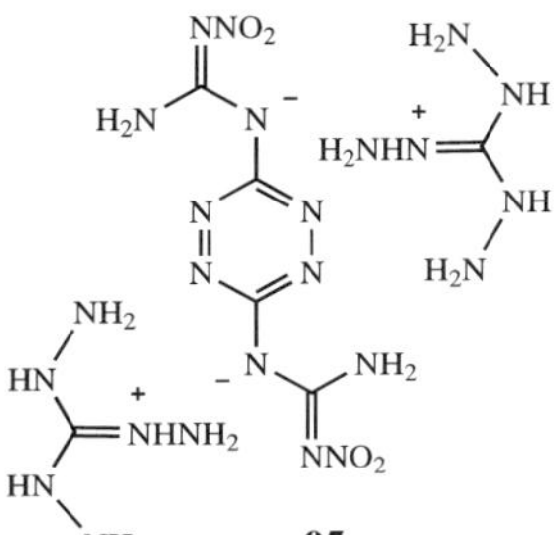

95

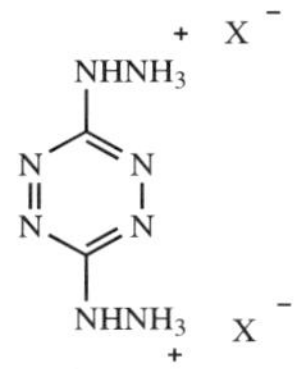

96a-c
a: X = ClO_4
b: X = NO_3
c: X = $N(NO_2)_2$

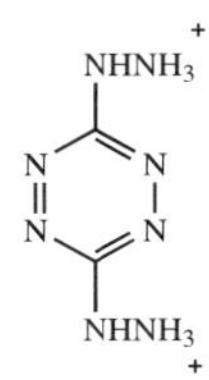

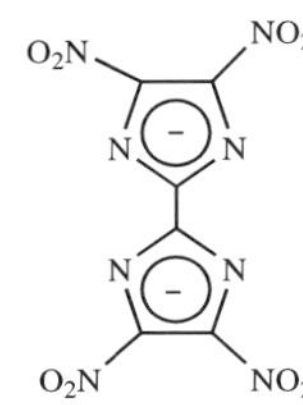

97

Fig. 5 Tetrazine-based energetic salts

Energetic salts **92a,b**, **93a,b**, and **96a,b** were synthesized by the reaction of nitric acid and perchloric acid with 3,6-diguanidino-1,2,4,5-tetrazine, 3,6-diguanidino-1,2,4,5-tetrazine-1,4-di-N-oxide, and 3,6-dihydrazino-1,2,4,5-tetrazine, respectively, whereas **96c** was made by the reaction of dinitramide with 3,6-dihydrazino-1,2,4,5-tetrazine.

Synthesis of **94** and **95** involved the reaction of the disodium salt of 3,6-bisnitroguanyl tetrazine with $AgNO_3$ and triaminoguanidinium hydrochloride, respectively.

Poly-rho tests, which are single shot experiments to determine detonation velocity as a function of density, were performed on **92a,b**. When **92a** is formulated with 5 weight percent of Kel-F 800 binder, a maximum pellet density of 1.79 g/cm^3 was obtained. At this density, the detonation velocity measured was 8.07 km/s. The dinitrate derivative (**92b**) was formulated with 3 weight percent Estane binder and 3 weight percent nitro-plasticizer. At a maximum pellet density of 1.60 g/cm^3, a detonation velocity of 7.31 km/s was obtained. Salt **93b** was found to be unstable based on DSC analysis, whereas salt **93a** showed improved thermal stability but less than **93a** [95]. The crystal density of **95** was reported to be 1.61 g/cm^3, and the heat of formation was 300 ± 2 kcal/mol [97]. The densities of **96a–c**, **97** are reported to be in the range of 1.80 to 1.96 g/cm^3 [96]. When these salts were heated over the temperature range 40–500 °C at a scan rate of 20 °C/min, **96a–c** left no residue, whereas **97** left an orange powder. In DSC studies, **96a, 96c** exhibited their major exotherm between 152 and 164 °C, respectively. Salt **96b** exhibited an exotherm at 190 °C, which is about 30 °C higher than **96a, 96c**. Salt **97** showed the highest exotherm at 220 °C [96]. All tetrazine-based salts seem to have interesting explosive performance and extraordinary combustion properties [95–98].

The synthesis and properties of several high-nitrogen materials with 3-amino-6-nitroamino-tetrazine (ANAT) as the anion are reported (Scheme 23). These salts (**99, 100, 103a,b**) provide a new and easy approach to highly energetic salts. All these salts are relatively dense (> 1.55 g/cm^3) and exhibit good thermal stability (T_d > 148 °C). The calculated detonation velocities and detonation pressures are comparable to those of explosives such as Tetryl, PETN, TATB, and RDX. A combination of theoretical and empirical calculations shows that all these salts have high molar enthalpies of formation (Table 9) [99].

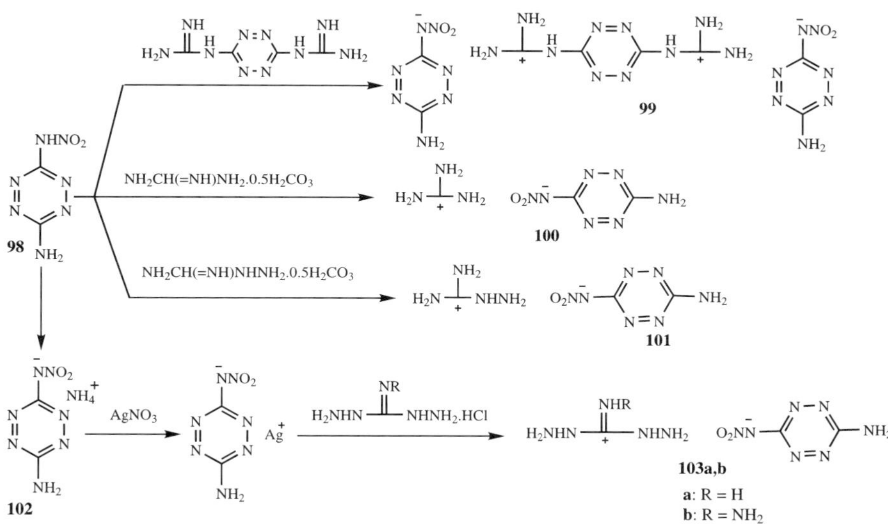

Scheme 23

The synthesis of 3,3′-azobis(6-amino-1,2,4,5-tetrazine) (DAAT) (**104**) is reported in a 20% yield after six reaction steps (Scheme 24). DAAT is a high nitrogen energetic material with remarkable thermal stability and insensitivity against friction and impact. DAAT decomposes at relatively high temperatures (> 250 °C) releasing a very large amount of heat. A high nitrogen content of 70% in combustion with a low oxygen balance of – 72.67% and a high heat of formation of + 1035 kJ/mol measured by combustion calorimetry are some of the promising properties of DAAT which

Table 9 Structure and properties of energetic salts with ANAT anion

Compound	Td[a] °C	Density g/cm^3	ΔH_m °[b] kJ/mol
98	164.0	1.82	441.0 [c]
99	232.3	1.56	1088.8
100	248.1	1.62	340.7
101	205.4	1.71	443.2
102	174.0	1.63	370.0
103a	147.7	1.56	564.2
103b	163.5	1.59	671.5

[a] thermal degradation temperature
[b] standard heat of formation (calculated by Gaussian 03)
[c] calculated heat of formation in the gas phase

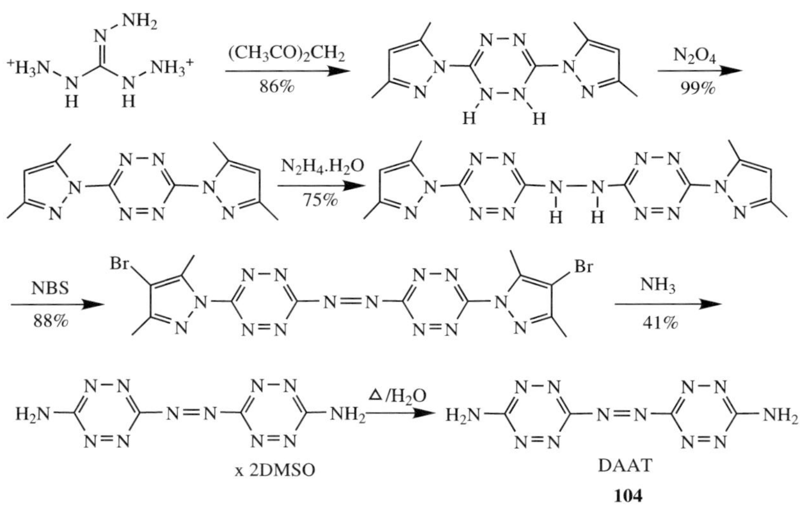

Scheme 24

make it interesting as a new energetic component. The density of DAAT is 1.76 g/cm^3 [100, 101].

The synthesis, characterization, and energetic properties of various heterocycles containing azide groups are reported. Compound **105** is prepared by the reaction of cyanuric chloride with sodium azide. The preparation of **106** utilizes the conversion of a pyrazole derivative by reaction with hydrazine to form the hydrazo intermediate which is subsequently reacted with $NaNO_2$. 2,5,8-Trichloro-s-heptazine is converted quantitatively into 2,5,8-triazido-s-heptazine (**107**) with trimethylsilyl azide. The nucleophilic reaction of hydrazine with 4,4′,6,6′-tetra(chloro)hydrazo-1,3,5-triazine gave 4,4′,6,6′-tetra(hydrazine)hydrazo-1,3,5-triazine which when diazotized gave **108**. Reaction of **108** with $CHCl_3/Cl_2$ in acetonitrile led to the formation of **109** (Scheme 25) [102–106]. The novel heterocyclic system with the phenanthrene-type skeleton **111**, in which the benzene ring is annulated with 1,2,3,4-tetrazene-1,3-di-N-oxide rings, is of considerable interest in the context of the high-nitrogen system stability and from a heteroaromaticity stand point. The step-by-step synthetic approach to this system involves treatment of **110** with N_2O_5 resulting in the first 1,2,3,4-tetrazene 1,3-dioxane ring. Subsequent repeated treatment with N_2O_5 led to the formation of the second 1,2,3,4-tetrazene 1,3-dioxane ring (Scheme 26). The structure of **111** was confirmed by ^{13}C and ^{14}NMR spectral studies. Compound **111** was also converted to **112** by the reaction with methylamine. Compound **111** begins to melt with decomposition at 140 °C, and above 210 °C; compound **112** begins to decompose without melting [107].

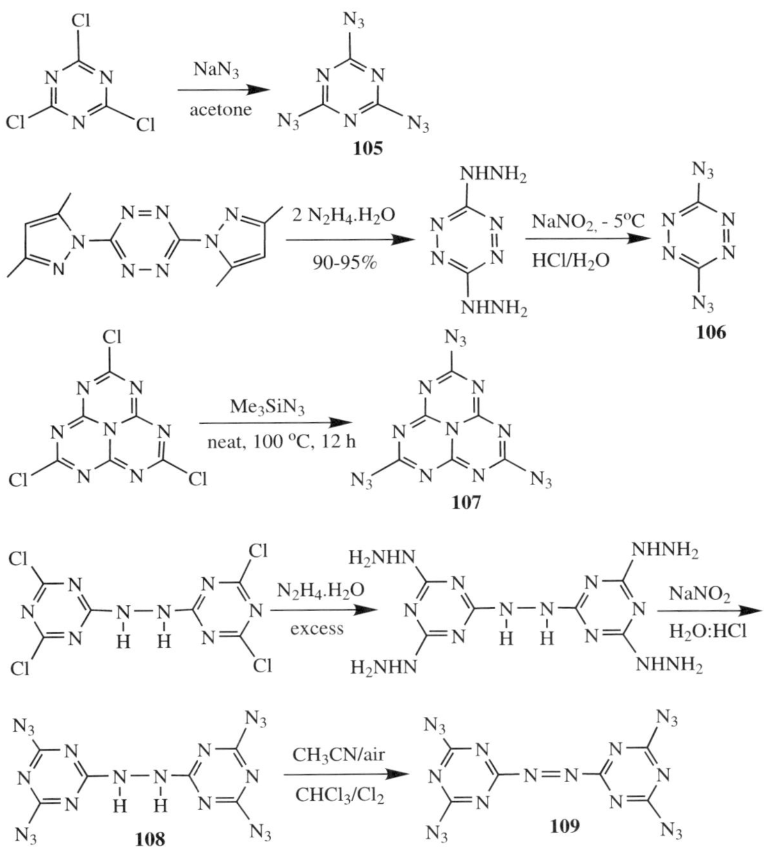

Scheme 25

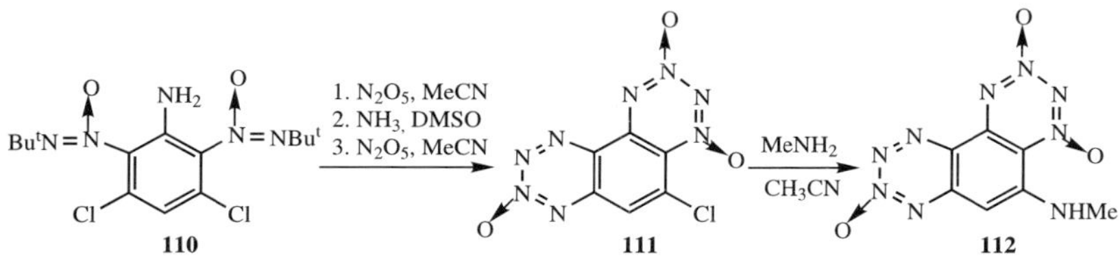

110 111 112

Scheme 26

6 Azetidinium-Based Heterocycles

Azetidine-based explosives, such as 1, 3, 3-trinitroazetidine [108, 109] demonstrate excellent performance partly because of the high strain associated with the four-membered ring. The basicity ($pKb = 6.5$) of 3,3-dinitroazetidine [110] allowed the preparation of a variety of solid energetic 3,3-dinitroazetidinium salts with high oxygen-balance [107], including 3,3′-dinitro-azetidinium nitrate (**113a**), 3,3′-dinitro-azetidinium 2,4-dinitroimidazolate (**113b**), 3,3′-dinitroazetidinium dinitramide (**113c**), 3,3′-dinitro-azetidinium 5-nitro-1,2,4-triazol-3-onate (**113d**), 3,3′-dinitro-azetidinium 3,5-dinitro-1,2,4-triazolate (**113e**), and 3,3′-dinitro-1-isopropylazedinium dinitramide (**114**) (Fig. 6). These salts were synthesized in 82–95% yields, either by mixing the free base 3,3′-dinitroazetidine with the appropriate acid or by metathesis of 3,3′-dinitroazetidinium trifluoromethanesulfonate [110] with ammonium salts of the acid. They were characterized by elemental analyses, IR, and ^{13}C NMR. Densities and thermal characteristics are tabulated in Table 10. All the salts were subjected to small-scale thermal and sensitivity tests [111].

113a-e

114

a: nitrate, **b**: 2, 4-dinitro imidazolate
c: dinitramide, **d**: 5-nitro-1,2,4-triazol-3-onate
e: 3, 5-dinitro-1, 2, 4-triazolate

Fig. 6 Azetidinium-based energetic salts

The single crystal X-ray structures of 3,3′-dinitroazetidinium dinitramide (**113c**) (orthorhombic, space group $C_{mc}2_1$) and 1-isopropyl-3,3-dinitroazetidinium dinitramide (**114**) (orthorhombic, space group Pbca) were also reported [112]. The latter was formed when an attempt to crystallize 3,3-dinitroazetidinium nitrate from acetone was being made. X-ray structures have confirmed that the conformation adopted by the dinitramide ions in **113c** is quite different from the bend, twist, and torsion angles for the

Table 10 Densities and thermochemical properties of azetidinium salts

Compound	Density g/cm^3	Tm [a] °C	ΔH_f° [b] kJ/mol
113a	1.76	142	-260 ± 8
113b	1.65	151	-75 ± 50
113c	1.79	139	-34 ± 16
113d	1.76	161	-201 ± 4
113e	1.70	148	$+84 \pm 8$
114	1.78	154	$+21 \pm 63$ [c]

[a] melting point
[b] [111] [c] [112]

dinitramide ion in **114**. The possible reason for these dramatic differences in the conformations observed for **113c** and **114** is claimed to be due to the different symmetries for this ion found in the two structures, as well as the absence of hydrogen bonding interaction in **114** [112]. Dehydration of 3,3-dinitroazetidinium nitrate with acetic anhydride provided an alternate route for the synthesis of 1,3,3-trinitroazetidine [111]. The synthesis and characterization of ^{15}N-labeled isomers of 3,3-dinitroazetidinium nitrate were also reported [113].

7
Picrate-Based Heterocycles

While anhydrous picric acid is unstable, and its impact and friction sensitivities are higher than those of trinitrotoluene, many organic and inorganic picrates have been reported [9]. The picrate anion, when combined with high-nitrogen azolium cations, formed energetic salts with rather high positive heats of formation. By taking advantage of the acidic properties of picric acid, energetic salts containing picrate and dipicrate anions have been prepared [114]. Scheme 27 depicts synthetic routes to salts composed of azolium cations with monopictrate as the anion, including the following picrate salts: 1,2,4-triazolium (**115a**), 1-methyl-1,2,4-triazolium (**115b**), 4-amino-1,2,4-triazolium (**116a**), and 5-amino-tetrazolium (**116b**); and 1,2,4-triazolyl-3-ammonium (**117**), 1,4-dimethyl-1,2,4-triazolium (**118a**), and 1,4-dimethyl-3-azido-1,2,4-triazolium (**118b**). In Scheme 28 are shown the synthetic routes to the energetic bisazolium dipicrate salts. These dipicrate salts are 1,1′-methylenebisimidazolium (**119**), 1,1′-methylene bis(3-methylimidazolium) (**120a**), and 1,1′-methylene bis(4-methyltriazolium) (**120b**) [114].

Triazolium or substituted triazolium picrates were first prepared via direct reaction of the triazole with picric acid in methanol or with silver picrate after

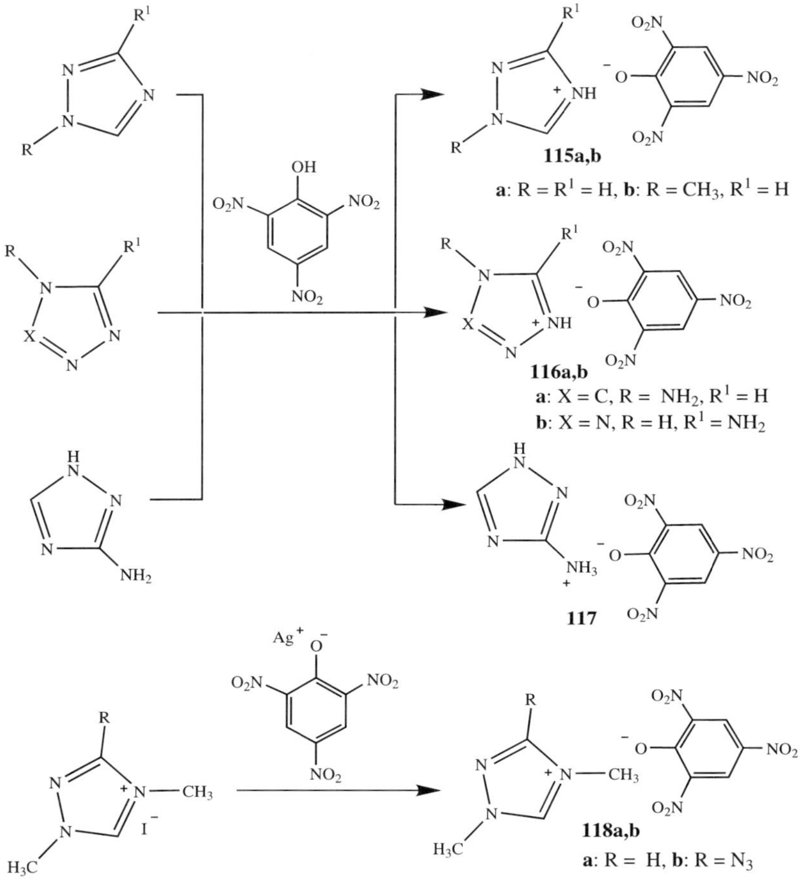

Scheme 27

quaternization with methyl iodide (Scheme 27). Bridged bis(imidazolium) or bis(tetrazolium)methane compounds were reacted with picric acid to form dipicrates, or they were quaternized and then metathesized to yield the desired salts. All of these salts were well-characterized, and X-ray crystal structures of **116a** and **116b** were also determined. Their physical characteristics and thermal properties along with those of other energetic materials which are included for comparison, are tabulated in Table 11.

Most of the picrates have good thermal stabilities and relatively high densities and oxygen balance. The bridged azolium picrates are thermally more stable than their monocationic picrate analogues. In general, the majority of these picrates were found to be more stable thermally than their nitrate analogues and less than the perchlorates. 5-Aminotetrazolium picrate (**116b**) was the densest of the picrate salts at 1.85 g/cm^3 (calc. X-ray), which

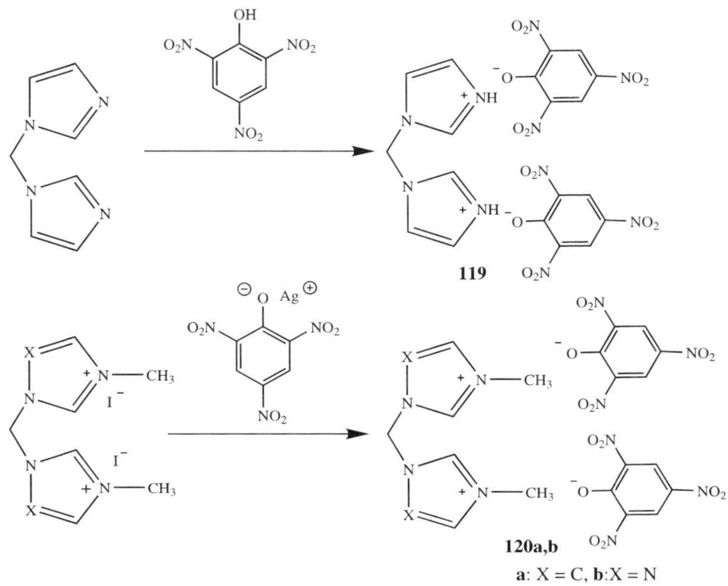

Scheme 28

Table 11 Physical characteristics and thermochemical data of picrate and dipicrate salts

Compound	Tm [a] °C	Td [b] °C	Density [c] g/cm^3	OB [d] %	ΔH_f° [e] kJ/mol
115a	169	196	1.77	- 67	259.0
115b	91	185	1.72	- 79	222.3
116a	197	228	1.64 [f]	- 66	375.0
116b	147	214	1.85 [f]	- 53	400.6
117	235	244	1.60	- 66	275.2
118a	141	271	1.80	- 91	189.5
118b	106	176	1.48	- 78	557.1
119	215	283	1.52	- 36	237.1
120a	184	313	1.63	- 93	147.4
120b	216	242	1.67	- 81	357.1
TAG-AT [g]	-	-	1.60	- 73	1075
HMX	-	-	1.90	- 21	75
Picric acid	-	-	1.77	- 42	- 213.6
RDX	-	-		- 22	83.8

[a] melting point;
[b] thermal decomposition temperature;
[c] gas pycnometer;
[d] oxygen balance = 1600 [(a+b/2-d)/FW for a compound with molecular formula CaHb-NcOd];
[e] standard enthalpy of formation (calculated by Gaussian 03);
[f] X-ray (calc);
[g] triaminoguanidinium azotetrazolate

places it between picric acid (1.77 cm/g^3) and HMX (1.90 g/cm^3). Unexpectedly protonation in **116a** occurred at one of the ring-nitrogen atoms rather than on the N-amino group. However, in contrast, in the reaction of picric acid with the C-amino triazole, a high-melting ammonium salt **(117)** was formed.

8
Imidazolium-Based Heterocycles

Nitro- and azido-substituted imidazoles, when paired with nitrate or perchlorates, form solid energetic salts in excellent yields [31]. These salts are 1,3-dimethyl-5-nitroimidazolium perchlorate **(121a)**, 1,3-dimethyl-5-nitroimidazolium nitrate **(121b)**, 1,2,3-trimethyl-5-nitroimidazolium perchlorate **(121c)**, 1,2,3-trimethyl-5-nitro-imidazolium nitrate **(121d)**, 1-ethyl-2,3-dimethyl-5-nitro-imidazolium perchlorate **(121e)**, 1-ethyl-2,3-dimethyl-5-nitroimidazolium nitrate **(121f)**, 2-azido-imidazolium perchlorate **(121g)**, and 2-azidoimidazolium nitrate **(121h)** (Fig. 7). Salts **(121a–f)** were made by the metathesis of the corresponding iodide salt with $AgClO_4$ and $AgNO_3$, whereas **121g–h** are formed by the reactions of 2-azidoimidazole with $HClO_4$ and HNO_3, respectively.

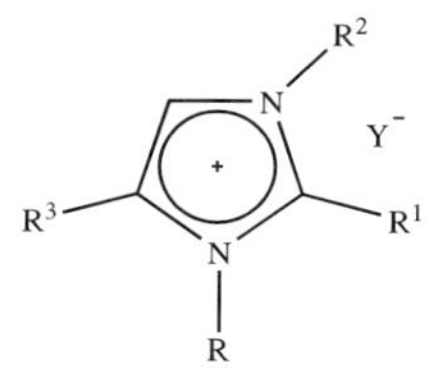

121a-h

a: R = CH_3, R^1 = H, R^2 = CH_3, R^3 = NO_2, Y = ClO_4
b: R = CH_3, R^1 = H, R^2 = CH_3, R^3 = NO_2, Y = NO_3
c: R = CH_3, R^1 = CH_3, R^2 = CH_3, R^3 = NO_2, Y = ClO_4
d: R = CH_3, R^1 = CH_3, R^2 = CH_3,R^3 = NO_2, Y = NO_3
e: R = C_2H_5, R^1 = CH_3, R^2 = CH_3, R^3 = NO_2, Y = ClO_4
f: R = C_2H_5, R^1 = CH_3, R^2 = CH_3, R^3 = NO_2, Y = NO_3
g: R = H, R^1 = N_3, R^2 = H, R^3 = H, Y = ClO_4
h: R = H, R^1 = N_3,R^2 = H, R^3 = H, Y = NO_3

Fig. 7 Imidazolium-based energetic ionic salts

All these compounds are well-characterized solids. In general, the nitrate salts have lower melting points and thermal stabilities than the corresponding perchlorates. Energetic salts containing imidazolium triazolate or tetrazolate, and tetrazolium imidazolate are described in Sect. 2.5.

9
Nitrocubane-Based Heterocycles

Highly nitrated cubanes were predicted to be shock-insensitive, very dense, high-energy compounds with great potential as explosives and propellants. Various nitrocubanes, such as penta, hexa, hepta, and octa are reported in the literature. Applying Kamlett–Jacobs equations to octanitrocubane using predicted values for density (1.9–2.2 $g\,cm^{-3}$) and ΔH_f (81–144 $Kcal\,mol^{-1}$) led to calculated detonation velocities and pressures higher than that of

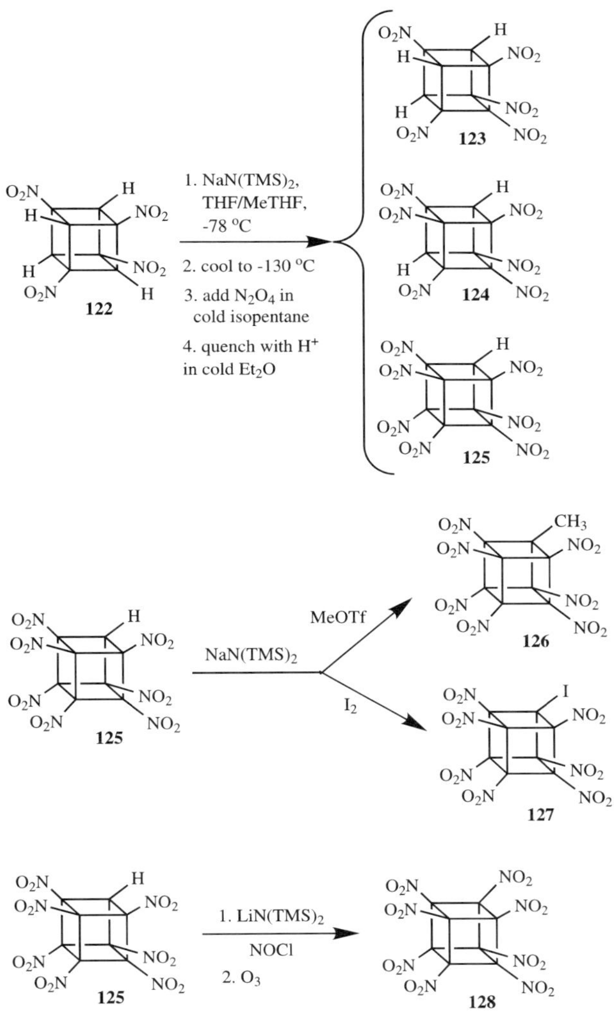

Scheme 29

trinitrotoluene (TNT), 1,3,5,7-tetranitro-1,3,5,7 tetraazacyclooctane (HMX), and 2,4,6,8,10,12-hexanitrohexaazaisowurtzitane (CL-20). The chemistry of mono- to tetranitrocubanes had been reported earlier. Eaton et al. showed that the reaction of tetranitrocubane under different reaction conditions produced penta (**123**), hexa (**124**), and hepta (**125**) nitrocubanes in variable yields (Scheme 29). Alkali metal salts of **125** were prepared at low temperature in dichloromethane by reacting with $NaN(TMS)_2$. When reacted with powerful electrophiles such as methyl triflate and iodine, methylheptanitrocubane (**126**) and iodoheptanitrocubane (**127**) were obtained in high yields. Octanitrocubane (**128**) is prepared by reacting **125** with $LiN(TMS)_2$ and excess nitrosyl chloride at – 78 °C followed by ozonation.

Heptanitrocubane (**125**) is easily soluble in polar solvents, such as acetone, THF, and dichloromethane. Single crystal X-ray analysis of **125** has provided an accurate density at 21 °C of 2.028 g cm^{-3} and showed many intermolecular contacts around each molecule. A solution of **125** in methanol turned yellow which may be due to the anion. Octanitrocubane (**128**) is a stable, white solid and somewhat soluble in hexane but readily soluble in polar organics. Characterization of **128** was based on single crystal X-ray analysis [115].

10 Miscellaneous

When 2,2,6,6-tetramethylpiperidino-boron diazide was reacted with HN_3, the salt 2,2,6,6-tetramethylpiperidinium tetraazido-borate **129** resulted (Scheme 30) [116–119].

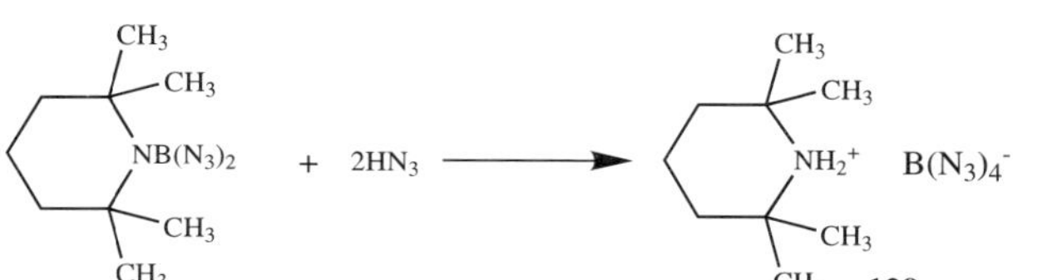

129

Scheme 30

Recently, the synthesis of a highly energetic, oxygen-balanced tetrazolium polynitratoaluminate (**130**) was reported (Fig. 8) [120]. The material is hydrolytically unstable. A microwave-assisted synthesis of an organic heterocycle azide (**131**) was reported in 88% yield (Scheme 31) [121].

Increasing the number of nitrogen atoms in heterocycles results in considerable gain in the standard enthalpy of formation in the resulting compounds. On the basis of this idea the synthesis and properties of cyclophosphazenes-containing azido and amino substituents on the phosphorus atoms were

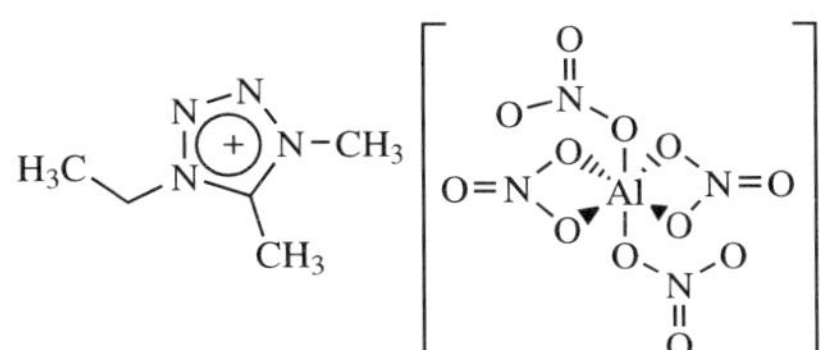

130

Fig. 8 Tetrazolium polynitratoaluminate (**30**)

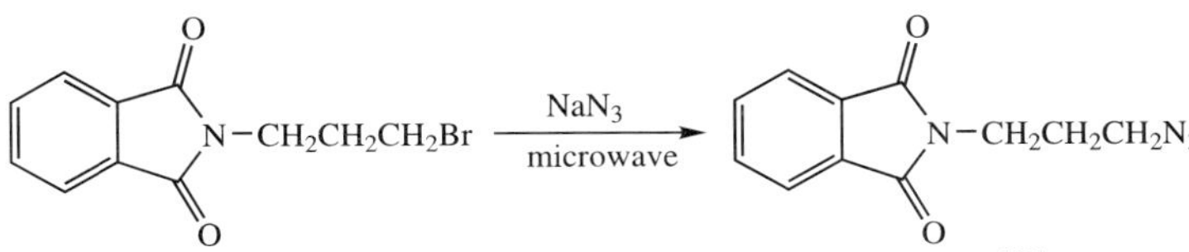

131

Scheme 31

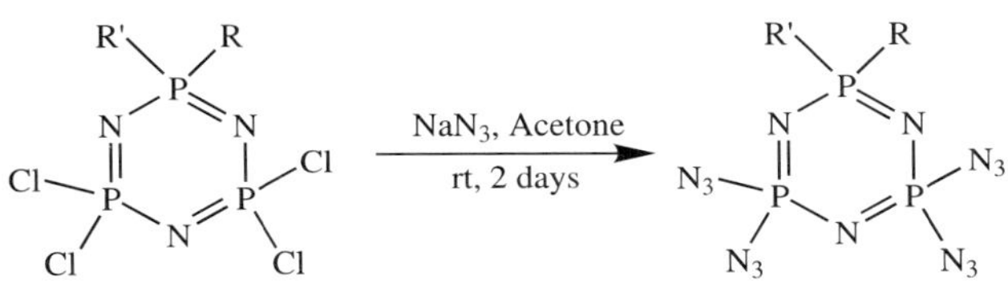

132a: R = R' = Cl
132b: R = Cl, R' = NMe_2
132c: R = R' = NMe_2

133a: R = R' = N_3
133b: R = N_3, R' = NMe_2
133c: R = R' = NMe_2

Scheme 32

studied. Reaction of **132a–c** with sodium azide in acetonitrile led to the formation of **133a–c** in very good yields (Scheme 32) [122, 123].

Compounds **133a** and **133b** are liquids, while **133c** is a solid at room temperature. Each was characterized by NMR, HRMS and IR spectroscopy. Compound **133a** is highly friction sensitive and is known to detonate. However, **133b** and **133c** are substantially more stable. The densities of **133b** and **133c** are 1.67 and 1.79 g/cm^3, respectively [122, 123]. Compound **133a** has been recently fully characterized and its single crystal X-ray structure determined [123].

Energetic heterocyclic salts containing SF_5-propyl substituents on imidazole, 4-amino-1,2,4-triazole (**134a–c, 135a–d**) and pyridine (**136**) moieties were synthesized (Scheme 33). These salts have moderate densities (> 1.4 g/cm^3). The calculated heats of formation for dinitramide-containing salts are higher than perchlorate and nitrate salts (Table 12) [124].

Reduction of 2,4,6,8,10,12-hexanitro-2,4,6,8,10,12-hexaazaisowurtzitane (**137**) with stannous chloride dihydrate is reported to give a mixture of isomeric 2,4,6,8,10- and 2,4,6,8,12-pentanitro-2,4,6,8,10,12-hexaazaisowurt-

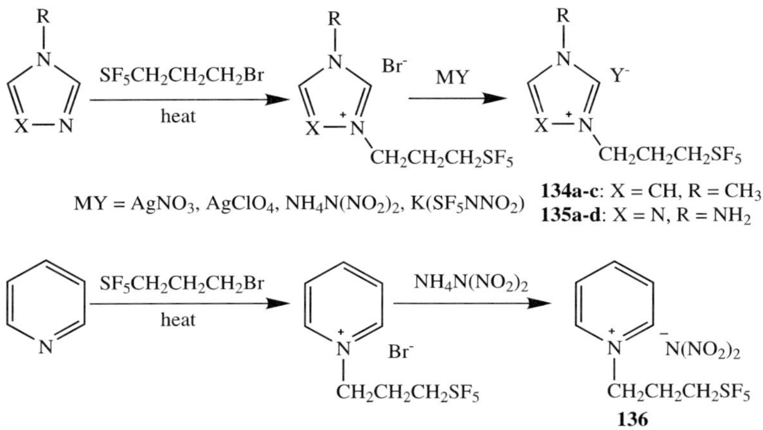

MY = $AgNO_3$, $AgClO_4$, $NH_4N(NO_2)_2$, $K(SF_5NNO_2)$

134a-c: X = CH, R = CH_3
135a-d: X = N, R = NH_2

Scheme 33

Table 12 Structure and properties of SF_5-containing salts

Cation	Anion	No	Tm [a] °C	Td [b] °C	d [d] g/cm^3	$\Delta_f H_m$ ° [d] kJ/mol
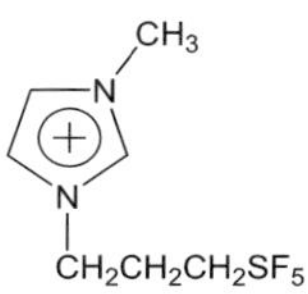	NO_3	**134a**	– 33.7	172.3	1.54	– 1030.9
	ClO_4	**134b**	– 68.0	247.1	1.46	– 982.4
	$[N(NO_2)_2]$	**134c**	– 50.7	201.7	1.52	864.0
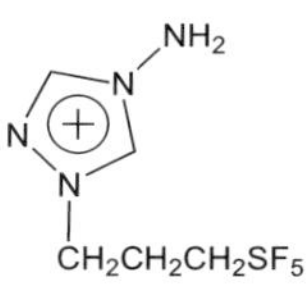	NO_3	**135a**	58.6	145.8	1.43	– 883.7
	ClO_4	**135b**	50.7	242.7	1.41	– 840.1
	$[N(NO_2)_2]$	**135c**	84.8	160.9	1.44	718.8
	$[SF_5NNO_2]$	**135d**	– 20.6	188.2	1.82	– 1755.2
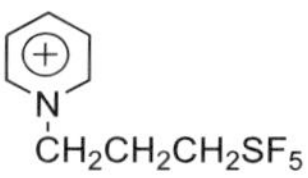	$[N(NO_2)_2]$	**136**	11.0	182.3	1.47	– 865.3

[a] melting point
[b] thermal degradation temperature
[c] density, gas pycnometer
[d] standard heat of formation (calculated by Gaussian 03)

zitanes (**138a,b**). The ratio of products depends on the solvents used. Treatment of **138a,b** with $NOBF_4$ gave isomeric pentanitroso-2,4,6,8,10,12-hexaazaisowurtzitanes (**139a,b**) which can be converted into the starting NH materials by treating with HCl (Scheme 34) [125]. Isomeric pentani-

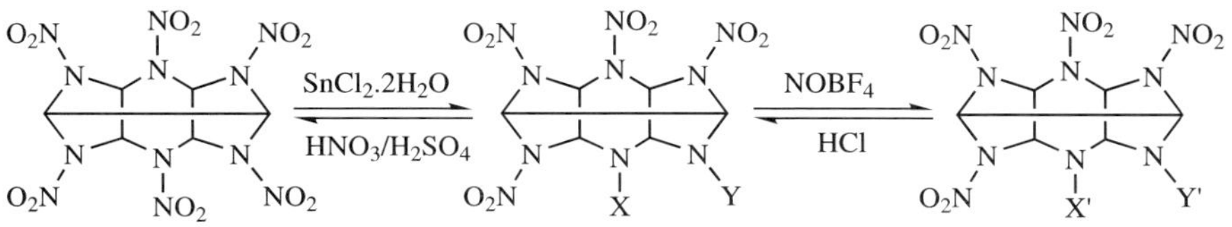

137 (CL-20)

138a: X = NO_2, Y = H
138b: X = H, Y = NO_2

139a: X' = NO_2, Y' = NO
139b: X' = NO, Y = NO_2

Scheme 34

trohexaazaisowurtzitanes (**138a,b**) are white powders which decompose at temperatures above 150 °C without melting. The nitroso derivatives (**139a,b**) are slightly yellowish powders, which decompose at temperatures below their melting points. Compound **138b** and **139b** are characterized by IR, NMR, and elemental analysis.

A nitroimidazole explosive molecule (**141**) has been synthesized by treating 2,2′-bi-1*H*-imidazole (**140**) with an excess of sodium nitrate (Scheme 35). This compound may serve as an energetic filler for high explosive formulations. The sensitivity results of **141** varied from "more sensitive than RDX" to substantially "less sensitive than RDX" according to the purity and condition of the test sample [126].

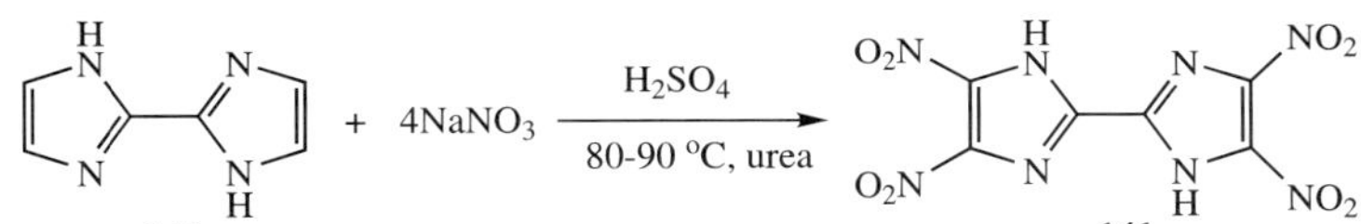

140 **141**

Scheme 35

11 Summary and Outlook

Many advantages accrue from nitrogen-rich heterocyclic compounds compared to traditional molecular energetic compounds. Utilization of heterocyclic nitrogen-containing cations and anions in energetic salts gives rise to lower vapor pressures, higher heats of formation, and higher densities. Additionally, smaller amounts of hydrogen and carbon contribute to a better oxygen balance than normally is found with their carbocyclic analogues. Nitrogen-rich compounds are promising high energetic materials that may be more acceptable than their alternatives for both industrial and military uses since a higher percentage of their decomposition products will be dinitrogen.

References

1. Gregory CE (1984) Explosives for North American Engineers, vol 5. Trans Tech Publications, Clausthal-Zellerfeld, Germany
2. Davis TL (1943) The Chemistry of Powder and Explosives, vol 2. Wiley, New York
3. Cook MA (1958) The Science of High Explosives. Reinhold, New York
4. Akhaven J (2005) Explosives and Propellants. In: Seidel A (ed) Encyclopedia of Chemical Technology. Wiley, Hoboken, N.J.
5. Urbanski T (1964) Chemistry and Technology of Explosives, vol 3. Pergamon, Oxford
6. Bailey A, Murray SG (1989) Explosives, Propellent and Pyrotechnics. Brassey's, Oxford
7. Urbanski T (1984) Chemistry and Technology of Explosives, vol 4. Pergamon, Oxford, p 202
8. Agrawal JP (1998) Prog Energy Combust Sci 24:1–30
9. Singh G, Kapoor IPS, Mannan SM, Kaur J (2000) J Hazard Mater A79:1–18
10. Fried LE, Manaa MR, Pagoria PF, Simpson RL (2001) Ann Rev Mater Res 31:291–321
11. Singh RP, Verma RD, Meshri DT, Shreeve JM (2006) Angew Chem 45:3584–3601
12. Shlyapochnikov VA, Tafipolsky MA, Tokmakov IV, Baskir ES, Anikin OV, Strelenko YA, Luk'yanov OA, Tartakovsky VA (2001) J Molec Struct 559:147–166
13. Pagoria PF, Lee GS, Mitchell AR, Schmidt RD (2002) Thermochim Acta 384:187–204
14. Lesnikovich AI, Ivashkevich OA, Levchik SV, Balabanovich AI, Gaponik PN, Kulak AA (2002) Thermochim Acta 388:233–251
15. Sikder AK, Sikder NJ (2004) J Hazard Mater A112:1–15
16. Bottaro J (2005) Ideas to Expand Thinking About New Energetic Materials. In: Shaw RW, Brill TB, Thompson DL (eds) Adv Ser Phys Chem, vol 16. World Scientific, Singapore, pp 473–501
17. Feuer H, Nielsen AT (1990) Nitro Compounds. VCH, New York
18. Nielsen AT (1995) Nitrocarbons. VCH, New York
19. Köhler J, Meyer R (1991) Explosivstoffe, 7th Ed. Wiley, Weinheim, Germany
20. Köhler J, Meyer R (1998) Explosivstoffe, 9th Ed. Wiley, Weinheim, Germany
21. Klapötke TM, Krumm B, Holl G, Kaiser M (2000) Energetic Materials: Modeling of Phenomena, Experimental Characterization, Environmental Engineering. Fraunhofer Institut, Karlsruhe, Germany
22. Simpson RL, Urtiew PA, Ornellas DL, Moody GL, Scribner KS, Hofman DM (1997) Propell Explos Pyrotech 22:249–255 and references cited therein
23. Krause HH (2005) New Energetic Materials. In: Teipal U (ed) Energetic Materials. Wiley, Weinheim, Germany
24. Hiskey MA, Chavez DE, Naud DL, Son SF, Berghout HL, Bolme CA (2000) Proc Int Pyrotech Semin 27:3–14
25. Pedley JB (1994) Thermochemical Data and Structure of Organic Compounds, vol I. Thermodynamic Research Center, College Station, TX
26. Jiminez P, Roux MV, Turrion CJ (1998) Chem Thermodyn 21:759–764
27. Drake GW, Hawkins T, Brand A, Hall L, McKay M, Vij A, Ismail I (2003) Propell Explos Pyrotech 28(4):174–180
28. Drake GW (2003) US Patent 6 509 473 B1
29. Schmidt MW, Gordon MS, Boatz JA (2005) J Phys Chem A 109:7285–7295
30. Xue H, Arritt SW, Twamley B, Shreeve JM (2004) Inorg Chem 43:7972–7977
31. Xue H, Gao Y, Twamley B, Shreeve JM (2005) Chem Mater 17:191–198

32. Drake GW, Hawkins T, Tollison K, Hall L, Vij A, Sobaski S (2005) In: Rogers RD, Seddon KR (eds) Ionic Liquids-III A: Fundamentals, Progress, Challenges, and Opportunities. ACS Symposium Series 902. ACS, Washington, DC, pp 259–302
33. Trohalaki S, Pachter R, Drake GW, Hawkins T (2005) Energ Fuel 19:279–284
34. Kaplan G, Drake G, Tollison K, Hall L, Hawkins T (2005) J Heterocycl Chem 42:19–27
35. Ostrovskii VA, Pevzner MS, Kofman TP, Tselinskii IV (1999) Targets Heterocycl Syst 3:467–526
36. Denault CC, Marx PC, Takimoto HH (1968) J Chem Eng Data 13:514–516
37. Xue H, Shreeve JM (2005) Adv Mater 17:2142–2146
38. Buscemi S, Pace A, Pibiri I, Vivona N, Spinelli D (2003) J Org Chem 68:605–608
39. Funabiki K, Noma N, Shibata K (1999) J Chem Res Synop pp 300–301, and references cited therein
40. Xue H, Twamley B, Shreeve JM (2004) J Org Chem 69:1397–1400
41. Mirzaei YR, Shreeve JM (2003) Synthesis 24–26
42. Mirzaei YR, Shreeve JM (2002) J Org Chem 67:9340–9345
43. Mirzaei YR, Xue H, Shreeve JM (2004) Inorg Chem 43:361–367
44. Shitov OP, Korolev VL, Bogdanov VS, Tartakovsky VA (2003) Russ Chem Bull Int Ed 52:695–699
45. Egashira M, Scrosati B, Armand M, Beranger S, Michot C (2003) Electrochem Solid State Lett 6(4):A71–A73
46. Katritzky A, Singh S, Kirichenko K, Holbrey JD, Smiglak M, Reichert MW, Rogers RD (2005) Chem Commun 868–870
47. Sitzmann MI (1978) J Org Chem 43:3389–3391
48. Kofman TP, Paketina EA (1997) Russ J Org Chem (Eng Trans) 33:1125–1132
49. Chernyshev VM, Zemlyakov ND, Il'in VB, Taranushich VA (2000) Zh Prikl Khim 73:791–793
50. Ogihara W, Yoshizawa M, Ohno H (2004) Chem Lett 33(8):1022–1023
51. Xue H, Gao Y, Twamley B, Shreeve JM (2005) Inorg Chem 44:5068–5072
52. Martin AR, Yallop HJ (1958) Trans Faraday Soc 54:257–267
53. Ostrovskii VA, Koldobskii GI, Shirokova NP, Poplavskii VS (1981) Khim, Grterotsikl Soedin 4:559–562
54. Krakovskii IM, MolChanova MS, Evtushenko AV, Shlyapochnikova VA (1998) Russ Chem Bull 47:1266–1273
55. Kozyro AA, Simirsky VV, Krasulin AP, Sevruk VM, Kabo GJ, Gopanik ML, Grigotiev YV (1990) Zh Fiz Khim 64:656–661
56. Gao A, Oyumi Y, Brill TB (1991) Combust Flame 83:345–352
57. von Denffer M, Klapötke TM, Kramer G, Spieß G, Welch JM, Heeb G (2005) Propell Explos Pyrotech 30:191–195
58. Klapötke TM, Mayer P, Schulz A, Weigand JJ (2005) J Am Chem Soc 127:2032–2033
59. Gálvez-Ruiz JC, Holl G, Karaghiosoff K, Klapötke TM, Löhnwitz K, Mayer P, Nöth H, Polborn K, Rohbogner CJ, Suter M, Weigand JJ (2005) Inorg Chem 44:4237–4253
60. Fisher G, Holl G, Klapötke TM, Weigand JJ (2005) Thermochim Acta 437:168–178
61. Geith J, Klapötke TM, Weigand J (2004) Propell Explos Pyrotech 29:3–8
62. Hammerl A, Klapötke TM, Nöth H, Warchhold M, Holl G (2003) Propell Explos Pyrotech 28:165–173
63. Hammerl A, Klapötke TM, Mayer P, Weigand JJ, Holl G (2005) Propell Explos Pyrotech 30:17–26
64. Klapötke TM, Mayer P, Verma V (2006) Propell Explos Pyrotech 31:263–268
65. Klapötke TM, Karaghiosoff K, Mayer P, Penger A, Welch JM (2006) Propell Explos Pyrotech 31:188–195

66. Karaghiosoff K, Klapötke TM, Mayer P, Piotrowski H, Polborn K, Willer RL, Weigand JJ (2006) J Org Chem 71:1295–1305
67. Hammerl A, Klapötke TM (2002) Inorg Chem 41:906–912
68. Thiele J (1892) Liebigs Ann 270:54–63
69. Thiele J, Marais JT (1893) Justus Liebigs Ann Chem 273:144–160
70. Thiele J (1893) Ber Dtsch Chem Ges 26:2645–2646
71. Thiele J (1898) Justus Liebigs Ann Chem 303:57–75
72. Hammerl A, Holl G, Klapötke TM, Mayer P, Noth H, Piotrowski H, Warchhold M (2002) Eur J Inorg Chem 834–845
73. Hiskey MA, Goldman N, Stine JR (1998) J Energ Mater 16:119–127
74. Peng Y, Wong C (1999) US Patent 5 877 300 CA 130:196656
75. Hammerl A, Klapötke TM, Nöth H, Warchhod M, Holl G, Kaiser M, Ticmanis U (2001) Inorg Chem 3570–3575
76. Hammerl A, Holl G, Kaiser M, Klapötke TM, Mayer P, Piotrowski H, Vogt M (2001) Naturforschung 847–856
77. Hammerl A, Holl G, Kaiser M, Klapötke TM, Mayer P, Nöth H, Piotrowski H, Suter M (2001) Naturforschung 857–870
78. Klapotke TM, Holl G (2001) Green Chem G75–G76
79. Hammerl A, Hiskey MA, Holl G, Klapötke TM, Polborn K, Stierstorfer J, Weigand J (2005) Chem Mater 17:3784–3793
80. Tappan BC, Ali AN, Son SF (2006) Propell Explos Pyrotech 31:163–168
81. Ye C, Xiao J-C, Twamley B, Shreeve JM (2005) Chem Commun 2750–2752
82. Hyoda S, Kita M, Sawada H, Nemugaki S, Otsuka S, Miyawaki Y, Ogawa T, Kubo Y (2000) US Patent 6 040 453, CA: 132:207845
83. Hyoda S, Kita M, Sugino A, Ueta T, Sato K (2001) EU Patent 1 162 198, CA: 136:20077
84. Hyoda S, Kita M, Swada H, Nemugaki S, Ueta T, Satoh K, Otsuka S, Miyawaki Y, Taniguchi H (2000) EU Patent Appl, EP 1 016 662, CA: 133:74021
85. Torii S, Tsuyama M, Miyawaki Y, Kubo Y, Ogawa T (2000) Jpn Kokai Tokkyo Koho, JP 2000281662, CA 133:252440
86. Friedrich M, Gálvez-Ruiz JC, Klapötke TM, Mayer P, Weber P, Weigand JJ (2005) Inorg Chem 44:8044–8052
87. Hammerl A, Holl G, Kaiser M, Klapötke TM, Piotrowski H (2003) Z Anorg Allg Chem 629:2117–2121
88. Kita M, Ueda T (2004) Jpn Kokai Tokkyo Koho, JP 2004067544, CA 140:217643
89. Naud DL, Hiskey MA (2003) US Patent Appl Publ, 2 003 060 634, CA:138:255236
90. Highsmith TK, Hajik RM, Wardle RB, Lund GK, Blau RJ (1995) US Patent 5 468 866
91. Gao Y, Ye C, Twamley B, Shreeve JM (2006) Chem-Eur J 12:9010
92. Foss ME, Hirst EL, Jones JKN, Springall HD, Thomas AT, Urbanski T (1950) J Chem Soc 624–628
93. Ang H-G, Fraenk W, Karaghiosoff K, Klapötke TM, Nöth H, Sprott J, Sutter M, Vogt M, Warchhold MZ (2002) Z Anorg Allg Chem 628:2901–2906
94. Xue H, Twamley B, Shreeve JM (2006) Eur J Inorg Chem 2959–2965
95. Chavez DE, Hiskey MA, Naud Di (2004) Propell Explos Pyrotech 29:209–215
96. Oxley JC, Smith JL, Chen H (2002) Thermochim Acta 91–99
97. Chavez DE, Hiskey MA, Gilardi RD (2004) Org Lett 6:2889–2891
98. Chavez DE, Tappan BC, Hiskey MA, Son SF, Harry H, Montoya D, Hagelberg S (2005) Propell Explos Pyrotech 30:412–417
99. Gao H, Wang R, Twamley B, Hiskey MA, Shreeve JM (2006) Chem Commun 4007–4009
100. Kerth J, Löbbecke S (2002) Propell Explos Pyrotech 27:111–118

101. Chavez DE, Hiskey MA, Gilardi RD (2000) Angew Chem Int Ed 39:1791–1793
102. Huynh MHV, Hiskey MA, Chavez DE, Naud DL, Gilardi RD (2005) J Am Chem Soc 127:12537–12543
103. Huynh MHV, Hiskey MA, Archuleta JG, Roemer EL, Gilardi R, Chavez DE, Naud DL, Gilardi RD (2004) Angew Chem Int Ed 43:5658–5661
104. Miller DR, Swenson DC, Gillan EG (2004) J Am Chem Soc 126:5372–5373
105. Huynh MHV, Hiskey MA, Hartline EL, Montoya DP, Gilardi R (2004) Angew Chem Int Ed 43:4924–4928
106. Huynh MHV, Hiskey MA, Pollard CJ, Montoya DP, Hartline EL, Gilardi R (2004) J Energ Mater 22:217–229
107. Frumkin AE, Churakov AM, Strelenko YA, Kachala VV, Tartakovsky VA (1999) Org Lett 1:721–724
108. Archibald TG, Gilardi R, Baum K, George C (1990) J Org Chem 55:2920–2924
109. Coburn MD, Hiskey MA, Archibald TG (1997) Waste Manag 17:143–146
110. Hiskey MA, Coburn MD, Mitchell MA, Benicewicz BC (1992) J Heterocycl Chem 29:1855–1856
111. Hiskey MA, Stincipher MM, Brown JE (1993) J Energ Mater 11:157–165
112. Gilardi RD, Butcher RJ (1998) J Chem Crystallogr 28(3):163–169
113. Coburn MD, Hiskey MA, Oxley JC, Smith JL, Zheng W, Rogers E (1998) J Energ Mater 16(2):73–99
114. Jin C-M, Ye C, Piekarski C, Twamley B, Shreeve JM (2005) Eur J Inorg Chem 3760–3767
115. Zhang M-X, Eaton PE, Gilardi R (2000) Angew Chem Int Ed 39(2):404–401
116. Chung G, Schmidt MW, Gordon MS (2000) J Phys Chem A 104:5647–5650
117. Fau S, Bartlett RJ (2001) J Phys Chem A 105:4096–4106
118. Nguyen NT (2003) Coordinat Chem Rev 244:93–113
119. Christe KO, Dixon DA, McLemore D, Wilson WW, Sheehy JA, Boatz JA (2000) J Fluor Chem 101:151–153
120. Jones CB, Haiges R, Schroer T, Christe KO (2006) Angew Chem Int Ed 45:4981–4984
121. Ju Y, Kumar D, Verma RS (2006) J Org Chem 71:6697–6700
122. Muralidharan K, Omotowa BA, Twamley B, Piekarski C, Shreeve JM (2005) Chem Commun 5193–5195
123. Göbel M, Karaghiosoff K, Klapötke TM (2006) Angew Chem Int Ed 45:6037–6040
124. Gao H, Ye C, Winter RW, Gard GL, Sitzmann ME, Shreeve JM (2006) Eur J Inorg Chem 3221–3226
125. Lukyanov OA, Shykova NI (2004) Russ Chem Bull Int Ed 53:566–568
126. Gyeong S, Cho JR, Goh EM, Kim J-K (2005) Propell Explos Pyrotech 30:445–449

Struct Bond (2007) 125: 85–121
DOI 10.1007/430_2007_057

Published online: 12 May 2007

New Nitrogen-Rich High Explosives

Thomas M. Klapötke

Ludwig-Maximilian University Munich, Chair of Inorganic Chemistry,
Energetic Materials Research, Butenandtstrasse 5–13 (D), 81377 Munich, Germany
tmk@cup.uni-muenchen.de

Abstract The possibility of new high explosives based on nitrogen-rich tetrazole building blocks is discussed. The expected advantages include gaseous products, high heats of formation, high propulsive/expolosive power, high specific impulse, and high flame temperatures. In addition, these new explosives do not have the toxicity and environmental activity of currently used organo-nitro explosives. The synthesis and characteristics of a series of neutral tetrazole compounds are looked at as well as the neutral nitramine, dinitrobiuret.

Keywords Dinitrobiuret · High energy density materials · Nitrogen-rich · Polynitrogen · Tetrazole

Abbreviations

BTNA	Bistrinitroethylamine
DAT	1,5-Diaminotetrazole
DNB	Dinitrobiuret
EI	energetic ingredient
NTNAP	1-Nitrotetrazolato-2-nitro-2-azapropane
TNE	Trinitroethanol
MMTHT	1-Methyl-5-(1-methyl-2-(2,2,2-trinitroethyl)hydrazinyl)-1*H*-tetrazole
MTHTE	2-(5-(1-methyl-2-(2,2,2-trinitroethyl)hydrazinyl)-1*H*-tetrazol-1-yl)ethanol

1 Introduction

Environmental contamination by nitro compounds is associated principally with the explosives industry and military testing of explosives [1, 2]. The current widely used nitro-explosives are TNT (trinitrotoluene), RDX (Royal demolition explosive), and HMX (high melting explosive). The nitro-explosives per se as well as their environmental transformation products are toxic, with symptoms of exposure that include methemoglobinemia, kidney trouble, jaundice etc. For HMX, anaerobic or anoxic degradations have been described in many studies [3]. Explosives released into the environment at production and processing facilities, as well as through field use, may be toxic at relatively low concentrations to a number of ecological receptors [4]. Toxicity studies on soil organisms using endpoints such as microbial processes (potential nitrification activity, dehydrogenase activity, substrate-induced respiration, basal respiration), plant and seedling growth (*Lactuca sativa* and *Hordeum vulgare*), and earthworm (*Eisenia andrei*) growth and reproduction were carried out at contaminated sites. Results showed that HMX was the principal polynitro-organic contaminant measured in soils. Soils from the contaminated site showed decreased microbial processes and earthworm reproduction. However, plant growth was not significantly reduced [4]. Monocyclic nitramine explosives such as RDX and HMX are toxic to a number of ecological receptors, including earthworms [5–7]. The results of recent investigations suggest that organisms exposed to explosives at contaminated sites show hormetic growth enhancement at concentrations less than 25 $mg\,kg^{-1}$ and at higher concentrations through increased mortality [8]. Factors effecting leaching and transport, microbial degradation, phytotoxicity and plant uptake, and toxicity to invertebrates and vertebrates will determine the ultimate environmental fate and hazard potential [9]. It is clear that the release of explosives into the environment at production and processing facilities, as well as through field use, is a major point of concern.

Since the toxicity and environmental activity of organo-nitro explosives (TNT, RDX, HMX, see Fig. 1) are usually related to the presence of organo-nitro ($C-NO_2$), organo-nitroso ($C-NO$), or organo-nitrito ($C-ONO$) groups either in the explosive itself or its degradation products [10], the development of novel energetic ingredients (EIs) that lack the environmental hazards of the organo-nitro explosives is of great interest. The presently used EIs with the highest performance (RDX, HMX) as well as new, even more powerful EIs that are still under investigation (CL-20, hepta- and octacubane, see Fig. 1) all belong to the class of typical organic ring and cage molecules.

In this chapter we want to discuss the possibility of new high explosives that are neutral compounds and are based on nitrogen-rich tetrazole building blocks.

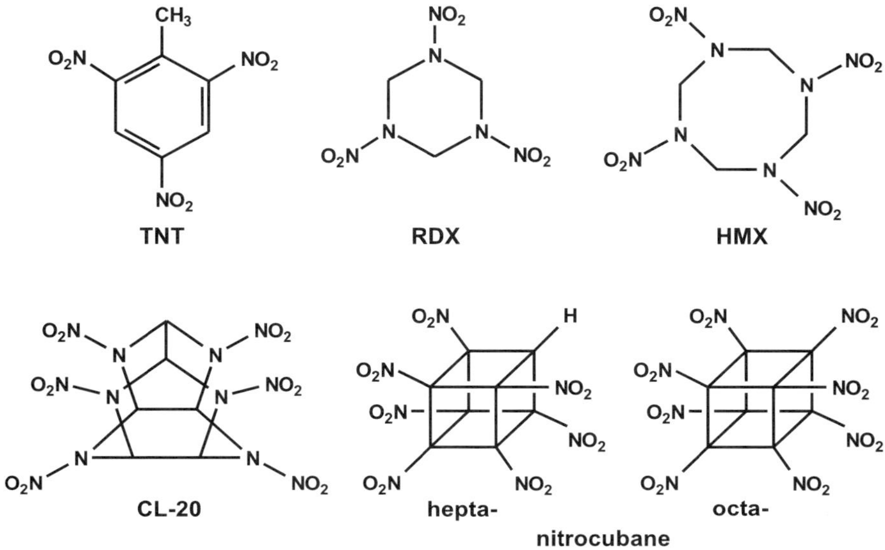

Fig. 1 Structures of TNT, RDX, HMX, CL-20, hepta-, and octa-cubane

The expected advantages of nitrogen-rich compounds include:

- Only gaseous products
- High heats of formation
- High propulsive/explosive power
- High specific impulse
- High flame temperatures

Also, recent modeling has shown that the presence of high concentrations of nitrogen species in the combustion products of propellants can reduce gun barrel erosion by promoting the formation of iron nitride rather than iron carbide on the interior surface of the barrel [11].

2 Strategies

Modern high-energy-density materials (HEDM) derive most of their energy from either:

1. Oxidation of the carbon backbone, as with traditional energetic materials
2. Ring or cage strain
3. Their very high positive heat of formation

Examples for the first class are traditional explosives such as TNT, RDX, HMX, and the recently reported 1,4-diformyl-2,3,5,6-tetranitropiperazine (Fig. 2). Modern nitro compounds such as CL-20 or the recently reported

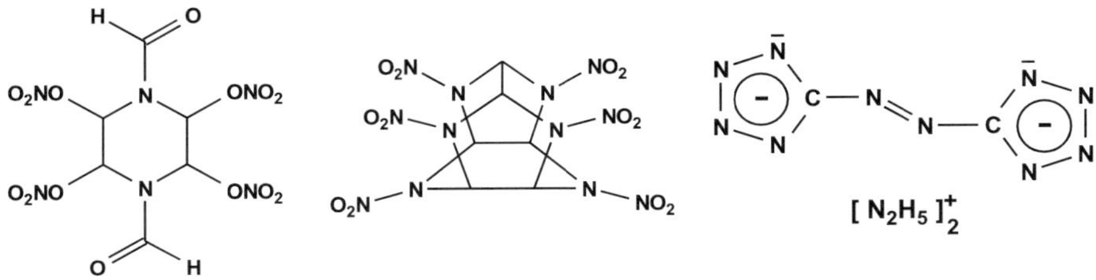

Fig. 2 Examples for EM that (i) show oxidation of the carbon backbone (*left*) [12], (ii) possess ring or cage strain (*center*) [13], or (iii) have very high positive heats of formation (*right*) [14]

hepta- and octanitrocubanes possess very high densities and enhance the energies utilizing substantial cage strain and therefore belong to the second class of explosives. Members of the third class of compounds are 5,5′-azotetrazolate salts, which show the desired remarkable insensitivity to electrostatic discharge, friction, and impact while having a very high (positive) heat of formation.

Nitrogen is unique amongst all other elements of the periodic table in so far that the bond energy per two-electron bond increases from a single over a double to a triple bond (Fig. 3). For carbon the situation is the opposite and one might expect acetylene ($H-C\equiv C-H$) to polymerize in an exothermic reaction whereas N_2 ($N\equiv N$) is more stable than any other polynitrogen species [15–18]. One may ask what the difference between formally sp-hybridized nitrogen in N_2 and sp-hybridized carbon in HCCH is. Indeed the bond situation is very similar and only an NBO analysis can help to explain the difference:

$N\equiv N$	$\sigma(N-N)$	64% s	36% p
	σ– LP(N)	34% s	66% p
$H-C\equiv C-H$	$\sigma(C-C)$	49% s	51% p
	$\sigma(C-H)C$	45% s	55% p
	H	100% s	

Whereas in acetylene the $C-C$ σ bond is indeed essentially made up of sp-hybrid orbitals, in N_2 the σ bond orbitals have two-third s and only one-third p character and form therefore a much stronger bond than the carbon atoms do in HCCH.

With respect to all-nitrogen species (N_x), there are only a very few species known that have been isolated and characterized (Table 1).

Quite recently, researchers from the Max Plank Institute for Chemistry and the Geophysical Laboratory at the Carnegie Institution of Washington reported the polymerization of nitrogen in sodium azide [16–18]. With increasing pressure on sodium azide, the sample became completely opaque above

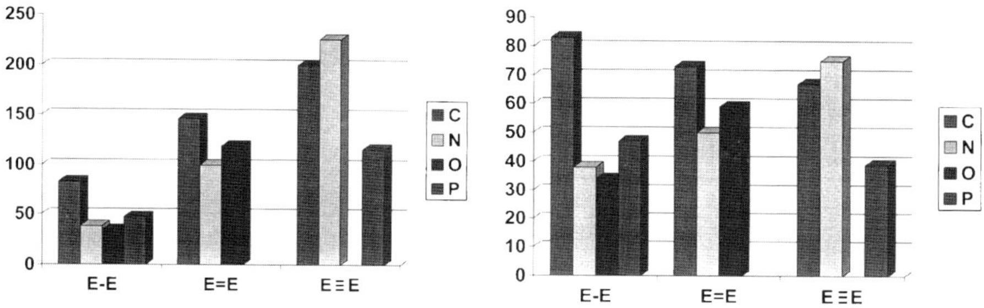

Fig. 3 Bond energies (in kcal mol^{-1}) for single, double and triple bonds (*left*) and per two-electron bond (*right*) [15]

Table 1 Isolated all-nitrogen species

All-nitrogen species	Name	Comment	Refs.
Neutral compounds			
N_2	Dinitrogen	Thermodynamic stable form of nitrogen, not energetic	[15]
cg-N_x	Cubic polynitrogen	Prepared from N_2 in diamond anvil cell at 200 K and 110 GPa; metastable only above 42 GPa	[16–18]
Ionic species			
N_3^-	Azide anion	Suitable energetic anion (or covalent azide group R-NNN)	[15]
N_5^+	Pentanitrogen cation	Metastable salts only with large counter anions ($Sb_2F_{11}^-$ etc.)	[19–21]
N_5^-	Pentanitrogen anion	Only detected in the gas phase (mass spectrometry)	[22, 23]

120 GPa, evidencing completion of the transformation to a non-molecular nitrogen state with amorphous-like structure that crystallized after laser heating to 3300 K. This change was interpreted in terms of a transformation of azide ions to larger nitrogen clusters and subsequent formation of a polynitrogen network. The polymeric form was preserved on decompression in the diamond anvil cell, but transformed back to the starting azide form under ambient conditions.

In an another publication [17], Eremets et al. reported a single-bonded cubic form of nitrogen. This material was synthesized directly from molecular nitrogen at temperatures above 2000 K and pressures above 110 GPa, again in a laser-heated diamond-cell. From X-ray and Raman scattering the substance was identified as the long-sought-after polymeric nitrogen with the theoretically predicted cubic gauche structure (cg-N). In this compound, each nitrogen atom is connected to three neighbors by three single covalent bonds (Fig. 4). This is a new member of a class of single-bonded nitrogen materials with unique properties such as an energy capacity more than five times that of most powerful non-nuclear energetic materials.

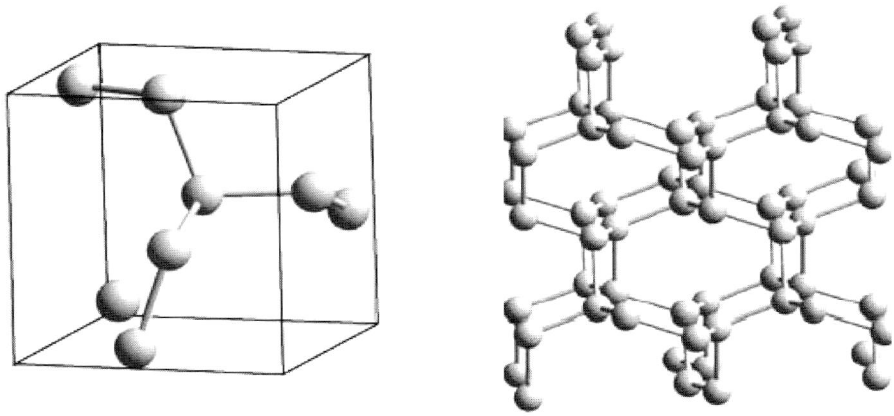

Fig. 4 cg-N structure. Each N atom is connected to three neighbors by three single covalent bonds. The primitive cell is shown on the *left*, and an extended structure of the polymeric N is shown on the *right*

The authors confirmed calculations that predicted that the cg-structure is the most energetically favorable. Their findings confirm the existence of a new allotrope of nitrogen where atoms are connected with single covalent bonds (cf. Fig. 4). The predicted wide band-gap of 8.1 eV (or 3.75 eV at 240 GPa) correlates well with the observed colorless transparent phase. Although theory also predicts that the cg-N could be metastable at atmospheric pressures, Eremets et al. found that at room temperature cg-N is metastable at pressures above 42 GPa. At this pressure, it transforms to a molecular phase under weak laser illumination. If cg-N is not metastable at ambient pressure, the authors hope it can be stabilized in compounds with other elements or by introducing impurities.

Only further experiments can show whether this new form of polynitrogen may ever be suitable for use as a high energy density material (HEDM), due to (i) its very high expected specific impulse (for propulsion) or, less likely, (ii) expected high explosive power (for use as a secondary explosive or high explosive). Whether polynitrogen finds application or not, the authors ought to be congratulated on their outstanding and break-through discovery.

In addition, there are numerous polynitrogen species that have been computed in the gas phase (Fig. 5) [24–30], however, none of these compounds has yet been prepared in the laboratory, not even on a milligram scale.

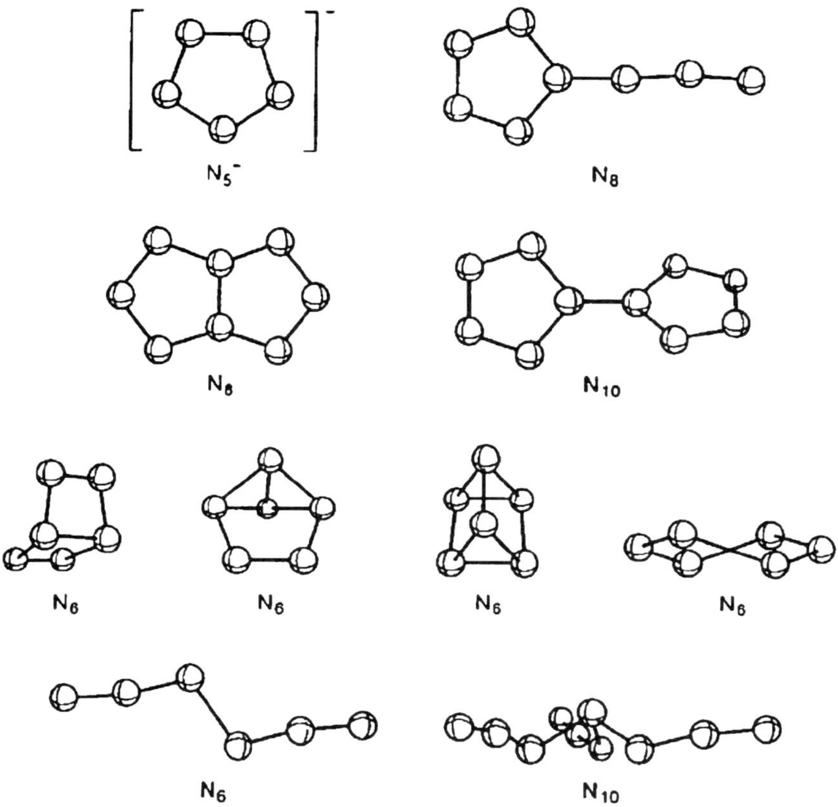

Fig. 5 Structure of N_5^- and some computationally known (but not yet synthesized) neutral polynitrogen species [24–30]

In the opinion of the author of this chapter high-nitrogen compounds (in contrast to all-nitrogen compounds) clearly have great potential as new energetic materials. Whereas high-nitrogen compounds without additional oxidizing groups may be good candidates for use as propellant charges for high explosives, the combination of high-nitrogen compounds with strongly oxidizing nitro, nitrato, nitramino, or nitrimino groups should be advantageous in delivering the explosive energy required (Fig. 6).

In order to stabilize nitrogen-rich molecules, π aromatic delocalization may be used. In this context five-membered heterocycles with three N atoms – triazoles –, four N atoms – tetrazoles –, and five N atoms – pentazoles – play an important role (Fig. 7). While triazoles often do not possess enough intrinsic energy to be suitable candidates for high explosives, pentazole derivatives are usually either far too sensitive and unstable for any application or need to be kinetically stabilized by bulky groups, which reduce the nitrogen con-

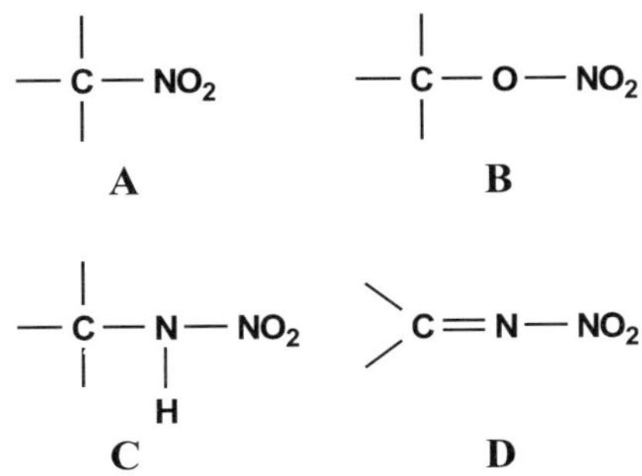

Fig. 6 Nitro (*A*), nitrato (*B*), nitramino (*C*), and nitrimino (*D*) groups

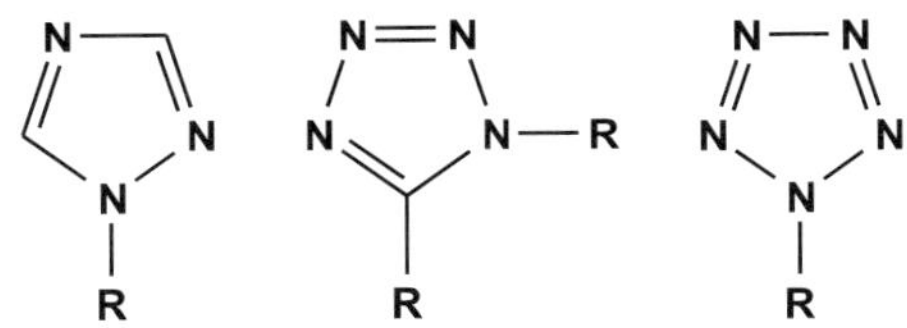

Fig. 7 Neutral triazole, tetrazole, and pentazole

tent substantially. Therefore, tetrazole compounds seem to be a good choice for combining high-nitrogen content with reasonable kinetic stability.

Energetic salts are important systems for the development of high energy density materials since salts are intrinsically non-volatile, are typically thermally stable under normal conditions, and are more dense [31, 32]. However, due to the fact that salts tend to have undesirable octanol/water partition coefficients and may therefore find their way into the groundwater, in this review we therefore want to focus on neutral species.

3 Neutral Tetrazole Compounds

3.1 1,5-Diaminotetrazole (DAT)

Because of the difficult accessibility of 1,5-diaminotetrazole (DAT, Fig. 8), only few synthetic methods for its preparation have been described in the literature. This compound, as all aminotetrazoles, has the highest content of nitrogen among organic substances (about 84%) and due to the aromaticity, it exhibits a relatively high thermal stability.

In addition, DAT presents a high energy of formation; for this reason it can be used as a valuable intermediate in the preparation of high-energy-density materials [33–35].

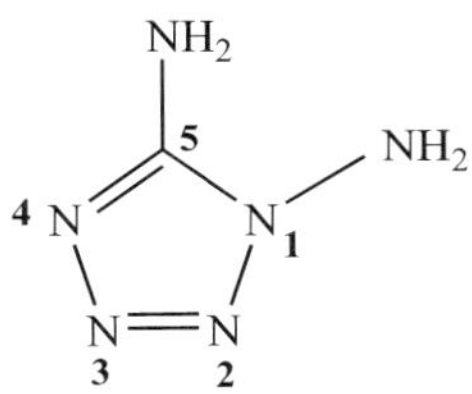

Fig. 8 Structure of 1,5-DAT

A milestone in the synthesis of DAT was described by Raap [36] who reacted the sodium salt of 5-aminotetrazole with hydroxylamine-*O*-sulfonic acid and obtained a mixture of 1,5-diaminotetrazole (1,5-DAT) and 2,5-diaminotetrazole (2,5-DAT), where the first constitutes 8.5% of the reaction yield (Fig. 9).

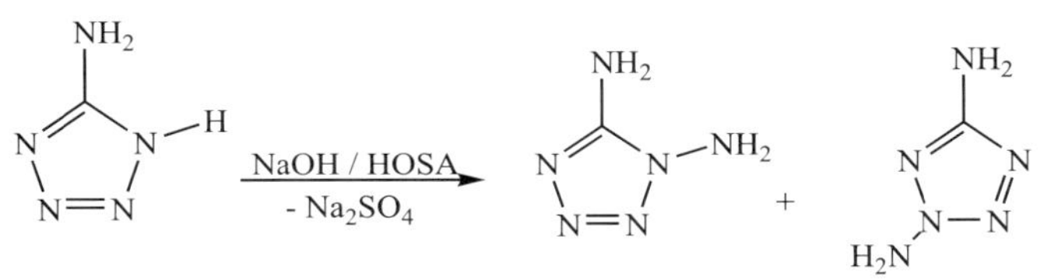

5-AT 1,5-DAT (8,5%) 2,5-DAT

Fig. 9 Synthesis of 1,5-DAT and 2,5-DAT according to Raap [36]

Before Raap, Gaponik et al. [37] improved the synthesis reported by Stolle et al. [38]. By reacting thiosemicarbazide with lead oxide and sodium azide in a CO_2 atmosphere, a carbodiimide intermediate is formed and reacts in situ with HN_3 to lead to 1,5-DAT (Fig. 10). Unfortunately, this reaction leads to large amounts of lead azide as the side-product, which makes this synthesis problematic for an industrial scale.

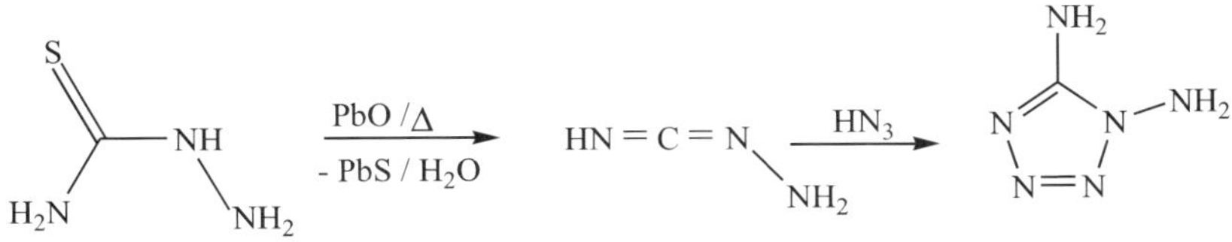

Fig. 10 Synthesis of 1,5-DAT using PbO [37, 38]

The most recent approach to the synthesis of DAT was made at LMU Munich [39, 40] and eliminates the formation of highly explosive lead azide

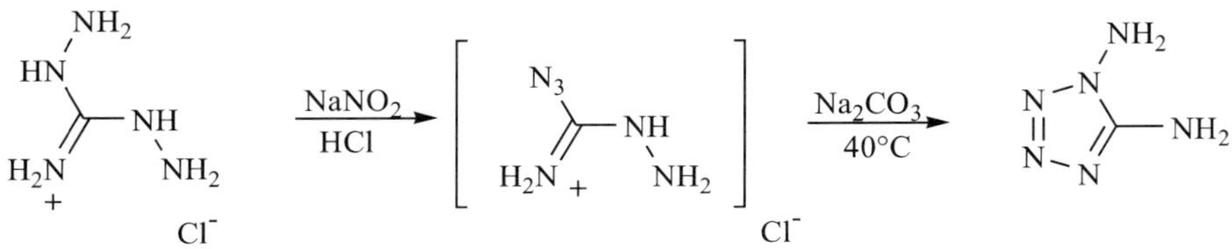

Fig. 11 Improved synthesis of 1,5-DAT without using PbO [39–41]

and produces a substantially greater yield. After diazotation of diaminoguanodinium chloride, the reaction mixture is brought to pH 8 to deprotonate the intermediate formed, which cyclizes yielding 1,5-DAT in 58% yield (Fig. 11). Nevertheless, the synthesis conditions must be perfectly controlled because the reaction of nitrous acid with aminoguanidinium is strongly dependent on the pH value as well as on the amounts of reactants, otherwise it might lead to the azide derivative, which is a very explosive byproduct [41].

3.2 1-Nitrotetrazolato-2-nitro-2-azapropane (NTNAP)

The reaction of 1-chloro-2-nitro-2-azapropane with silver nitrotetrazolate yielded 1-nitrotetrazolato-2-nitro-2-azapropane (NTNAP) as a white solid in over 90% yield (Fig. 12) [42].

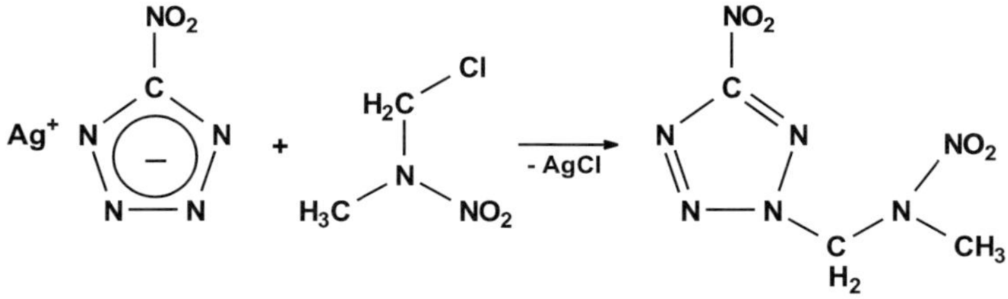

Fig. 12 Synthesis of NTNAP

The molecular structures of NTNAP in the crystalline state was determined by low-temperature X-ray crystallography (Fig. 13). The aromatic tetrazolium ring is, as expected, planar with the following torsion angles: C1 – N1 – N2 – N3 0.2(2)°, N2 – N1 – C1 – N4 – 0.4(2)°, N1 – N2 – N3 – N4 0.0(2)°. The crystalline density was determined to be 1.735 g cm^{-3}.

NTNAP was fully characterized by IR and Raman spectroscopy and multinuclear NMR spectroscopy (^{1}H, ^{13}C, ^{14}N). The nitrotetrazolium compound shows a rather complex ^{14}N NMR spectrum, however, the two nitro (– NO_2) resonances are well resolved at – 26.2 and – 28.7 ppm. In the ^{13}C NMR spec-

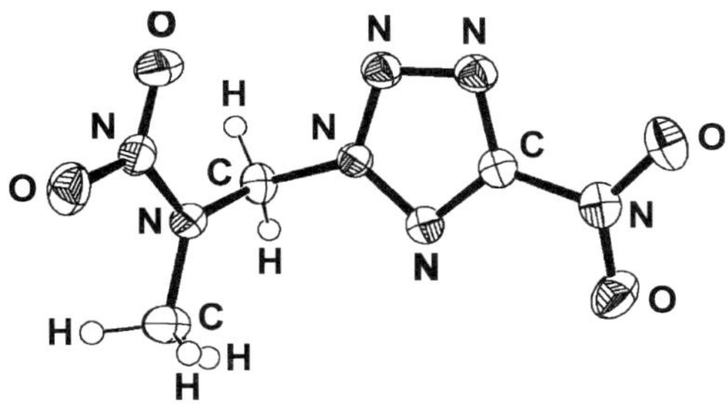

Fig. 13 Molecular structure of NTNAP in the crystalline state. Displacement ellipsoids are shown at the 50% probability level

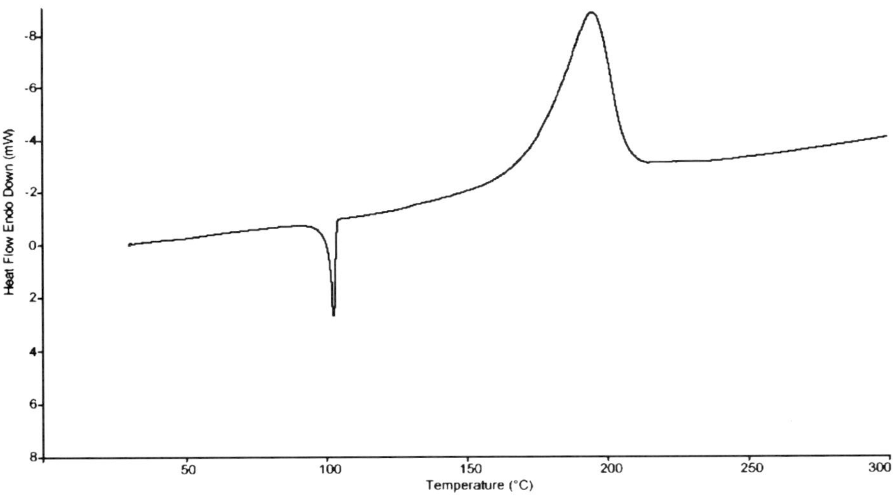

Fig. 14 DSC plot of NTNAP, heating rate 2 °C min^{-1}

trum the methyl ($-CH_3$, 3.55 ppm) and methylene ($-CH_2-$, 6.85 ppm) protons show singlet resonances clearly shifted, as expected, to low field.

The DSC plot of NTNAP was recorded with a heating rate of 2 °C min^{-1} (Fig. 14). The covalent compound is thermally reasonably stable, melts without decomposition at 100 °C, and shows decomposition with an onset of about 180 °C. Therefore, the compound possesses properties that may make it suitable for melt-casting processes.

The sensitivity and detonation parameters of NTNAP are summarized in Table 2.

Table 2 Energetic properties of NTNAP

	NTNAP
Appearance	White powder
Stability	Hydrolytically stable
Deflagration	Explodes when thrown in a flame
Oxygen balance [%]	– 35.5
Impact sensitivity [J] [18, 19]	6
Friction sensitivity [N]	60

3.3 *N*-Trinitroethyl Derivatives of Nitrogen Containing Compounds

Although compounds carrying the *N*-trinitroethyl fragment have been reported, relevant data were published mainly in the patent literature, often without giving information about synthetic procedures or specifying the physicochemical characteristics of the compounds obtained. In the course of investigations into high energy density materials (HEDM) at LMU Munich, we recently focused our attention on new derivatives of energetic materials that combine both the advantages of the tetrazole, as well as the trinitroethyl, moiety. The tetrazole unit with its high nitrogen content together with its endothermic character is remarkably thermodynamically stable and the trinitroethyl fragment contributes to a positive oxygen balance.

Trinitroethanol (TNE) and bistrinitroethylamine (BTNA) can easily be synthesized according to the literature [43, 44]. The best synthetic approach for the synthesis of 1-methyl-5-(1-methyl-2-(2,2,2-trinitroethyl)hydrazinyl)-1*H*-tetrazole (MMTHT) and 2-(5-(1-methyl-2-(2,2,2-trinitroethyl)hydrazinyl)-1*H*-tetrazol-1-yl)ethanol (MTHTE) utilizes a condensation of the starting amino derivative with 2,2,2-trinitroethanole (Fig. 15) [45].

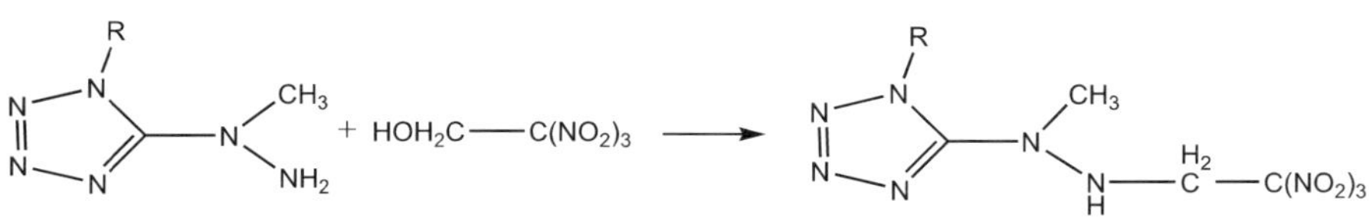

R = CH_3, CH_2CH_2OH

Fig. 15 Synthesis of MMTHT and MTHTE

All compounds were characterized using vibrational spectroscopy. In addition, TNE, MMTHT, and MTHTE were characterized using ^{1}H, ^{13}C, and

Table 3 Properties of MMTHT

Sum formula	$C_5H_9N_9O_6$
Formula weight [g mol^{-1}]	291.21
TMD [g cm^{-3}]	1.63
$T_{decomposition, onset}$ (DSC, 2 K min^{-1}) [°C]	82.5
– $\Delta_c U$ [cal g^{-1}]	2936
Impact sensitivity [J]	> 30
Friction sensitivity [N]	108 (visible flame)
Oxygen balance [%]	– 46.7
– $\Delta_f U$ [kJ kg^{-1}]	+ 1889
Q_V [kJ kg^{-1}]	– 6368
T_{ex} [K]	4404
P [kbar]	277
D [m s^{-1}]	8307
V_0 [L kg^{-1}]	783

Table 4 Selected structural data of TNE, BTNA, MMTHT, and MTHTE

Compound	TNE	BTNA	MMTHT	MTHTE
Formula	$C_2H_3N_3O_7$	$C_4H_5N_7O_{12}$	$C_5H_9N_9O_6$	$C_6H_{11}N_9O_7$
Formula weight [g mol^{-1}]	181.07	343.12	291.21	321.208
Temperature [K]	200	200	100	100
Crystal system	Monoclinic	Orthorhombic	Triclinic	Monoclinic
Space group	$P2_1/c$ (no. 14)	P bca (no. 61)	P-1	$P2_1/n$ (no. 14)
a [Å]	6.1218(4)	12.8996(6)	7.2651(13)	13.0419(4)
b [Å]	18.8120(12)	11.7753(5)	7.5773(16)	7.3020(2)
c [Å]	11.7391(8)	16.1577(7)	11.695(7)	14.8002(5)
α [°]	90	90	102.89(3)	90
β [°]	104.997(4)	90	103.82(3)	112.118(4)
γ [°]	90	90	99.387(17)	90
Volume [Å^3]	1305.87(15)	2454.30(19)	593.3(4)	1305.73(7)
Z	8	8	2	4
Absorption coefficient [mm^{-1}]	0.191	0.188	0.146	0.147
Density exptl. [g cm^{-3}]	1.842	1.857	1.630	1.634
F(000)	736	1392	300	664
2θ [°]	51.0	51.5	52.0	54.0
Reflections collected	12556	22836	5227	9034
Reflections unique	2424	2330	2308	2831
R_{int}	0.0631	0.0494	0.0225	0.0237
Parameters	237	228	217	243
GOOF	1.223	1.294	0.979	1.061
R_1/wR_2 [I > 2σ(I)]	0.0651/0.1048	0.0802/0.1766	0.0300/0.0730	0.0297/0.0740
R_1/wR_2 (all data)	0.0825/0.1135	0.0831/0.1787	0.0432/0.0772	0.0438/0.0806

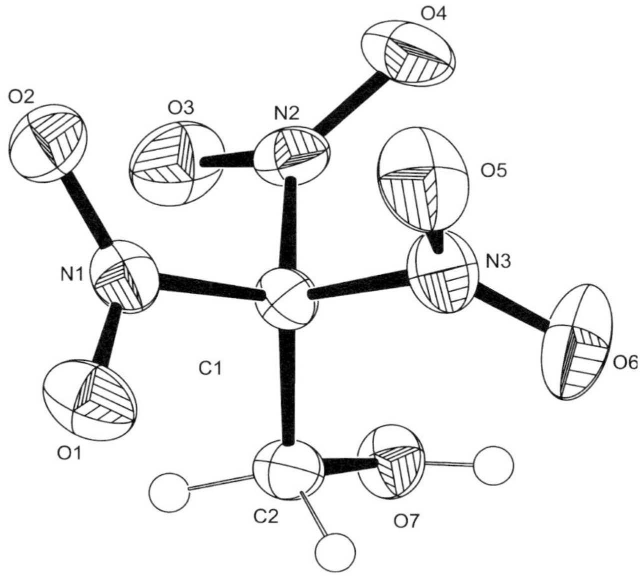

Fig. 16 ORTEP representation of the molecular structure of TNE in the crystalline state. The thermal ellipsoids are shown at the 50% probability level

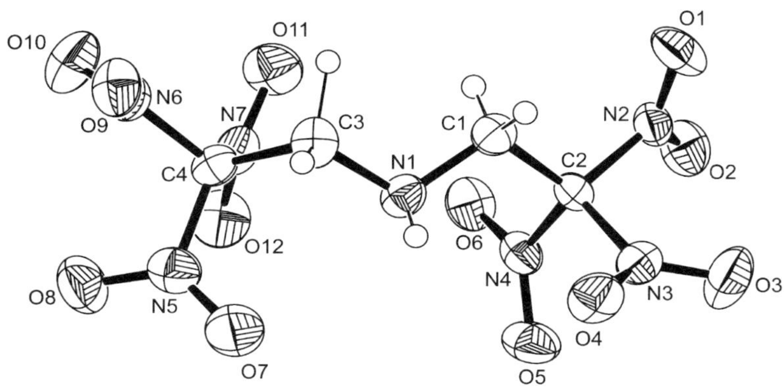

Fig. 17 ORTEP representation of the molecular structure of BTNA in the crystalline state. The thermal ellipsoids are shown at the 50% probability level

$^{14/15}N$ NMR spectroscopy, mass spectrometry, and elemental analysis. Their impact, friction, and electrostatic sensitivity data were measured in order to establish safe handling procedures for these compounds (Table 3). Furthermore, all compounds have been structurally fully characterized using single crystal X-ray diffraction methods (Table 4, Figs. 16–19).

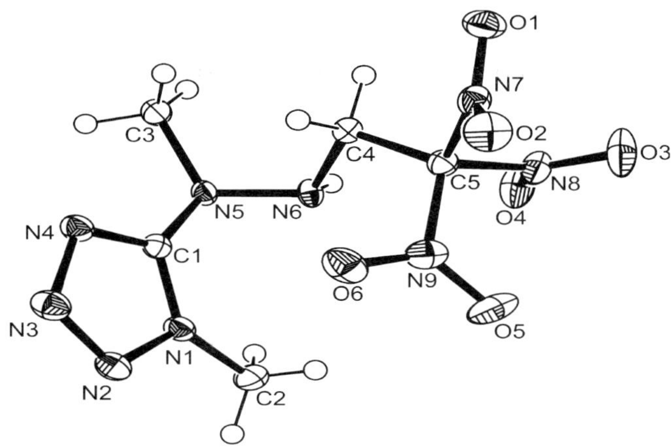

Fig. 18 ORTEP representation of the molecular structure of MMTHT in the crystalline state. The thermal ellipsoids are shown at the 50% probability level

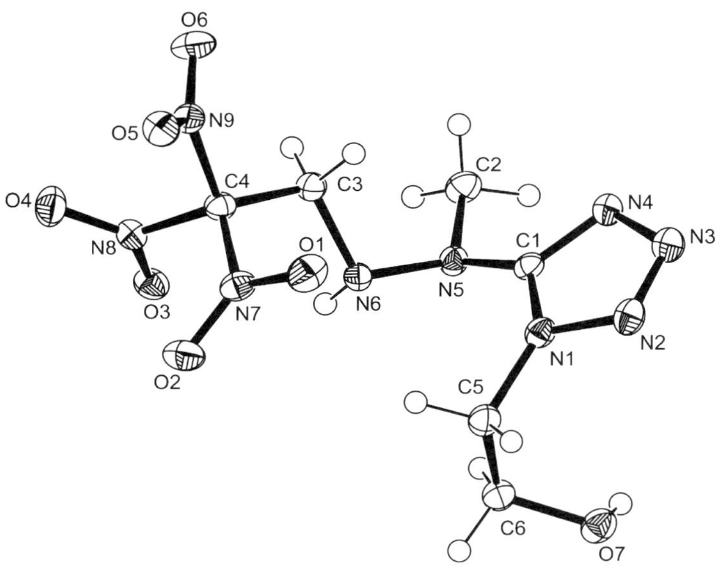

Fig. 19 ORTEP representation of the molecular structure of MMTHT in the crystalline state. The thermal ellipsoids are shown at the 50% probability level

3.4 Nitrated Aminotetrazoles

The combination of a tetrazole ring with energetic groups containing oxygen such as nitro groups ($R-NO_2$) [46], nitrate esters ($R-O-NO_2$) [47], or nitramines (R_2N-NO_2) [48] is of particular interest. Energetic materials based on tetrazoles show the desirable properties of high nitrogen contents on the one hand, and astonishing kinetic and thermal stabilities due to aromaticity on the other. Compounds with a high nitrogen contend are potential candidates for replacing common secondary explosives like RDX [49] (hexahydro-1,3,5-trinitro-*S*-triazine) and HMX [50] (octahydro-1,3,5,7-tetranitro-1,3,5,7-tetrazine), or in high-tech propellants when combined with a suitable oxidizer [51]. Nitroaminotetrazoles are of special interest because they combine both the oxidizer and the energetic nitrogen-rich backbone in one molecule.

The nomenclature of nitroaminotetrazoles (also referred to as nitramino- or nitriminotetrazoles) is usually inconsistent in literature [52] as a result of incomplete characterization of the previously reported compounds. Therefore, a complete characterization of three well-known nitroaminotetrazoles is given in this chapter (see also [53]). The crystal structures show the first examples of neutral 5-aminotetrazoles that have been nitrated at the primary NH_2 group. For 5-nitroaminotetrazole only the cell parameters have been previously published [54], while several examples of 5-methyl-nitraminotetrazoles have been structurally characterized and reported in the literature [55]. On the basis of the crystal structures obtained, the nitration product of 5-aminotetrazole (**1**) is now referred to as 5-nitriminotetrazole (**2**).

Nitriminotetrazoles and the corresponding metal nitramino-tetrazolates salts [56–58] have been known for a long time since they are cheap and easy to manufacture via various routes. However there are two main methods. The first synthesis uses protonation of 5-aminotetrazole (**1**) [59] using warm concentrated HNO_3 to form 5-aminotetrazole nitrate, followed by dehydration with concentrated sulfuric acid [60, 61] to form **2**. Another synthetic route is based on the cyclization of nitroguanylazide [62–64] (also known as nitroazidoformamidine). A further method is the N-nitration of aminotetrazoles using tetranitromethane [65, 66].

The investigated compounds **2**, **5**, and **6** are prepared starting from the corresponding 5-amino-1*H*-tetrazoles. 5-Aminotetrazole (**1**, 5-AT), first described by Thiele in 1892 [67], can be easily alkylated forming the methyl derivatives [68, 69]. The investigated and probably simplest way for preparing 5-nitriminotetrazole is the treatment of 5-aminotetrazole with fuming HNO_3 (Fig. 20).

The single crystal X-ray diffraction data were collected using an Oxford Xcalibur3 diffractometer with a Spellman generator (voltage 50 kV, current 40 mA) and a KappaCCD detector. The data collections were undertaken using the CrysAlis CCD software [70] and the data reductions were performed with the CrysAlis RED software [71]. The structures were solved with SIR-92 [72] and refined with SHELXL-97 [73] and finally checked using PLATON [74]. In all structures the hydrogen atoms were found and refined. The absorptions of **5** and **6** were corrected using the SCALE3 ABSPACK multiscan

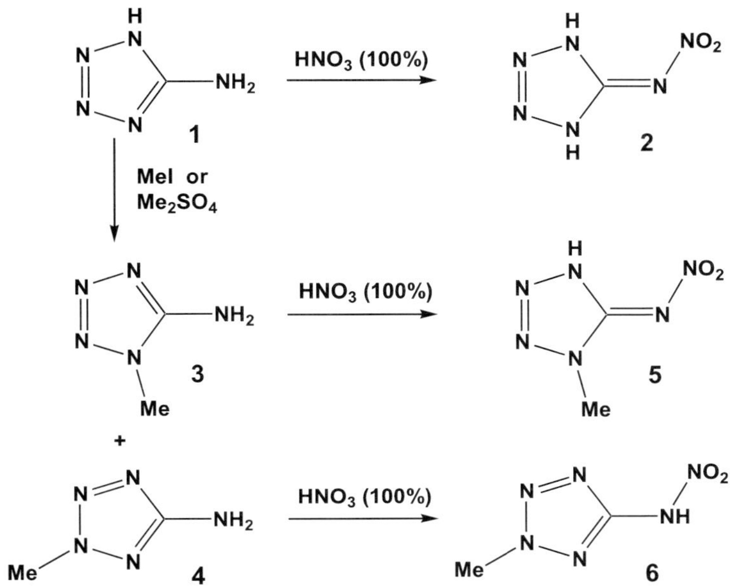

Fig. 20 The reaction of 5-aminotetrazole (**1**), 1-methyl-5-aminotetrazole (**3**), and 2-methyl-5-aminotetrazole (**4**) with 100% HNO_3

method [75]. In the chiral space group $P2_12_12_1$ the Friedel pairs were merged. Selected data from the X-ray data collection and refinements are given in Table 5. Further information regarding the crystal structure determinations has been deposited with the Cambridge Crystallographic Data Centre [76] as supplementary publication numbers 635164 (**2**), 635163 (**5**), and 635160 (**6**).

Compound **2** was characterized as a dibasic acid with pK_a values of 2.5 and 6.1 [60]. Previously it was not possible to locate the hydrogen atoms and it was assumed that the more acid hydrogen is located at the nitramino group [77]. However, using single crystal diffraction both hydrogen atoms were located on the tetrazole ring at nitrogen atoms N1 and N4. The ^{15}N NMR spectra shows only four signals, which suggests the same connectivity present in solution. DFT calculations show that this isomer is the one calculated lowest in energy even in the gas phase. The molecular unit can be seen in Fig. 21.

The geometry of the tetrazole ring in **2** can be compared with that of 5-aminotetrazole monohydrate (5-AT) [78]. The bond lengths in **2** are about 1.0 Å shorter than in 5-AT, with the shortest distance in **2** between the atoms N2 and N3 with 1.278(1) Å and the longest between N1 and N2 with 1.358(1) Å. The C1 – N5 bond length is 1.341(1) Å, which is closer to a C = N double bond (1.28 Å) than a C – N single bond (1.46 Å), whereas the nitramine bond N5 – N6 is considerably longer (1.363(1) Å). The angles in **2** differ from those in 5-AT. The N4 – C1 – N1 angle in **2** (103.9(1)°) is smaller than in 5-AT (107.9(1)°), which can be explained using the VSEPR

Table 5 Crystallographic data for compounds **2**, **5**, and **6**

	2	**5**	**6**
Formula	$CH_2N_6O_2$	$C_2H_4N_6O_2$	$C_2H_4N_6O_2$
Form. weight [$g\,mol^{-1}$]	130.09	144.11	144.11
Crystal system	Monoclinic	Orthorhombic	Monoclinic
Space Group	$P2_1/c$ (14)	$P2_12_12_1$ (19)	$P2_1/c$ (14)
Color/habit	Colorless cuboids	Colorless rods	Colorless blocks
Size [mm]	0.18 × 0.13 × 0.08	0.19 × 0.16 × 0.08	0.24 × 0.18 × 0.15
a [Å]	9.4010(3)	6.6140(1)	9.5278(9)
b [Å]	5.4918(1)	8.5672(2)	7.7308(7)
c [Å]	9.3150(3)	19.2473(4)	8.4598(9)
α [°]	90.0	90.0	90.0
β [°]	105.762(3)	90.0	112.875(9)
γ [°]	90.0	90.0	90.0
V [Å^3]	462.84(2)	1090.62(4)	574.1(1)
Z	4	8	4
ρ_{calc} [$g\,cm^{-3}$]	1.867	1.755	1.667
μ [mm^{-1}]	0.169	0.153	0.145
F(000)	264	592	296
$\lambda_{MoK\alpha}$ [Å]	0.71073	0.71073	0.71073
T [K]	100	100	200
θ min, max [°]	4.3, 32.1	3.7, 32.1	4.6, 26.0
Dataset	– 13 : 13; – 8 : 8; – 13 : 13	– 9 : 9; – 12 : 12; – 28 : 27	– 11 : 11; – 9 : 9; – 10 : 10
Reflections collected	6375	15881	5625
Independent reflections	1537	2104	1128
R_{int}	0.037	0.034	0.034
Observed reflections	1050	1616	1087
No. parameters	90	213	107
R_1 (obs)	0.0343	0.0297	0.0440
wR_2 (all data)	0.0960	0.0696	0.0950
GooF	1.00	0.99	1.21
Weighting scheme	0.05710	0.04220	0.03930 0.16880
Resd. Density [e/ Å^3]	– 0.23, 0.31	– 0.32, 0.21	– 0.22, 0.17
Device type	Oxford Xcalibur3 CCD	Oxford Xcalibur3 CCD	Oxford Xcalibur3 CCD
Solution	SIR-92	SIR-92	SIR-92
Refinement	SHELXL-97	SHELXL-97	SHELXL-97
Absorption correction	None	Multiscan	Multiscan
CCDC	635164	635163	635160

model [79] in which a double bond requires more space. A condition of the 6π-Hückel-aromaticity is a planar ring system, which can be seen at the torsion angle between N1 – N2 – N3 – N4 of 0.5(1)°. The nitramine

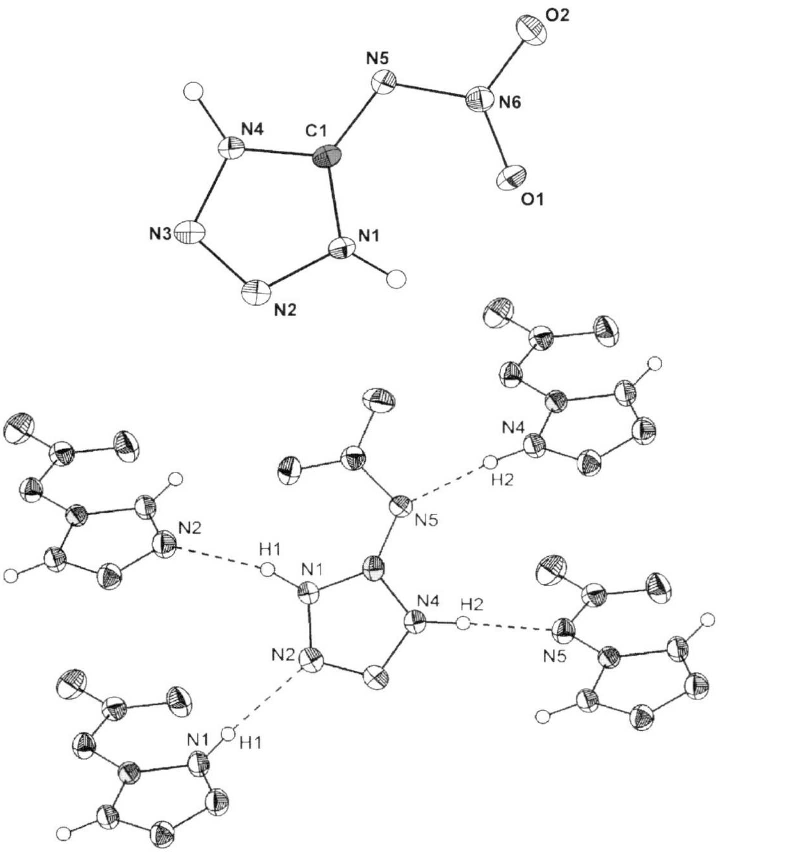

Fig. 21 *Left*: View of the molecular structure of **2** with thermal ellipsoid drawn at the 50% probability level. Hydrogen atoms are shown as small spheres of arbitrary radii. *Right*: H-bonding in **2** (N1 – H1 – N2^{i}: D – H 0.89(2) Å, H – A 2.11(2) Å, D – A 2.948(1) Å, D – H – A 156 (1)°; N4 – H4 – N5^{ii}: 0.98(2) Å, 1.83(2) Å, 2.850(2) Å, 175 (2)°; N4 – H4 – O2^{ii}: 0.98(2) Å, 2.48(2) Å, 3.146(1) Å, 125 (1)°; N4 – H4 – N6^{ii}: 0.98(2) Å, 2.58(2) Å, 3.4748(14) Å, 153 (1)°; ((i) 1 – *x*, 0.5 + *y*, 1.5 – *z* (ii) 2 – *x*, – 0.5 + *y*, 1.5 – *z*))

group also shows only slight derivations from the planarity (torsion angle C1 – N5 – N5 – O1 1.8(1)°), and is stabilized via a intramolecular hydrogen bond between N1 – H1 – O1. Bond lengths for the non-hydrogen atoms in **2** are given in Table 6. Each 5-nitriminotetrazole molecule participates in two strong intermolecular hydrogen bonds, as is illustrated in Fig. 21. The strong hydrogen bonds appear to be the most important reason for the high density of 1.867 g cm^{-3}.

Accordingly, the densities of **5** and **6** decrease because of the presence of the methyl groups. The molecular unit of **5** (Fig. 22) shows a similar con-

Table 6 Bond lengths of **2**, **5**, and **6**

Atoms	Bond length d(1–2) [Å]		
	2	**5**	**6**
O2 – N6	1.234(1)	1.234(2)	1.224(2)
O1 – N6	1.237(1)	1.266(2)	1.217(2)
N1 – N2	1.358(1)	1.355(2)	1.327(2)
N1 – C1	1.341(1)	1.345(2)	1.325(2)
N4 – C1	1.336(1)	1.338(2)	1.339(2)
N4 – N3	1.352(1)	1.364(2)	1.321(2)
N2 – N3	1.278(1)	1.284(2)	1.318(2)
N5 – N6	1.363(1)	1.338(2)	1.379(2)
N5 – C1	1.341(1)	1.346(2)	1.397(2)
N1(2) – C2		1.455(2)	1.459(2)

nectivity to that of **2**, whereby the N1 – H group in **2** has been substituted by a methyl group at atom N1. Again, in **5** the nitro group is tilted sideways where the hydrogen atom is located, forming an intramolecular hydrogen bond. In **5**, the nitrimine unit is also found to lie in the plane of the tetrazole ring (torsion angle O1 – N6 – N5 – C1 of 4.3(2)°), whereby the tetrazole ring in **5** shows comparable bond lengths with those observed in **2**. The most significant difference is the nitramine bond between N5 and N6, which is shorter in compound **5** (d (N5 – N6) = 1.338(2) Å). Finally, there is no significant differences for the angles observed in compounds **2** and **5**. The structure observed for **5** in the crystalline state is again influenced by several strong intermolecular hydrogen bonds, which are illustrated in Fig. 22.

The structure of **6** is considerably different to the structures of **2** and **5** discussed previously. The methyl group at the nitrogen atom N2 directs the remaining proton to the nitrogen atom N5, building a nitraminotetrazole (Fig. 23). In **6**, the C1 – N5 bond length is crucially longer with a distance of 1.397(2) Å and the nitramine unit does not lie in the plane of the tetrazole ring, as is clearly shown by the C1 – N5 – N6 – O1 torsion angle of – 19.2(2)°. The N6 – N5 – C1 angle (117.9(1)°) is larger than is observed in **2** and **5**, and the N5 – N6 nitramine bond of 1.379(2) Å is the longest in this series of compounds and can be seen as contributing to the low density of 1.667 g cm^{-3}, which is the lowest among the three structures discussed in this work. Further reasons for the lower observed density of **6** are the absence of strong intramolecular hydrogen bonds and the presence of only two moderately strong hydrogen bonds, as illustrated in Fig. 23. Relevant structural parameters are listed in Tables 6 and 7.

Vibrational spectroscopy is a suitable method for identifying nitraminotetrazoles. The IR and Raman (Fig. 24) spectra of compounds **2**, **5**, and **6** show

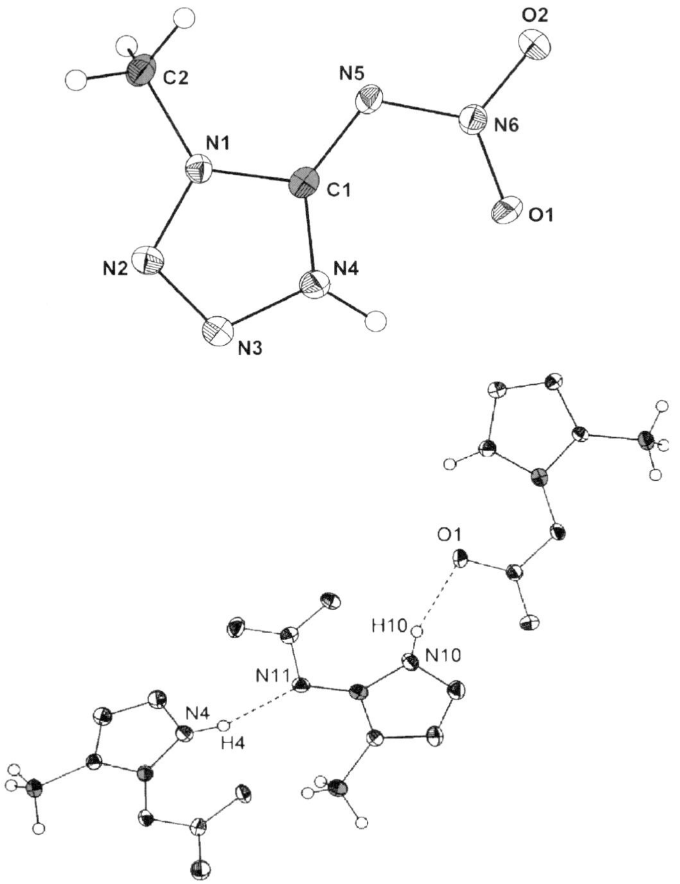

Fig. 22 *Left*: View of the molecular structure of **5**, representing the half of the asymmetric unit. Thermal ellipsoid are drawn at the 50% probability level and hydrogen atoms are shown as small spheres of arbitrary radii. *Right*: H-bonding in **5** (N4 – H4 – N11^i: D – H 0.93(2) Å, H – A 1.96(2) Å, D – A 2.848(2) Å, D – H – A 158 (2)°; N4 – H4 – O4: 0.93(19) Å, 2.59(2) Å, 3.098(2) Å, 115.3(15)°; N10 – H10 – O1^i: 0.80(2) Å, 2.10(2) Å, 2.874(2) Å, 163 (2)°; ((i) – *x*, 0.5 + *y*, 0.5 – *z*))

the vibrations expected from comparison with similar compounds in the literature [80, 81]. The N – NO_2 groups result in the strong absorptions in the 1280–1300 ($\nu_{sym}(NO_2)$) and 1560–1620 ($\nu_{asym}(NO_2)$) cm^{-1} regions as well as to a weak band at 945–970 (ν(N – N)) cm^{-1} [82]. The IR and Raman spectra of compounds **2**, **5**, and **6** show further characteristic absorption bands in the regions listed below: 3250–3100 cm^{-1} [ν(N – H)], 3000–2850 [ν(C – H) **5**, **6**], 1680–1550 [ν(N – H)], 1550–1350 [ν, tetrazole ring, $\nu_{as}(CH_3)$], ~ 1380 [$\nu(CH_3)$], 1350–700 [ν(N1 – C1 – N4), ν(N – N), ν(N – H), ν, tetrazole ring], < 700 [$\nu\delta_{oop}$(N – H)].

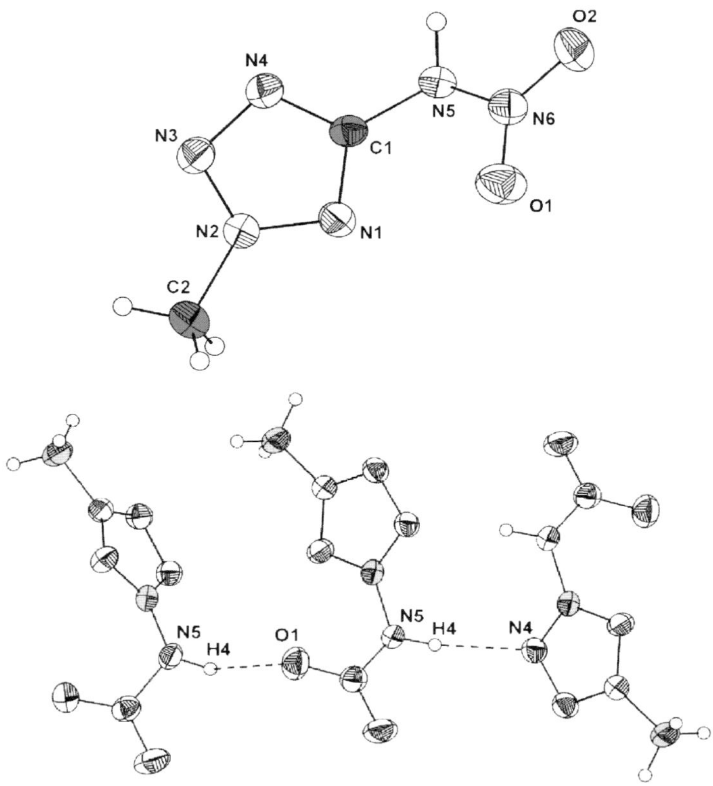

Fig. 23 *Left*: View of the molecular structure of **6**. Thermal ellipsoid are drawn at the 50% probability level and hydrogen atoms are shown as small spheres of arbitrary radii. *Right*: H-bonding in **6** (N5 – H1 – N4^{i}: D – H 0.86(2) Å, H – A 2.13(2) Å, D – A 2.965(2) Å, D – H – A 163 (2)°; N5 – H1 – O1^{ii}: 0,86(2), 2.66(2), 3.070(2), 111.0(18) ((i) $-x$, $-y$, $2-z$; (ii) x, $0.5-y$, $0.5+z$))

The ^{13}C and ^{15}N NMR chemical shifts and the ^{15}N – ^{1}H coupling constants are presented in Table 8. For all compounds the proton coupled as well as the proton decoupled ^{15}N NMR spectra (with full NOE) were recorded. The assignments are given based on the values of the ^{15}N – ^{1}H coupling constants and on comparison with the literature [55]. The chemical shifts are given with respect to CH_3NO_2 (^{15}N) and TMS (^{13}C) as external standards. In the case of ^{15}N NMR, negative shifts are upfield from CH_3NO_2. In all cases d_6-DMSO was used as the solvent.

DSC measurements to determine the decomposition temperatures of **2**, **5**, and **6** were performed in covered Al-containers with a nitrogen flow of 20 mL min^{-1} on a PerkinElmer Pyris 6 DSC [83], calibrated by standard pure indium and zinc at a heating rate of 5 °C min^{-1}. The DSC plots in Fig. 25 show the thermal behavior of 1.0 mg of **2**, **5**, and **6** in the 50–300 °C tem-

Table 7 Bond angles of **2**, **5**, and **6**

Atoms	Bond angles (1–2–3) [°] 2	5	6
N2 – N1 – C1	109.87(9)	110.4(1)	101.0(1)
C1 – N4 – N3	110.5(1)	110.2(1)	105.8(1)
N1 – N2 – N3	107.97(9)	107.8(1)	114.1(1)
N6 – N5 – C1	115.43(9)	115.7(1)	117.9(1)
N2 – N3 – N4	107.73(9)	107.7(1)	106.0(1)
N1 – N2 – C2		122.0(1)	
C1 – N1 – C2		127.3(1)	
N2 – N3 – C2			122.1(1)
N1 – N2 – C2			123.8(1)
O1 – N6 – O2	123.5(1)	121.6(1)	126.1(2)
O1 – N6 – N5	122.07(9)	121.9(1)	118.2(1)
O2 – N6 – N5	114.44(9)	116.5(1)	115.6(2)
N4 – C1 – N1	103.9(1)	103.9(1)	113.2(2)
N4 – C1 – N5	121.4(1)	136.9(1)	122.3(1)
N1 – C1 – N5	134.6(1)	119.2(1)	124.5(2)

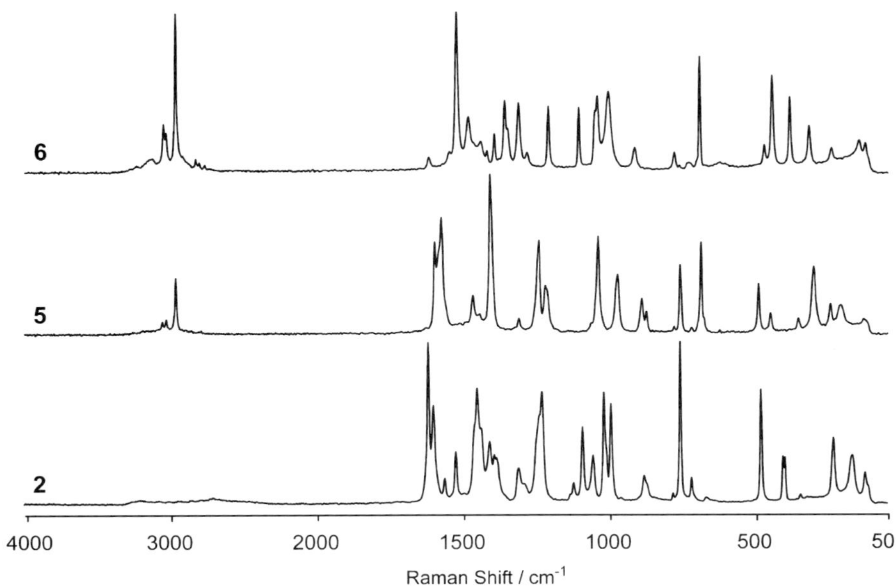

Fig. 24 Solid state Raman spectra of compounds **2** (*bottom*), **5** (*middle*), and **6** (*top*)

perature range. Compound **2** shows the lowest decomposition point at 122 °C, whereby compounds **5** and **6** decompose during melting at 125 °C and 123 °C, as shown by the curve of Fig. 5.

Table 8 ^{15}N NMR and ^{13}C NMR chemical shifts and ^{15}N–^{1}H coupling constants

Compound	NMR chemical shift [ppm]						
	δ(N1)	δ(N2)	δ(N3)	δ(N4)	δ(N4)	δ(N6)	δ(C1)
2	- 144.6	- 24.6	- 24.6	- 144.6	- 174.9	- 24.6	152.6
5	- 177.4^{a}	- 26.8^{b}	- 29.8	- 159.2	- 159.3	- 18.8	150.3
6	- 83.4^{b}	- 103.6^{c}	- 0.3^{b}	- 58.3	- 209.3	- 35.0	157.6

^{a} $^{2}J_{N-H}$ =2.1 Hz
^{b} $^{3}J_{N-H}$ =1.9 Hz
^{c} $^{2}J_{N-H}$ =2.2 Hz

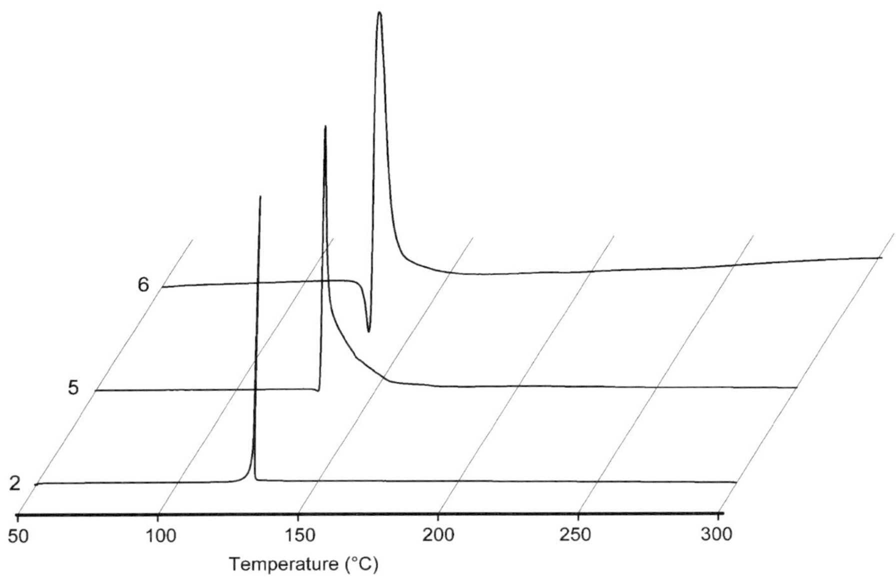

Fig. 25 DSC plot (endo up) of compounds **2**, **5**, and **6** (5 °C min^{-1}). T_{onset}: **2** 122 °C; **5** 125 °C; **6** 123 °C

To determine the heats of decomposition, a Linseis DSC PT10 [84] was used. Three samples (ca. 1 mg) were heated with a heating rate of 2 °C min^{-1} and a fixed nitrogen flow of 5 L h^{-1} over the decomposition peaks. The surface was integrated using the Linseis software and the average of three measurements was calculated to yield heats of decomposition $\Delta_{dec}H^{\circ}_{m}$ of 2638 J g^{-1} (**2**), 1685 J g^{-1} (**5**), and 2158 J g^{-1} (**6**).

For initial safety testing, the impact and friction sensitivities were tested according to BAM methods [85, 86] with the "BAM Fallhammer" and "BAM friction tester". Compound **2** is very sensitive towards impact (< 1.5 J) and friction (< 8 N) and since the value is comparable with lead azide, it should be considered to be a primary explosive, and should therefore only be handled with appropriate precautions. Compound **5** is moderately sensitive towards impact (< 12.5 J) and friction (< 160 N). However, **6** shows increased sensitivities comparing to **5** (impact: < 3.0 J, friction: < 145 N). Accordingly, **5** and **6** fall into the group of compounds described as "sensitive".

The heats of combustion for the compounds **2**, **5**, and **6** were determined experimentally, using a Parr 1356 bomb calorimeter (static jacket) equipped with a Parr 1108CL oxygen bomb [87]. To ensure better combustion, the samples (ca. 200 mg) were pressed with a defined amount of benzoic acid (ca. 800 mg) to form a tablet and a Parr 45C10 alloy fuse wire was used for ignition. In all measurements, a correction of 2.3 cal cm^{-1} wire burned has been applied, and the bomb was examined for evidence of noncombusted carbon after each run. A Parr 1755 printer was furnished with the Parr 1356 calorimeter to produce a permanent record of all activities within the calorimeter. The reported values are the average of three separate measurements. The calorimeter was calibrated by combustion of certified benzoic acid (SRM, 39i, NIST) in an oxygen atmosphere at a pressure of 3.05 MPa. The experimental results of the constant volume combustion energy ($\Delta_c U$) of the salts are summarized in Table 9. The standard molar enthalpy of combustion ($\Delta_c H^\circ$) was derived from $\Delta_c H^\circ = \Delta_c U + \Delta nRT$ ($\Delta n = \Delta n_i$(products, g) – Δn_i(reactants, g); Δn_i is the total molar amount of gases in the products or reactants). The enthalpy of formation, $\Delta_f H^\circ$, for each of the salts was calculated at 298.15 K using Hess' law and the following combustion reactions:

$$\textbf{2}: \quad CH_2N_6O_2 + 1/2O_2 \rightarrow CO_2 + H_2O + 3N_2 \tag{1}$$

$$\textbf{5,6}: \quad C_2H_4N_6O_2 + 2O_2 \rightarrow 2CO_2 + 2H_2O + 3N_2\,. \tag{2}$$

Table 9 shows that **2**, **5**, and **6** are strongly endothermic compounds ($\Delta_f H^\circ$ **2**: + 264 kJ mol^{-1}, **5**: + 260 kJ mol^{-1}, **6**: + 380 kJ mol^{-1}). The enthalpies of energetic materials are governed by the molecular structure of the compounds, and therefore, heterocycles with a higher nitrogen content (e.g., imidazole $\Delta_f H^\circ_{cryst} = 14.0$ kcal mol^{-1} [88]; 1,2,4-triazole $\Delta_f H^\circ_{cryst} = 26.1$ kcal mol^{-1}; tetrazole $\Delta_f H^\circ_{cryst} = 56.7$ kcal mol^{-1} [89, 90]) show higher heats of formation. From the experimentally determined heats of formation and densities obtained from single crystal structure X-ray diffraction, various thermochemical properties have been calculated using the EXPLO5 software and are summarized in Table 9.

The detonation parameters were calculated using the program EXPLO5 V5.02 [91, 92] and are summarized in Table 9. The program is based on the steady-state model of equilibrium detonation and uses BKW E.O.S [93] for gaseous detonation products and Cowan–Fickett E.O.S. for solid carbon. The calculations were performed using the maximum densities according to the crystal structures and the BKWN set of constants was used. **2** shows a very

Table 9 Physico-chemical properties of **2**, **5**, and **6**

	2	**5**	**6**
Formula	$CH_2N_6O_2$	$C_2H_4N_6O_2$	$C_2H_4N_6O_2$
Molecular mass	130.09	144.11	144.11
Impact sensitivity [J][a]	1.5	12.5	3.0
Friction sensitivity [N][b]	8	160	145
Electrical discharge	No reaction	No reaction	No reaction
N [%][c]	64.61	58.32	58.32
Ω [%][d]	- 12.30	- 44.41	- 44.41
Combustion	Yes	Yes	Yes
T_{dec} [°C][e]	122	125	122
Density [$g\,cm^{-3}$][f]	1.87	1.76	1.67
$-\Delta_c U$ [$cal\,g^{-1}$][g]	1750	2700	2902
$-\Delta_c H°$ [$kJ\,mol^{-1}$][h]	944	1619	1740
$\Delta_f H^{\circ}_m$ [$kJ\,mol^{-1}$][i]	264	260	380
$\Delta_{dec} H^{\circ}_m$ [$J\,g^{-1}$][j]	2638	1685	2158
$-\Delta_E U^{\circ}_m$ [$J\,g^{-1}$][k]	5326	5235	5998
T_E [K][l]	4309	3824	4283
P [kbar][m]	363	295	289
D [$m\,s^{-1}$][n]	9173	8433	8434
Gas volume [$mL\,g^{-1}$][o]	404	395	413

[a] BAM methods, insensitive > 40 J, less sensitive ≥ 35 J, sensitive ≥ 4 J, very sensitive ≤ 3 J [85, 86];
[b] BAM methods, insensitive > 360 N, less sensitive = 360 N, sensitive < 360 N to > 80 N, very sensitive ≤ 80 N, extremely sensitive ≤ 10 N [85, 86];
[c] Nitrogen content;
[d] Oxygen balance;
[e] Decomposition temperature from DSC ($\beta = 5\,°C$);
[f] Estimated from X-ray diffraction;
[g] Experimental (constant volume) combustion energy;
[h] Experimental molar enthalpy of combustion;
[i] Molar enthalpy of formation;
[j] Experimental enthalpy of decomposition using DSC;
[k] Energy of explosion, EXPLO5 V5.02;
[l] Explosion temperature, EXPLO5 V5.02;
[m] Detonation pressure, EXPLO5 V5.02;
[n] Detonation velocity, EXPLO5 V5.02;
[o] Assuming only gaseous products, EXPLO5 V5.02

high calculated detonation pressure of 363 kbar and a detonation velocity of 9173 $m\,s^{-1}$ higher than TNT ($P = 202$ kbar, $D = 7150\,m\,s^{-1}$) [91]. Even compound **5** also shows promising values for the detonation pressures (295 kbar) and explosion velocity (8433 $m\,s^{-1}$). The influence of the density on the properties of energetic materials is clearly shown by **6**, which shows the lowest detonation pressure of 289 kbar in spite of the highest positive heat of formation due to its low density.

In order to determine the thermal decomposition products, the compounds were heated in a evacuated steel tube for ca. 30 s at a temperature of 250 °C and the gaseous products were transferred into an evacuated gas IR cell. Figure Fig. 26 shows the gas phase IR spectra of the decomposition products of **2**, **5**, and **6**.

The thermal decomposition of **2** results in the formation of only two main products, which could be identified using IR spectroscopy, namely CO_2 [94] and CO [95]. In addition trace amounts of HCN and CH_4 were visible in the gas-phase IR spectrum, however, no evidence for the formation of water vapor was found. In the methylated compounds, many more decomposition products were observed using gas-phase IR spectroscopy. Besides CO_2 and CO, larger amounts of CH_4 [96] and HCN were found in the decomposition of **5** and **6**, in comparison with **2**. In contrast to **5**, where bigger amounts of expected NH_3 were detected, the thermal decomposition of **6** shows only traces of NH_3 but moderate amounts of N_2O [97].

The electrostatic potentials were illustrated after computing a optimal geometry at the B3LYP/6-31G* level of theory using the program package HyperChem 7.52 [98]. Figure 27 shows the 0.001 electron bohr^{-3} 3D isosurface of electron density for **2**, **5**, and **6**. In these diagrams a Jorgensen–Salem

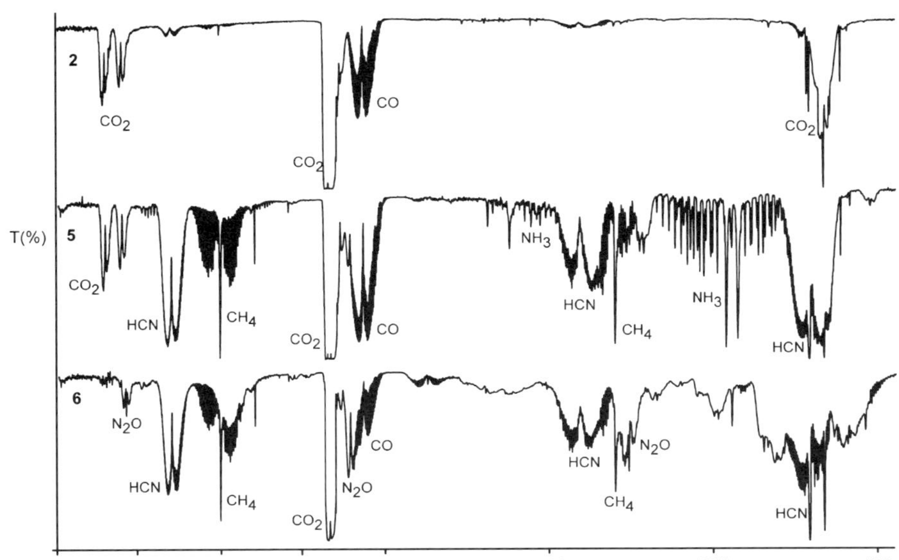

Fig. 26 IR spectra showing the gas-phase decomposition products of **2** (*top*), **5** (*middle*), and **6** (*bottom*)

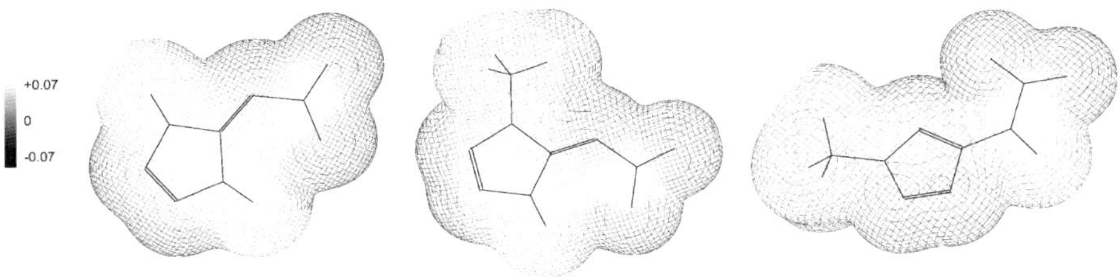

Fig. 27 Calculated (B3LYP /6-31G*) electrostatic potential of **2** (*left*), **5** (*middle*), and **6** (*right*). The *dark regions* represent electron-rich regions, the *light regions* electron-deficient regions

representation was chosen with an electrostatic potential contour value of 0.07 hartree. The dark regions represent extremely electron-rich regions ($V(r) < 0.07$ hartree) and the light regions extremely electron-deficient regions ($V(r) > 0.071$ hartree). In general, the patterns of the calculated electrostatic potentials of the surface of molecules can be related to the impact sensitivities [99–106]. In contrast to non-energetic organic molecules where the positive potential is larger but weaker in strength, in nitro and azo compounds usually more extensive regions with larger and stronger positive potentials are observed, which can be related to the increased impact sensitivities.

Long-term stability tests were performed using a Systag FlexyTSC [107] (thermal safety calorimetry) in combination with a RADEX V5 oven and the SysGraph software. The tests were undertaken as long-term isoperibolic evaluations in glass test vessels at atmospheric pressure with ca. 200 mg of the compounds. It was shown that tempering a substance for 48 h at 40 °C under the decomposition point results in a storage period of 58 years at room temperature. In our cases we chose a temperature of 80 °C and investigated the possible occurrence of exo- or endothermic behavior over a period of 48 h (Fig. 28). **2** and **5** were completely stable for 48 h, while **6** showed negligible minimal exothermic steps in the first 6 h. It can therefore be reasoned, that all

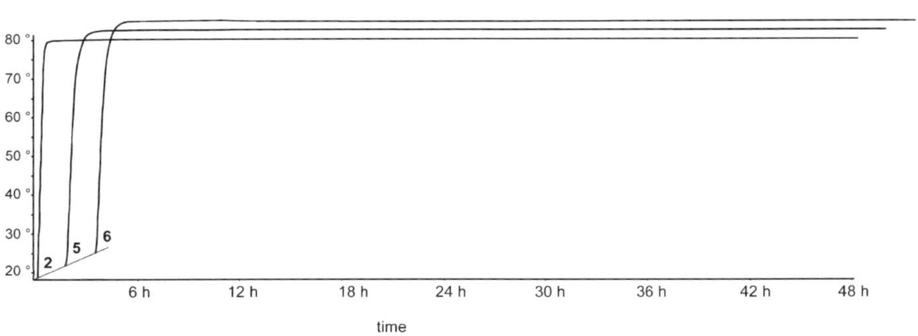

Fig. 28 Long-term stability screen of **2**, **5**, and **6** using a FlexyTSC (80 °C, 48 h)

three compounds show long-term stability, which is a basic requirement for possible applications.

From the combined experimental and computational study discussed in this chapter the following conclusions can be drawn:

- 5-Nitriminotetrazole (**2**), 1-methyl-5-nitriminotetrazole (**5**), and 2-methyl-5-nitraminotetrazole (**6**) can be synthesized in high yields and purity from aminotetrazole and 1- and 2-methyl-5-aminotetrazole, respectively, in simple one-step syntheses by reaction with fuming HNO_3.
- The crystal structures of **2**, **5**, **6** were determined using low-temperature single crystal X-ray diffraction. In the structure of **2** both hydrogen atoms could be located at the tetrazole ring forming a nitriminotetrazole. A similar structure can be observed for **5**, whereas **6** corresponds to a nitraminotetrazole, where the hydrogen atom is located at the nitrogen atom of the nitramine group. All of the compounds are stabilized in the crystalline state because of the presence of strong hydrogen bonds.
- Thorough characterization of the chemical, thermochemical, and energetic properties of **2**, **5** and **6** is reported. All compounds presented are energetic materials, showing increased sensitivities towards friction and impact and have long-term stability at room temperature. In the case of **2**, increased precautions should be undertaken when the compound is prepared on a larger scale.

3.5 1,6-Dimethyl-5-nitraminotetrazole

In an early publication [108] our research group inadvertently published a wrong value for the heat of formation determined from bomb-calorimetric measurements.

In the publication we described several nitroso- and nitraminotetrazoles, including 1,6-dimethyl-5-nitraminotetrazole (Fig. 29), as products of the nitration of 1,6-dimethyl-5-aminotetrazole using a mixture of nitric acid (100%) and trifluoroacetic anhydride (Fig. 30).

In the thermochemistry part of the earlier publication [108] a heat of formation of + 2.8 kcal mol^{-1} was reported. According to recent calculations by Rice et al. (see chapter in this volume), however, the heat of forma-

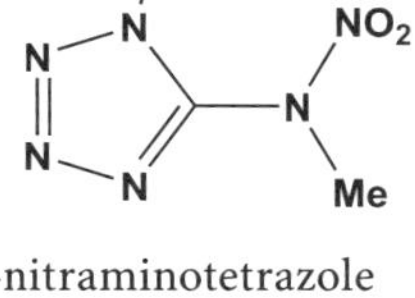

$C_3H_6N_6O_2$

Mol. Wt.: 158,12

C, 22.79; H, 3.82;
N, 53.15; O, 20.24

Fig. 29 1,6-Dimethyl-5-nitraminotetrazole

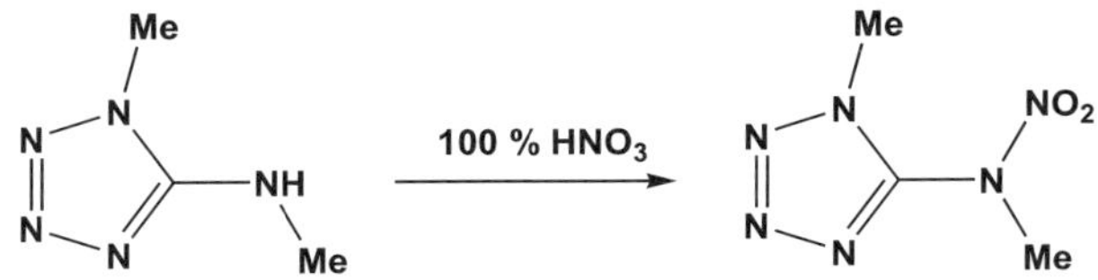

Fig. 30 Synthesis of 1,6-dimethyl-5-nitraminotetrazole

tion was expected to be 70 kcal mol^{-1}. This prompted us to synthesize 1,6-dimethyl-5-nitraminotetrazole again and to re-investigate the heat of combustion (Table 10). From the re-determined experimental heat of combustion of 564 kcal mol^{-1}, a heat of formation of 75 kcal mol^{-1} was then calculated, which is in much better agreement with the predicted value of 70 kcal mol^{-1}.

The calorimetric measurements were performed using a Parr 1356 bomb calorimeter (static jacket) equipped with a Parr 1108CL oxygen bomb for the combustion of highly energetic materials [87]. The samples (ca. 200 mg) were pressed with a well-defined amount of benzoic acid (ca. 800 mg) forming a tablet and a Parr 45C10 alloy fuse wire was used for ignition. In all measurements a correction of 2.3 (IT) cal cm^{-1} wire burned has been applied, and the bomb was examined for evidence of noncombusted carbon after each run. A Parr 1755 printer was furnished with the Parr 1356 calorimeter to produce a permanent record of all activities within the calorimeter. The reported values are the average of three single measurements. The calorimeter was calibrated by combustion of certified benzoic acid (SRM, 39i, NIST) in an oxygen atmosphere at a pressure of 3.05 MPa. The standard molar enthalpy of combustion ($\Delta_c H^\circ$) was derived from $\Delta_c H^\circ = \Delta_c U + \Delta nRT$ ($\Delta n = \Delta n_i$(products, g) – Δn_i(reactants, g); Δn_i is the total molar amount of gases in the products or reactants). The enthalpy of formation, $\Delta_f H^\circ$, for each of the salts was calculated at 298.15 K using the Hess' law and the following combustion reaction:

$$C_3H_6N_6O_2 + 3.5O_2 \rightarrow 3CO_2(g) + 3H_2O(l) + 3N_2(g) \quad \Delta n = 2.5 .$$

The heats of formation of the combustion products were obtained from the literature (NIST) [96].

Table 10 Thermodynamic values of 1,6-dimethyl-5-nitraminotetrazole

	Correct values	Wrong published values
$\Delta_c U$ [cal g^{-1}]	3565	3119
$\Delta_c U$ [kcal mol^{-1}]	564	490
$\Delta_f H^\circ$ [kcal mol^{-1}]	75	2.8
Calculated $\Delta_f H^\circ$	70	

4 Neutral Nitramine Compounds

4.1 Dinitrobiuret (DNB)

The LMU Munich research group has also been looking at tetrazole-free nitrogen-rich compounds that contain oxidizing groups such as nitramine functionalities. In this context, the preparation and structural characterization of dinitrobiuret (DNB) (Fig. 31) was carried out [109, 110]. The high chemical and thermal stability of DNB and the determined critical diameter of 6 mm for DNB (Figs. 32, 33) in the Koenen test (steel shell test) is comparable to the values reported for HMX (8 mm), RDX (8 mm), or PETN (6 mm) and prompted us to obtain the thermodynamic data and detonation pressures and velocities for DNB in a combined experimental and theoretical study.

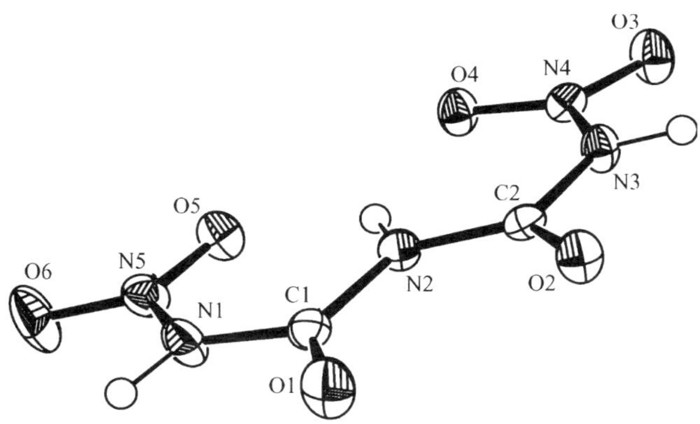

Fig. 31 Molecular structure of DNB in the crystalline state

The heat of combustion ($\Delta_c H$) of dinitrobiuret (DNB) was determined experimentally using oxygen bomb calorimetry: $\Delta_c H$(DNB) = 5195 ± 200 kJ kg^{-1}. The standard heat of formation ($\Delta_f H°$) of DNB was obtained on the basis of quantum chemical computations at the electron-correlated ab initio MP2 (second order Møller–Plesset perturbation theory) level of theory using a correlation consistent double-zeta basis set (cc-pV-DZ): $\Delta_f H°$(DNB) =– 353 kJ mol^{-1}, – 1829 kJ kg^{-1} (Fig. 34). The detonation velocity (D) and detonation pressure (P) of DNB was calculated using the empirical equations by Kamlet and Jacobs: D(DNB) = 8.66 mm µs^{-1}, P(DNB) = 33.9 GPa.

From this combined experimental and theoretical study the following conclusions could be drawn:

1. Dinitrobiuret (DNB) is a very powerful and promising new explosive

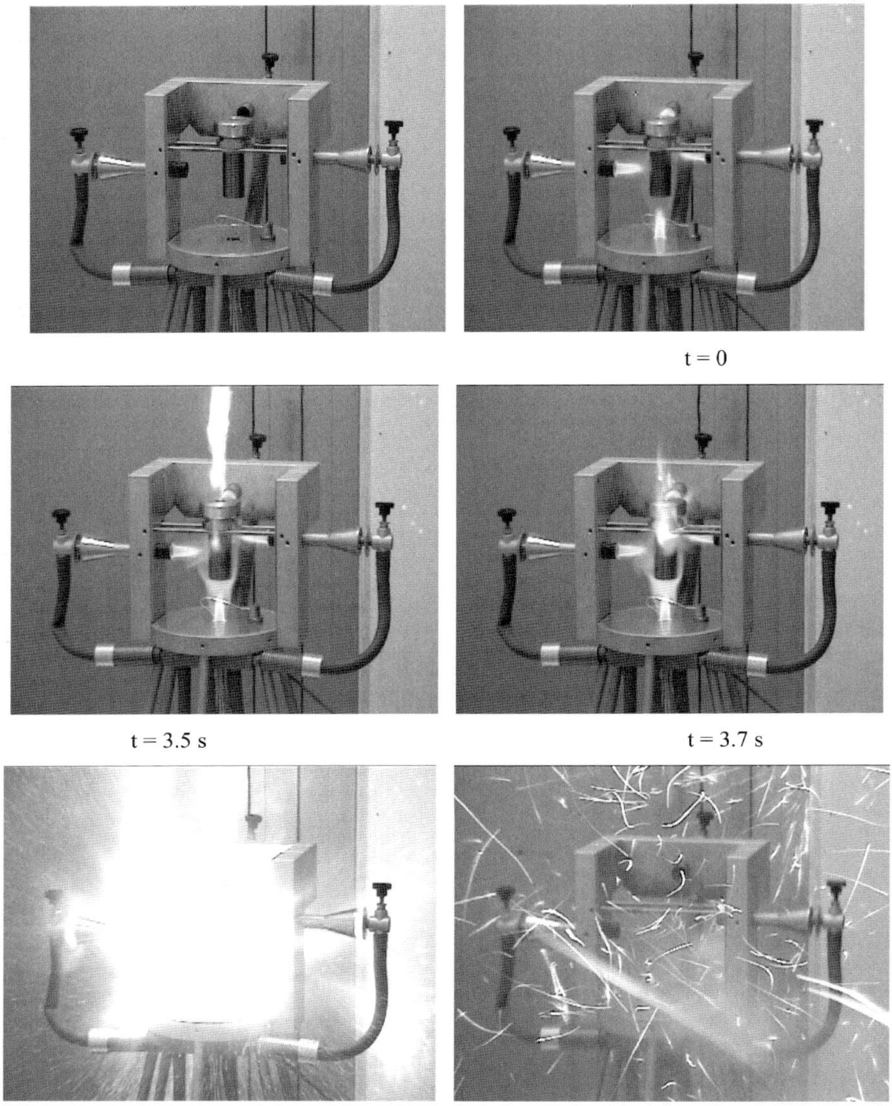

t = 0

t = 3.5 s

t = 3.7 s

t = 4.0 s

t = 4.1 s

Fig. 32 Steel sleeve test of DNB (10 g, 6 mm)

2. DNB has an almost neutral, even slightly positive oxygen balance of + 4.1%
3. DNB shows a detonation velocity and detonation pressure similar to well-established energetic materials such as PETN, RDX, or HMX

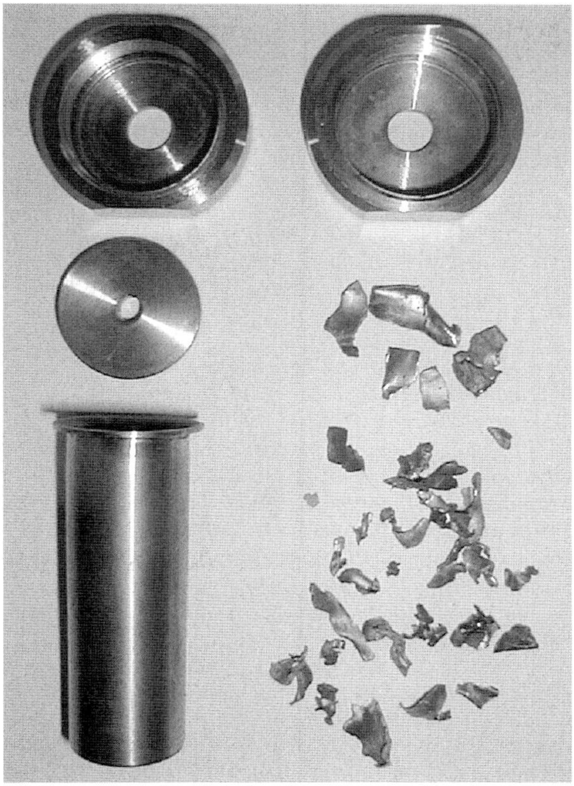

Fig. 33 Steel sleeve test of DNB (10 g, 6 mm) before (*left*) and after test (*right*)

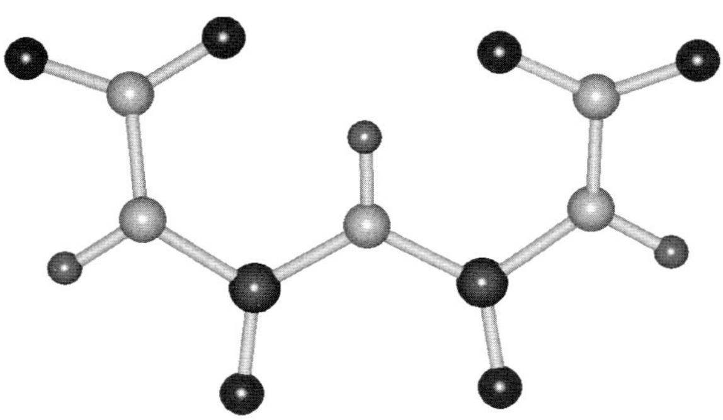

Fig. 34 Ab initio (MP2/cc-pVDZ) computed molecular structure of DNB

Acknowledgements Financial support of this work by the University of Munich (LMU), the Fonds der Chemischen Industrie (FCI) and the European Research Office (ERO) of the US Army Research Laboratory (ARL) under contract nos. 9939-AN-01 & N 62558-

05-C-0027 and the Bundeswehr Research Institute for Materials, Explosives, Fuels and Lubricants (WIWEB) under contract nos. E/E210/4D004/X5143 and E/E210/7D002/4F088 is gratefully acknowledged. The work presented in this chapter would not have been possible without the help of several outstanding coworkers in the LMU research group. The author especially thanks Prof. Konstantin Karaghiosoff, Dr. habil. Margaret-Jane Crawford, Dr. Jan J. Weigand, Dipl.-Chem. Michael Göbel and Dipl.-Chem. Jörg Stierstorfer for their help and support. Ms. Carmen Nowak is thanked for drawing all structures and preparing the figures and Ms. Irene Scheckenbach for her help with a difficult manuscript.

References

1. Klapötke TM, Holl G (2001) Green Chem 3:G75
2. Klapötke TM, Holl G (2002) Chem Aust, p 11
3. Kanekar P, Dautpure P, Sarnaik S (2003) Ind J Exp Biol 41:991
4. Robidoux PY, Gong P, Sarrazin M, Bardai G, Paquet L, Hawari J, Dubois C, Sunahara GI (2004) Can Ecotoxicol Environ Safe 58:300
5. Robidoux PY, Sunahara GI, Savard K, Berthelot Y, Dodard S, Martel M, Gong P, Hawari J (2004) Can Environ Toxicol Chem 23:1026
6. Simini M, Checkai RT, Kuperman RG, Phillips CT, Kolakowski JE, Kurnas CW, Sunahara GI (2003) Pedobiologia 47:657
7. Robidoux PY, Hawari J, Bardai G, Paquet L, Ampleman G, Thiboutot S, Sunahara GI (2002) Can Arch Environ Contam Toxicol 43:379
8. Steevens JA, Duke BM, Lotufo GR, Bridges TS (2002) Environ Toxicol Chem 21:1475
9. Pennington JC, Brannon JM (2002) Thermochim Act 384:163
10. Fournier D, Halasz A, Spain J, Spanggord RJ, Bottaro JC, Hawari J (2004) Appl Environment Microbiol 70:1123
11. Doherty RM (2003) In: De Luca LT, Galfetti L, Pesce-Rodriguez RA (eds) Novel energetic materials and applications. Proceedings of the 9th IWCP, Lerici, La Specia, Italy
12. Karaghiosoff K, Klapötke TM, Michailovski A, Nöth H, Suter M (2003) Prop Explos Pyrotech 28:1
13. Klapötke TM, Krumm B, Holl G, Kaiser M (1999) Proc of 30th int annual conference of ICT, June 29–July 2, Karlsruhe, Germany, p 120
14. Hammerl A, Klapötke TM, Nöth H, Warchhold M, Holl M, Kaiser M (2001) Inorg Chem 40:3570
15. Klapötke TM (2007) Nichtmetallchemie. In: Riedel E (ed) Moderne Anorganische Chemie, 3rd edn. Walter de Gruyter, Berlin
16. Eremets MI, Gavriliuk AG, Serebryanaya NR, Trojan IA, Dzivenko DA, Boehler R, Mao HK, Hemley RJ (2004) J Chem Phys 121:11296
17. Eremets MI, Gavriliuk AG, Trojan IA, Dzivenko DA, Boehler R (2004) Nat Mater 3:558
18. Eremets MI, Popov MY, Trojan IA, Denisov VN, Boehler R, Hemley RJ (2004) J Chem Phys 120:10618
19. Christe KO, Wilson WW, Sheehy JA, Boatz JA (1999) Angew Chem Int Ed 38:2004
20. Vij A, Wilson WW, Vij V, Tham FS, Sheehy JA, Christe KO (2001) J Am Chem Soc 123:6308
21. Klapötke TM (1999) Angew Chem 111:2694
22. Schroer T, Haiges R, Schneider S, Christe KO (2005) Chem Comm, p 1607

23. Vij A, Pavlovich JG, Wilson WW, Vij V, Christe KO (2002) Angew Chem Int Ed 41:3051
24. Lauderdale WJ, Stanton JF, Bartlett RJ (1992) J Phys Chem 96:1173
25. Perera SA, Bartlett RJ (1999) Chem Phys Lett 314:381
26. Tobita M, Bartlett RJ (2001) J Phys Chem A 105:4107
27. Glukhovtsev MN, Jiao H, Schleyer PVR (1996) Inorg Chem 35:7124
28. Glukhovtsev MN, Schleyer PVR (1992) Chem Phys Lett 198:547
29. Klapötke TM (2000) J Mol Struct (THEOCHEM) 499:99
30. Klapötke TM, Harcourt RD (2001) J Mol Struct (THEOCHEM) 541:237
31. Schmidt MW, Gordon MS, Boatz JA (2005) J Phys Chem A 109:7285
32. Wang R, Gao H, Ye C, Twamley B, Shreeve JM (2007) Inorg Chem 46:932
33. Geith J, Klapötke TM, Weigand JJ, Holl H (2004) Prop Explos Pyrotech 29:3
34. Klapötke TM, Mayer P, Schulz A, Weigand JJ (2005) J Am Chem Soc 127:2032
35. Fischer G, Holl G, Klapötke TM, Weigand JJ (2005) Thermochim Acta 437:168
36. Raap R (1969) Can J Chem 47:3677
37. Gaponik PN, Karavai VP (1984) Khim Geterotsikl Soedin 12:1683
38. Stolle R, Netz H, Kramer O, Rothschild S, Erbe E, Schick O (1933) J Prak Chem 138:1
39. Gálvez-Ruiz JC, Holl G, Karaghiosoff K, Klapötke TM, Löhnwitz K, Mayer P, Nöth H, Polborn K, Rohbogner CJ, Suter M, Weigand JJ (2005) Inorg Chem 44:4237
40. Darwich C, Klapötke TM (2006) New trends in research of energetic materials. Proceedings of the 9th seminar, Pardubice, Czech Republic, p 551
41. Gálvez-Ruiz JC, Holl G, Karaghiosoff K, Klapötke TM, Löhnwitz K, Mayer P, Nöth N, Polborn K, Rohbogner CJ, Suter M, Weigand JJ (2005) Inorg Chem 44:5192
42. Boese R, Klapötke TM, Mayer P, Verma V (2006) Prop Explos Pyrotech 31:263
43. Murray WM, Sauer WC (Arthur D, Little Inc) (1961) US Patent 3006957
44. Marans NS, Zelinski RP (1950) J Am Chem Soc 72:5329
45. Göbel M, Klapötke TM (2007) New trends in research of energetic materials. Proceedings of the 10th seminar, Pardubice, Czech Republic, p L13
46. Adam D, Karaghiosoff K, Holl G, Kaiser M, Klapötke TM (2002) Prop Explos Pyrotech 27:7
47. Karaghiosoff K, Klapötke TM, Michailovski A, Nöth H, Suter M (2003) Prop Explos Pyrotech 28:1
48. Klapötke TM, Mayer P, Verma V (2006) Prop Explos Pyrotech 31:263
49. Deal WE (1957) J Chem Phys 27:796
50. Mader CL (1963) Report LA-2900: Fortran BKW code for computing the detonation properties of explosives. Los Alomos Scientific Laboratory, NM
51. Urbanski T (1985) Chemistry and technology of explsoives. Pergamon, England
52. Astakhov AM, Vasilev AD, Molokeev MS, Revenko VA, Stepanov RS (2005) Russ J Org Chem 41:910
53. Klapötke TM, Stierstorfer J (2007) New trends in research of energetic materials. Proceedings of the 10th seminar, Pardubice, Czech Republic, p 35
54. Bryden JH (1953) Acta Cryst 6:669
55. Karahiosoff K, Klapötke TM, Mayer P, Piotrowski H, Polborn K, Willer RL, Weigand JJ (2005) J Org Chem 71:1295
56. Tappan BC, Beal RW, Brill TB (2002) Thermochim Act 288:227
57. Tappan BC, Incarnito CD, Rheingold AL, Brill TB (2002) Thermochim Act 384:113
58. Brill TB, Tappan BC, Beal RW (2001) New trends in research of energetic materials. Proceedings of the 4th seminar, Pardubice, Czech Rep, p 17
59. Thiele J (1892) Ann 270:1
60. Herbst RM, Garrison JA (1953) J Org Chem 18:941

61. Lieber E, Sherman E, Henry RA, Cohen J (1951) J Am Chem Soc 73:2327
62. Astachov AM, Nefedo AA, Vasiliev AD, Kruglyakova LA, Dyugaev KP, Stepanov RS (2005) Proc of 36th int annual conference of ICT, Jun 28–July 1, Karlsruhe, Germany, p 113
63. Lieber E, Sherman E, Henry RA, Cohen J (1951) J Am Chem Soc 73:2327
64. O'Connor TE, Fleming G, Reilly J (1949) J Soc Chem Ind (London) 68:309
65. Mayants AG, Klimenko VS, Erina VV, Pyreseva KG, Gordeichuk SS, Leibzon VN, Kuz'min VS, Burtsev UN (1991) Khim Geterot Soed 8:1067
66. Göbel M, Klapötke TM, Mayer E (2006) Z Anorg Allg Chem 2632:1043
67. Thiele J (1892) Ann 270:1
68. Henry RA, Finnegan WG (1954) J Am Chem Soc 76:923
69. Henry RA, Finnegan WG (1956) J Am Chem Soc 78:411
70. Oxford Diffraction (2005) CrysAlis CCD, Version 1.171.27p5 beta (release 01-04-2005 CrysAlis171.NET)
71. Oxford Diffraction (2005) CrysAlis RED, Version 1.171.27p5 beta (release 01-04-2005 CrysAlis171.NET)
72. Altomare A, Cascarano G, Giacovazzo C, Guagliardi A (1993) SIR-92, A program for crystal structure solution. J App Cryst 26:343
73. Sheldrick GM (1994) SHELXL97 program for the refinement of crystal structures. University of Göttingen, Germany
74. Spek AL (1999) PLATON, a multipurpose crystallographic tool. Utrecht University, Utrecht, The Netherlands
75. Oxford Diffraction (2005) SCALE3 ABSPACK (1.0.4,gui:1.0.3) (C)
76. Cambridge Crystallographic Data Centre (2007) http://www.ccdc.cam.ac.uk/, last visited: 30 Mar 2007
77. Lieber E, Patinkin T (1951) J Am Chem Soc 73:1792
78. Bray DD, White JG (1979) Acta Cryst B 35:3089
79. Riedel E (1999) Anorganische Chemie, 4th edn. Walter de Gruyter, Berlin, p 134
80. Lieber E, Levering DR, Patterson LJ (1951) Anal Chem 23:1594
81. Weigand JJ (2005) Dissertation, Ludwig Maximilian University Munich
82. Daszkiewicz Z, Nowakowska EM, Preździo WW, Kyzioł JB (1995) Pol J Chem 69:1437
83. PerkinElmer (2007) http://www.perkinelmer.com, last visited: 30 Mar 2007
84. Linseis (2007) http://www.linseis.com, last visited: 30 Mar 2007
85. United Nations Economic Commission for Europe (2005) UN recommendations on the transport of dangerous goods, 14th edn. http://www.unece.org/trans/danger/publi/unrec/rev14/14files_.html, last visited: 30 Mar 2007
86. Bundesanstalt für Materialforschung und -prüfung (2007) http://www.bam.de, last visited: 30 Mar 2007
87. Parr Instrument Company (2007) http://www.parrinst.com, last visited: 30 Mar 2007
88. West RC, Selby SM (eds) (1967–1968) Handbook of chemistry and physics, 48th edn. CRC, Cleveland, OH
89. McEwan WS, Rigg MW (1951) J Am Chem Soc 73:4725
90. Ostrovskii VA, Pevzner MS, Kofman TP, Tselinskii IV (1999) Targets Heterocycl Syst 3:467
91. Suceska M (1999) Proc of 30th int annual conference of ICT, June 29–July 2, Karlsruhe, Germany, p 50
92. Suceska M (2001) EXPLO5.V2: computer program for calculation of detonation parameters. Proc of 32nd int annual conference of ICT, July 3–6, Karlsruhe, German, p 110
93. Suceska M (1991) Prop Explos Pyrotech 16:197

94. Mecke R, Langenbucher F (1965) Infrared spectra. Heyden, London, Serial no. 6
95. Shimanouchi T (1972) Tables of molecular vibrational frequencies consolidated, vol II. J Phys Chem Ref Data 6:993
96. National Institue of Standards and Technology (2007) Vibrational energy search http://webbook.nist.gov/chemistry/vib-ser.html, last visited: 30 Mar 2007
97. Nakamoto K (1986) Infrared and Raman Spectra of inorganic and coordination compounds, 4th edn. Wiley, New York
98. Hypercube (2002) HyperChem 7.52: Molecular visualization and simulation program package. Hypercube, Gainsville, FL
99. Murray JS, Lane P, Politzer P (1995) Mol Phys 85:1
100. Murray JS, Lane P, Politzer P (1998) Mol Phys 93:187
101. Politzer P, Murray JS (1999) Computational characterization of energetic materials. In: Maksic ZB, Orville-Thomas WJ (eds) Pauling's legacy: modern modelling of the chemical bond. Theor Comput Chem 6:347
102. Politzer P, Murray JS, Seminario JM, Lane P, Grice ME, Concha MC (2001) J Mol Struct (THEOCHEM) 573:1
103. Rice BM, Chabalowski CF, Adams GF, Mowrey RC, Page M (1991) Chem Phys Lett 184:335
104. Rice BM, Hare JJ (2002) J Phys Chem A 106:1770
105. Rice BM, Sahu S, Owens FJ (2002) J Mol Struct (THEOCHEM) 583:69
106. Rice BM (2005) Adv Ser Phys Chem 16:33
107. Systag (2007) Process development and safety http://www.systag.ch, last visited: 30 Mar 2007
108. Karaghiosoff K, Klapötke TM, Mayer P, Piotrowski H, Polborn K, Willer RL, Weigand JJ (2005) J Org Chem 71:1295
109. Geith J, Klapötke TM, Weigand JJ, Holl G (2004) Prop Explos Pyrotech 29:3
110. Geith J, Holl G, Klapötke TM, Weigand JJ (2004) Combust Flame 139:358

Struct Bond (2007) 125: 123–151
DOI 10.1007/430_2007_058

Published online: 12 May 2007

Organic Difluoramine Derivatives

Robert D. Chapman

Chemistry Branch (Code 498200D),
Research Division, Research & Engineering Sciences Department,
Naval Air Warfare Center Weapons Division, Naval Air Systems Command,
China Lake, CA 93555-6106, USA
robert.chapman@navy.mil

Abstract This article reviews recent developments in synthesis and characterizations of organic difluoramine (NF_2) derivatives since the 1980s. Specific classes of chemical compounds are covered: cyclic difluoramino-nitramines, especially *gem*-bis(difluoramino)-substituted analogs of conventional nitramines; products from the relatively new transformation of electrophilic difluoramination; other classes of difluoramines such as polymers, plasticizers, and compounds not otherwise categorized. Also covered are aspects of other recently developed synthetic methodology to prepare compounds in this general class, as well as purely theoretical treatments of the chemistry and properties of this class.

Keywords Difluoramine · Difluoroamine · Difluoramination · HNFX · Nitramines

Abbreviations

BNMO 3,3-Bis(nitratomethyl)oxetane
HMX Octahydro-1,3,5,7-tetranitro-1,3,5,7-tetrazocine

HNFX	3,3,7,7-Tetrakis(difluoramino)octahydro-1,5-dinitro-1,5-diazocine
NMMO	3-Methyl-3-(nitratomethyl)oxetane
PETN	Pentaerythrityl tetranitrate
RDX	Hexahydro-1,3,5-trinitro-1,3,5-triazine
RNFX	5,5-Bis(difluoramino)hexahydro-1,3-dinitropyrimidine
TNFX	3,3-Bis(difluoramino)octahydro-1,5,7,7-tetranitro-1,5-diazocine

1 Introduction

This review will emphasize recent developments in synthesis and characterizations of organic difluoramine (NF_2) derivatives[1]. Organic monofluoramines have been recently reviewed [1–3], as many of them tend to be suitable as electrophilic fluorinating reagents. Early [4–7] and recent [8] reviews of inorganic N–F derivatives have appeared.

Interest in organic difluoramines has been predominantly due to their potential as energetic materials in rocket propellant and explosives formulations. This potential derives from the intrinsically high energy of the difluoramino (NF_2) group relative to other substituents of oxidizing capability in energetic ingredients, such as nitro (NO_2). While this difference in energy content is not conspicuously apparent in thermodynamic properties such as the heats (enthalpies) of formation, which tend to be more negative than those of analogous nitro derivatives [9], the energetics of processes of interest (e.g., combustion) manifest this difference distinctly, as will be seen in discussions of these properties for certain compounds of interest. Thus, quantitative measures of realistic chemical processes (such as specific impulse of rocket propellant combustion) offer a more valid comparison of the energetics of difluoramino derivatives.

An attractive feature of fluorinated oxidizers is that their combination with metal fuel ingredients (e.g., aluminum or boron) offers a propellant performance advantage of producing more moles of more-volatile, lower-molecular-weight combustion products, such as BOF (O=B–F) rather than simple oxides (B_2O_3) produced by solely oxygenated oxidizers [10, 11]. Difluoramino derivatives should also offer a kinetically more available form of fluorine (to achieve higher combustion or explosion efficiency) than the "stabler" fluoroalkyl (C–F) derivatives that have been employed in energetic formulations.

The difluoramino (NF_2) substituent is also more dense than NO_2, contributing favorably to explosive performance and to the propellant property density×(specific impulse). According to Ammon's linear vol-

[1] This review will use the colloquial American English term "difluoramino" rather than the (admittedly more systematic) equivalent "difluoroamino", which is more commonly used in non-English publications and English translations of non-English (especially Russian) publications.

ume additivity method [12], NF_2 has a substituent density of 2.303 g cm^{-3} (52.0035 amu/37.504 Å^3) vs. an NO_2 density of 2.166 g cm^{-3} (46.005 amu/35.265 Å^3).

2
Historical Development

Although an organic difluoramine derivative (CF_3NF_2 by-product formed by direct fluorination of silver cyanide) had been reported as early as 1936 [13], the impetus for the major development of this class of compound was the recognition by the US Advanced Research Projects Agency and Department of Defense that various N–F compounds, especially organic difluoramines, offered the prospect of being superior propellant oxidizers, for reasons alluded to above. Thus, Project Principia was carried out over the period 1958–1965 to explore the possibilities offered by these materials [14]. While research projects funded by this program ended in 1964–1965, open-literature publications of their results appeared for several years afterward. Hundreds of organic difluoramines (even hundreds with the *gem*-bis(difluoramino) linkage, $C(NF_2)_2$) were prepared under the aegis of Project Principia and of foreign efforts along the same lines. Several good reviews [15–25] over subsequent years have described the early organic difluoramine chemistry of this era.

Project Principia failed to produce a difluoramine-based oxidizer candidate sufficiently superior to replace nitro derivatives that were the technological standards. Almost all difluoramine derivatives at that time had inadequate stability, insensitivity, or physical properties to serve as such replacements. Organic difluoramine research languished for a couple of decades afterward, though results were published throughout the 1970s and 1980s by a few researchers, especially Fokin and coworkers at the Nesmeyanov Institute of Heteroorganic Compounds (Moscow).

3
A New Hope

A resurgence in interest in organic difluoramines occurred in the USA in the late 1980s, based on support by the Office of Naval Research [26], with additional support soon received from the Ballistic Missile Defense Organization (manager of the Strategic Defense Initiative program). Efforts were made to pursue a specific new class of compounds proposed by Baum and Archibald of Fluorochem [27]: cyclic difluoramino-nitramines that were analogs of conventional nitramines (e.g., HMX and RDX) but with *gem*-bis(difluoramino) groupings replacing some of the nitramine groups. Also in the late 1980s, Zheng et al. (Xi'an Modern Chemistry Institute, China) alluded, in a conference proceeding, to preparing at least one member of this class [28]: "For example, the difluoramination of cyclic derivatives of 1,3-diaminoacetone (the substituents on the nitrogen atoms were $-NO_2$... etc.) is easier to be carried out than that of the chain derivatives." However, no specific example was described, and subsequent journal reports [29, 30] of other chemistry from this conference proceeding made no mention of cyclic derivatives.

This section will review a variety of experimental results arising from this renaissance period in difluoramine chemistry.

3.1 Cyclic Difluoramino-nitramines

Although nitramines containing (difluoramino)alkyl substituents had been prepared as early as 1963 and published by 1972 [31, 32], they were *vic*-bis(difluoramino)-substituted primary nitramines and *N*-alkyl-*N*-(difluoraminomethyl)nitramines with insufficient stability and difluoramine content to constitute attractive candidates for formulations. In contrast, the newly proposed *gem*-bis(difluoramino)-substituted cyclic nitramines would have high difluoramine content and should exhibit high densities compared to the analogous conventional nitramines. The favorable effect of cyclic (or even cage) structures in imparting desirable higher density and better oxygen balance was better appreciated by this time. It would also be expected that difluoramine derivatives with a nitramine component would exhibit better physical properties (e.g., melting point) than the earlier simple (difluoramino)alkanes. However, cyclic nitramines with β,β-bis(difluoramino) substituents relative to nitramine components posed a likely synthetic challenge: conventional methodology for preparation of *gem*-bis(difluoramino)alkanes required strongly acidic conditions (such as anhydrous sulfuric acid, fuming sulfuric acid, difluorosulfamic acid, or fluorosulfonic acid) with which most nitramines are incompatible. Alternatively, conventional nitrolyzably protected heterocyclic nitrogens tended not to survive such difluoramination conditions either. So, finding methodology to efficiently prepare this new class of organic difluoramine would be a daunting task.

3.1.1 HNFX

On the basis of estimations (personal communications from Baroody E of the Naval Ordnance Station, Indian Head, MD; Ammon HL of the University of Maryland, College Park; and Adolph HG, Koppes WM, and Lawrence GW of the Naval Surface Warfare Center, Silver Spring, MD, cited in [33]) of physical and performance properties of several *gem*-bis(difluoramino)-substituted

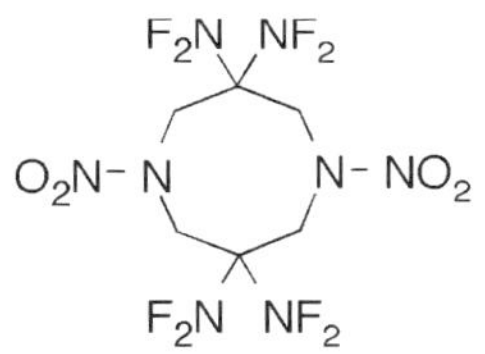

1

analogs of known cyclic nitramines, 3,3,7,7-tetrakis(difluoramino)octahydro-1,5-dinitro-1,5-diazocine (HNFX, **1**) was seriously pursued early on as the most attractive candidate member of this class. (The acronym HNFX [26] signifies the HMX analog of the class of NF_2-modified cyclic nitramine explosives.)

Synthesis

In the first synthesis of HNFX, the use of a nitramine reactant, 5-(difluoramino)-3,7-dinitro-9-oxa-3,7-diazabicyclo[3.3.1]nonan-1-ol (**2**), in a typical difluoramination reaction (difluoramine–difluorosulfamic acid–sulfuric acid) produced HNFX in only ~ 1% yield [34] (Scheme 1).

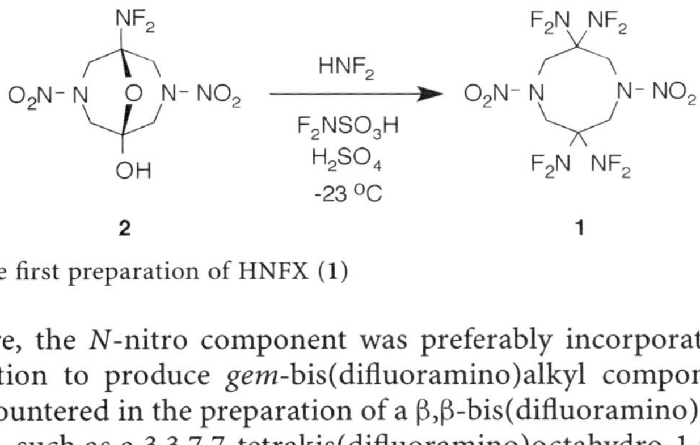

2 **1**

Scheme 1 The first preparation of HNFX (**1**)

Therefore, the *N*-nitro component was preferably incorporated after difluoramination to produce *gem*-bis(difluoramino)alkyl components. A dilemma encountered in the preparation of a β,β-bis(difluoramino)-substituted heterocycle, such as a 3,3,7,7-tetrakis(difluoramino)octahydro-1,5-diazocine, is that the nitrogen in most aminoacetone derivatives would be more basic than the ketone carbonyl (and oxygen in hemiaminal intermediates). This would deactivate difluoramination via difluoramino-carbocations unless the nitrogen is protected with a sufficiently electronegative protecting group to favorably affect this basicity. The dilemma was resolved by employment of a strong-acid-stable N-protecting group, 4-nitrobenzenesulfonyl, allowing acceptably efficient difluoramination of the ketone reactant tetrahydro-1,5-bis(4-nitrobenzenesulfonyl)-1,5-diazocine-3,7(2*H*,6*H*)-dione (**3**) while keeping the *N*-nosyl group intact [35] (Scheme 2). Also, trichlorofluoromethane ($CFCl_3$) was utilized as an inert cosolvent for difluoramine in order to alleviate the known hazards of condensed-phase neat HNF_2. The next dilemma

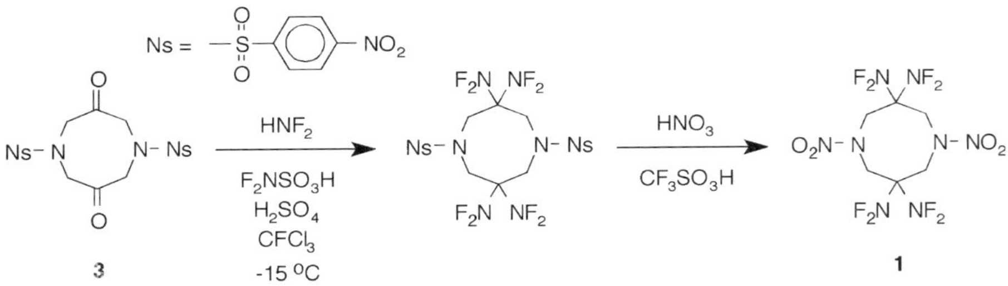

Scheme 2 An improved preparation of HNFX (**1**): *p*-nosyl N-protection during difluoramination plus nitrolysis with protonitronium ion

encountered in this approach, however, was the known reluctance of more-electronegatively substituted amides and of sterically hindered amides to undergo direct nitrolysis. The *N*-nosyl protecting group could be directly nitrolyzed by use of the protonitronium (NO_2H^{2+}) reagent formed from a nitronium source in the presence of a superacid, such as the system nitric acid–trifluoromethanesulfonic acid [36].

Although this transformation worked well enough in small reactions, it proved inefficient upon scale-up. A superior, scalable one-pot process (Scheme 3) from diazocinedione to HNFX was developed more recently [37]. Here, efficient protolytic N-denosylation is effected by trifluoromethanesulfonic acid added to the crude 3,3,7,7-tetrakis(difluoramino)octahydro-1,5-bis(4-nitrobenzenesulfonyl)-1,5-diazocine intermediate (**4**), still contained in the acid system used for difluoramination after expulsion of volatiles such as excess difluoramine and trichlorofluoromethane. The denosylated intermediate, a 3,3,7,7-tetrakis(difluoramino)octahydro-1,5-diazocinium salt (**5**), is straightforwardly nitrated to HNFX by added nitric acid.

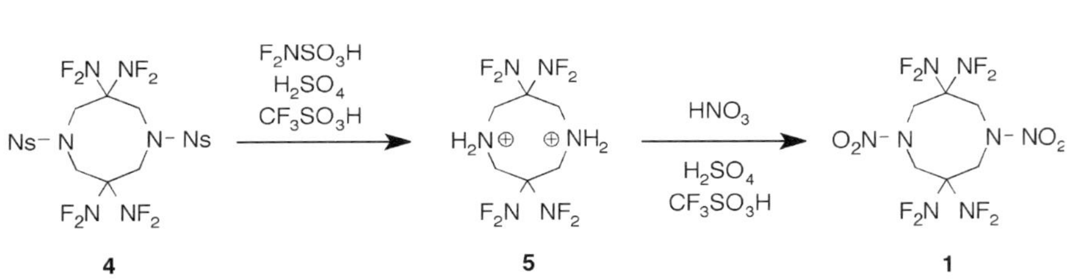

Scheme 3 Scalable preparation of HNFX (**1**): protolytic N-denosylation plus simple nitration

Chapman et al. [36] had suggested that solvents with less ozone-depleting potential than $CFCl_3$ could be used, and a preliminary demonstration of difluoramination with dichloromethane solvent was described. Adolph and Stern more recently chose *n*-pentane as an example of an alternative cosolvent (with $CFCl_3$) for this purpose [38].

Structure

X-ray crystallographic analysis of HNFX recrystallized from a variety of solvent systems showed an interesting, unpredicted feature (Fig. 1): channels, with a threefold axis of symmetry surrounded by HNFX molecules, passing through each unit cell [36]. Analysis sometimes showed the presence of disordered mass due to recrystallization solvent(s), though solvent-free crystals can be prepared by driving out solvent (e.g., by heating under vacuum).

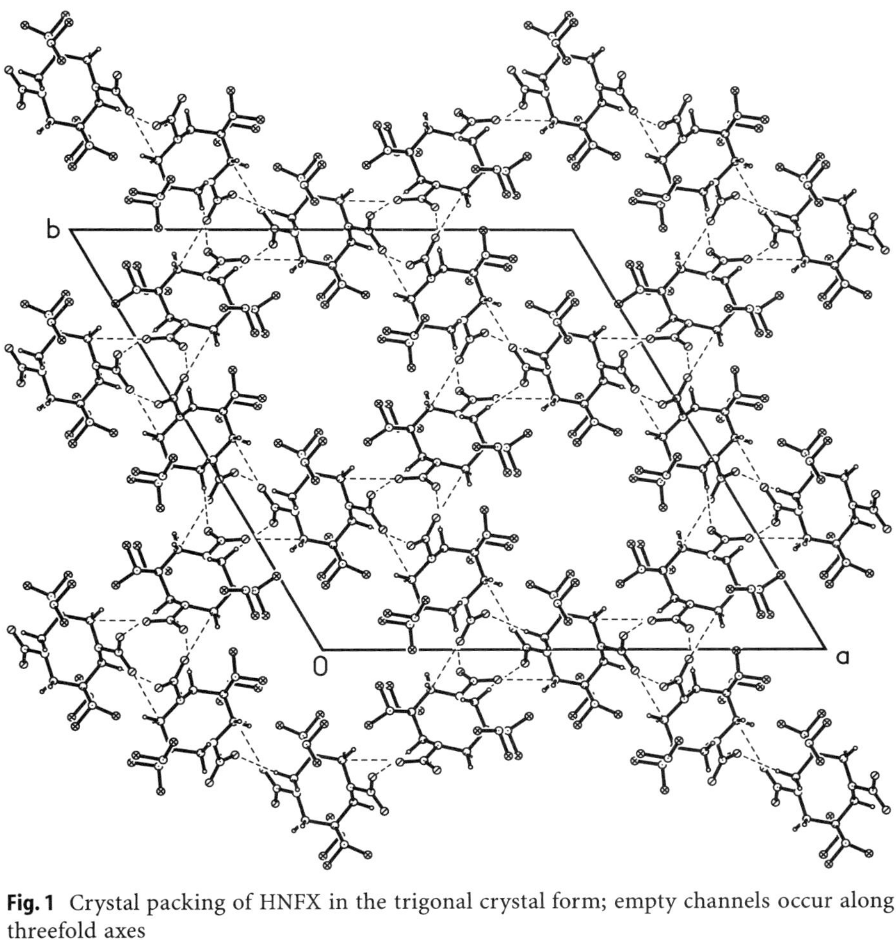

Fig. 1 Crystal packing of HNFX in the trigonal crystal form; empty channels occur along threefold axes

The Manifest Density of HNFX

The crystal density of the initial form is 1.807 g cm^{-3} (21 °C), assuming a vacuum (i.e., zero mass) in the channels. This contrasts significantly with a density of 2.027 g cm^{-3} recently predicted by Ammon [39] from its possible crystal packings. However, cyclic nitramines, especially those incorporating eight-membered rings, are well known to exhibit polymorphism [40, 41]: HMX has four known polymorphs of different densities. The phenomenon of solvent inclusion observed in HNFX crystals is also qualitatively similar to that found in the first polymorph of hexanitrohexaazaisowurtzitane (α-CL-20) [42], which is the lowest-density form. With the promise of preparing a higher-density polymorph of HNFX, which would dramatically improve its attractiveness as a munitions ingredient, several technical approaches have been undertaken to effect such a phase transition. Peiris monitored HNFX for phase transitions during compression in a diamond anvil cell at pressures up to 5 GPa but observed none [43]. Nicol and coworkers subjected HNFX to pressures up to 30 GPa in a diamond anvil cell and studied it by synchrotron radiation X-ray diffraction and Raman and infrared spectroscopy at ambient temperature, also without observing evidence of a polymorphic transition [44]. More recently, a combined experimental and modeling study of the vibrational spectra of solid HNFX was reported [45]. Kalyon and coworkers have performed a systematic study of recrystallizations of HNFX with several different solvent systems [46], similarly without evidence of a new polymorph, though they independently computationally corroborate the feasible existence of a higher-density form [47]. While these failures to achieve formation of a new polymorph of HNFX might indicate that its manifest density is that observed in the crystals prepared to date, they may instead reflect the recognized poor understanding, in general, of how to predict and prepare new stable polymorphs [48].

Hazard Properties

Sensitivity measurements on HNFX (the "polymorph" tentatively called α-HNFX) show it to be comparable to CL-20 and PETN on impact but more sensitive by friction and electrostatic discharge [49]. Impact sensitivity (Explosives Research Laboratory impact tester, modified Type 12 tooling, 2.5-kg weight): HNFX $h_{50} = 8$ cm; CL-20 ~ 8 cm; PETN = 10–13 cm; RDX = 16–17 cm. Friction sensitivity (Allegany Ballistics Laboratory test, per MIL-STD-1751, threshold initiation value): HNFX fired at 72 lbf; PETN low-fires ranged 159–316 lbf. Electrostatic sensitivity: HNFX fired at 0.05 J; PETN showed 10/10 no-fires at 0.25 J. DSC (2 °C min^{-1} per ASTM E537) onset/peak: 221/231 °C.

However, it is also well known that sensitivities of the polymorphs of another eight-membered-ring nitramine, HMX, correlate inversely with density [50–52]: the lowest-density δ form exhibits the highest sensitivity. If this correlation transfers to HNFX, a higher-density "β-HNFX" would exhibit lower sensitivity than "α-HNFX".

3.1.2
4,4-Bis(difluoramino)-1-nitropiperidine

Although HNFX was expected to be the most attractive target difluoramino-nitramine in terms of performance, other examples of the class were desired in order to characterize the behavior of the mixed functionalities present in the compounds. For this reason, a synthetically more straightforward example was prepared [53]: 4,4-bis(difluoramino)-1-nitropiperidine (**6**). As an N-protected γ-aminoketone, the starting material, 1-acetyl-4-piperidinone, would have fewer complications in undergoing difluoramination than β-aminoketones used for more-energy-dense heterocycles. The acetyl N-protecting group would also undergo nitrolysis more readily than *N*-nosyl-protected precursors, such as for HNFX.

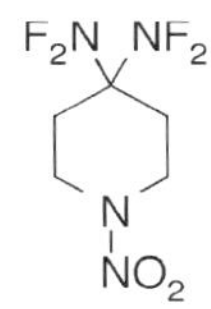

6

Mass spectral fragmentation pathways of several cyclic nitro and difluoramino compounds, including HNFX and 4,4-bis(difluoramino)-1-nitropiperidine, were studied by Smith and coworkers [53], and their thermal decomposition behavior was reported by Oxley et al. [54]. Kuo and Young studied the burning behavior of formulated propellants (including boronized and aluminized compositions) containing 4,4-bis(difluoramino)-1-nitropiperidine in comparison to simple nitro derivatives 1,4-dinitropiperazine and 1,4,4-trinitropiperidine [55]. They determined that a non-metallized propellant based on 4,4-bis(difluoramino)-1-nitropiperidine offered an enhancement of specific impulse by 7.5% over that of a formulation based on 1,4-dinitropiperazine. The difluoramine-containing propellants had burning rates significantly higher than those of the nitramines. However, they exhibited a drastic and uncontrollable increase in burning rate at an initial temperature of 50 °C. This was believed to be caused by melting of the 4,4-bis(difluoramino)-1-nitropiperidine at 40 ∼ 50 °C in the formulations (even though the pure ingredient melts at 71–72 °C), producing a burn rate as high as 508 cm s^{-1}.

3.1.3
RNFX

Another attractive example of the class of difluoramino-nitramines was pursued by Chapman and Nguyen [56]: RNFX, 5,5-bis(difluoramino)hexahydro-

1,3-dinitropyrimidine (**7**). (RNFX is an RDX analog in the series of NF_2-modified cyclic nitramine explosives.) A key step of the synthesis (Scheme 4) was mechanistically similar to that of HNFX: tetrahydro-1,3-bis(4-nitrobenzenesulfonyl)pyrimidin-5(4*H*)-one (**8**) served as an *N*-nosyl-protected reactant for difluoramination of the carbonyl group. However, subsequent conversion to a heterocyclic nitramine product differed significantly from the route used to prepare HNFX. Rather than nitrolytic or protolytic denosylation of the hexahydropyrimidine intermediate with protonitronium ion or superacid, the hexahydropyrimidine ring was intentionally cleaved at the methylene bridge by treatment with ~ 98% nitric acid. The resulting initial product, 2,2-bis(difluoramino)-*N*-(nitratomethyl)-*N*′-nitro-*N*,*N*′-bis(4-nitrobenzenesulfonyl)-1,3-propanediamine (**9**), underwent further nitrolysis to 2,2-bis(difluoramino)-*N*,*N*′-dinitro-*N*,*N*′-bis(4-nitrobenzenesulfonyl)-1,3-propanediamine (**10**), which was adventitiously N-denosylated by water contained in the nitric acid, forming 2,2-bis(difluoramino)-*N*,*N*′-dinitro-1,3-propanediamine (**11**). This did not accumulate but combined spontaneously with the formaldehyde equivalent generated in the nitrolysis, methylene dinitrate, which recyclized the primary nitramine termini to form RNFX.

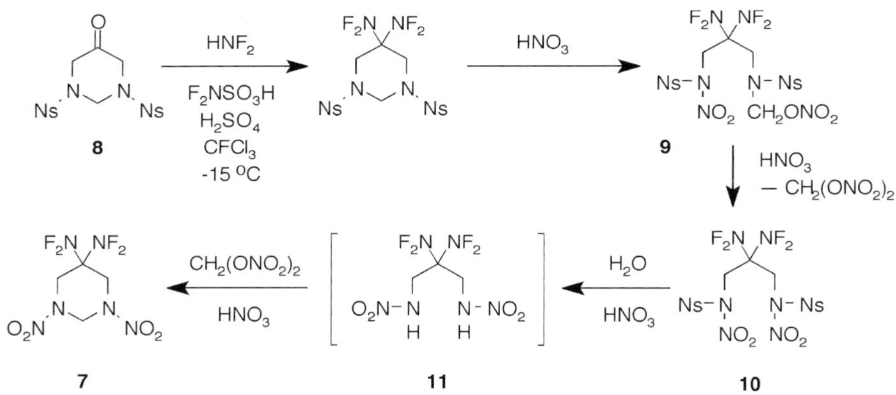

Scheme 4 The first preparation of RNFX (7)

Nucleophilic N-desulfonation of *N*-nitrosulfonamides is a well documented transformation [57–60], and N-denosylation of the 2,2-bis(difluoramino)-*N*,*N*′-dinitro-*N*,*N*′-bis(4-nitrobenzenesulfonyl)-1,3-propanediamine intermediate (**10**) was relatively facile, employing only the minor water content of the nitric acid medium. Beyond definitive identification of its preparation, properties of RNFX have not been experimentally well elucidated.

3.1.4
TNFX

Another early conceptual target member of this class was one in which one each of two nitramine groups of HMX was replaced by a *gem*-bis(difluoramino) grouping and a *gem*-dinitro grouping. This would provide a product of better oxygen balance than HNFX but still some NF_2 content. The introduction of asymmetric components into a 1,5-diazocine substrate suitable for the incorporation of three different energetic functionalities was a formidable challenge, which was undertaken by Axenrod et al. [61]; the successful synthetic sequence is shown below (Scheme 5). The ultimate nitrolysis step proved inordinately difficult because of steric and electronic deactivating effects of $C(NF_2)_2$ *and* $C(NO_2)_2$. Thus, the reaction required not only a superacid (trifluoromethanesulfonic acid) as a medium for nitric acid in order to generate protonitronium reagent, but also the addition of a strong Lewis acid, antimony pentafluoride, in order to increase the superacid strength to generate a higher concentration of NO_2H^{2+}. Upon its successful preparation, the new product (**12**) was named TNFX (a tetranitro derivative in the series of NF_2-modified cyclic nitramine explosives).

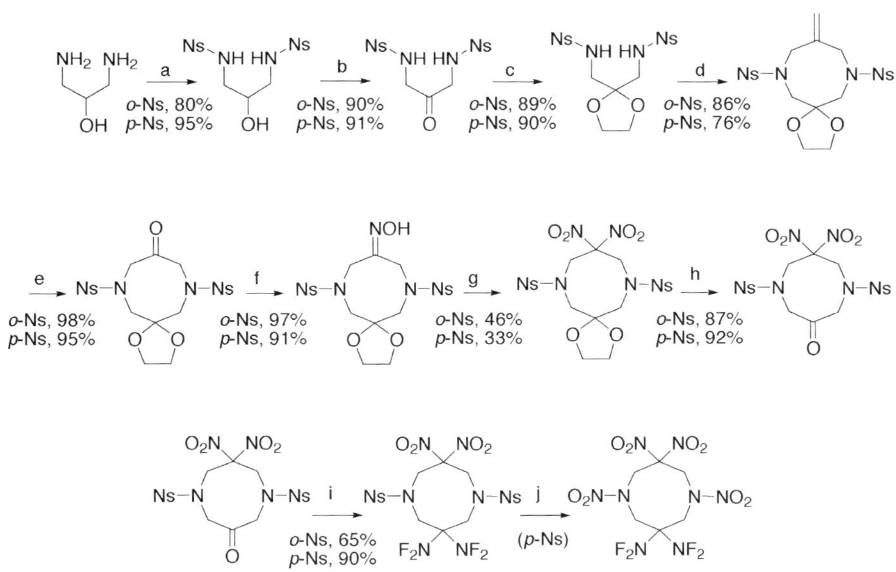

12

Scheme 5 The first preparation of TNFX (**12**). Reagents and conditions: *a* *o*- or *p*-nitrobenzenesulfonyl chloride, THF–H_2O, K_2CO_3, r.t.; *b* CrO_3, H_2SO_4, acetone, r.t.; *c* ethylene glycol, *p*-TsOH, toluene; *d* $CH_2{=}C(CH_2Br)_2$, K_2CO_3, acetone, reflux; *e*(i) O_3, CH_2Cl_2, – 78 °C, *e*(ii) Me_2S; *f* $NH_2OH{\cdot}HCl$, NaOAc, EtOH, reflux; *g* HNO_3, NH_4NO_3, urea; *h* conc. H_2SO_4, r.t.; *i* HNF_2–F_2NSO_3H–H_2SO_4–$CFCl_3$, – 15 °C; *j* HNO_3–CF_3SO_3H–SbF_5

Yoon and coworkers have explored alternative preparations (Scheme 6) of the asymmetric precursors to TNFX [62].

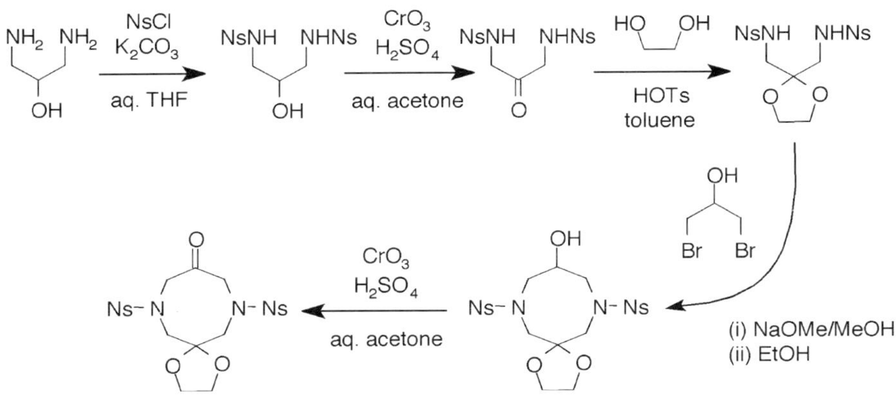

Scheme 6 An alternative preparation of precursors to TNFX or HNFX

TNFX proved structurally interesting in forming two crystallographically distinct polymorphs [63] upon crystallization from its first preparation (as HNFX was hoped to do): a low-density trigonal form of $\rho = 1.712\,\mathrm{g\,cm^{-3}}$ and a high-density orthorhombic form of $\rho = 1.904\,\mathrm{g\,cm^{-3}}$. Thus, the low-density trigonal form of HNFX ($\rho = 1.807\,\mathrm{g\,cm^{-3}}$) has an experimentally observed density enhancement of $0.095\,\mathrm{g\,cm^{-3}}$ imparted by replacement of $C(NO_2)_2$ in TNFX with $C(NF_2)_2$ in HNFX. A similar enhancement in a previously unobserved orthorhombic form of HNFX would provide an experimental density of $\rho = 1.999\,\mathrm{g\,cm^{-3}}$, coincidentally identical to the (still attractive) value initially estimated for HNFX by Ammon [26].

Beyond synthetic and structural elucidation, properties of TNFX have not been experimentally well characterized.

3.1.5 Properties and Applications

Solid-state heats of formation of some cyclic difluoramino-nitramines have been computationally estimated by Politzer and Lane [9]: HNFX, $-67\,\mathrm{kcal\,mol^{-1}}$; TNFX, $-39\,\mathrm{kcal\,mol^{-1}}$; RNFX, $-21\,\mathrm{kcal\,mol^{-1}}$. Application of these values to rocket engine specific impulse (I_{sp}) calculations (performed with a code such as the Air Force Specific Impulse Program [64]) allows ranking of the cyclic difluoramino-nitramines (as monopropellants) relative to state-of-the-art nitramines: β-HMX < ε-CL-20 < TNFX < RNFX < HNFX. The I_{sp} of CL-20 has been reported by Hong et al. as 272.6 s [65, 66]. An early estimate of the I_{sp} of RNFX was reported by Politzer et al. as being 108% of that of HMX, which is 265.1 s [67].

As reports of preparations of this new class of energetic material started appearing, their potential in ordnance applications started receiving notice overseas. Feng reviewed recent examples of renewed interest in organic difluoramine derivatives in Western research [68, 69], including early reports on the pursuit of HNFX. Tan reported the new preparation of HNFX [36] to the Xi'an Modern Chemistry Research Institute [70]; and Li et al. cited Tan's report in an outlook on the prospects of this and other classes of ingredients in advanced solid propellants [71, 72]. This new class of ingredient has received interest in Korea as well [73].

3.2 Electrophilic Difluoramination and Its Products

In the 1990s, researchers throughout the Russian Academy of Sciences reported syntheses and characterizations of a variety of compounds derived from a new transformation, electrophilic difluoramination, particularly of some compounds derived from difluoramination of various nitroalkyl reactants. Examples of such new products were first described by Pepekin and coworkers in terms of thermodynamic properties [74, 75] and physical properties [76, 77], which will be mentioned below.

3.2.1 Syntheses

Synthetic routes involving the new transformation started appearing in 1996. Shevelev and coworkers [78] prepared a variety of *N*-(difluoramino)azoles (Scheme 7) by reactions of electronegatively substituted azole anions with *N*,*N*-difluorohydroxylamine-*O*-sulfonyl fluoride ("*O*-fluorosulfonyl-*N*,*N*-difluorohydroxylamine"), F_2NOSO_2F, preferably in water–ethylene glycol–dichloroethane mixtures with a phase transfer catalyst such as polyethylene glycol (PEG-400). If the pK_a of an azole was < 5, there was no interference by formation of *N*-fluorosulfonylazoles. They later offered computational predictions of molecular and crystal parameters and of detonation and combustion parameters of five examples of this class of compound [79].

Fokin et al. used *N*,*N*-difluorohydroxylamine-*O*-sulfonyl fluoride for electrophilic difluoramination of anions of α,α-dinitroalkyl derivatives, such as

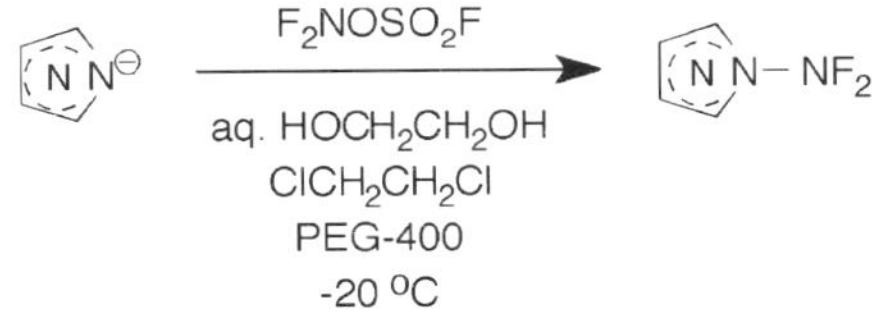

Scheme 7 Electrophilic difluoramination of azole (pyrazole, imidazole, triazole) anion [78]

1,1-dinitroethane and dinitroacetonitrile. Sodium dinitroacetonitrile underwent difluoramination in 70.8% yield, but the other compounds formed in 18.8–29.3% yields at the time of this report [80, 81]. However, yields of dinitroacetate esters were later improved to 50–52%, and other α-mononitro derivatives were prepared in 67–91% yields [82, 83] (Scheme 8).

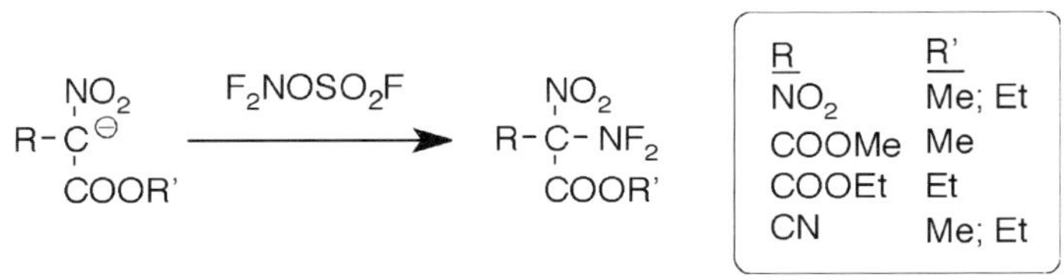

Scheme 8 Electrophilic difluoramination of nitroalkyl anion [82, 83]

The chemistry of (difluoramino)dinitroacetonitrile was further explored as well [84, 85]: it underwent common transformations such as hydrolysis to (difluoramino)dinitroacetamide; reaction with HCl in anhydrous methanol to form the *O*-methyl imidate hydrochloride; trimerization to 2,4,6-tris[(difluoramino)dinitromethyl]-1,3,5-triazine; and reaction with azides (NaN_3, HN_3) to make 5-[(difluoramino)dinitromethyl]tetrazoles.

Electrophilic difluoramination of certain (especially tertiary) alkoxide salts and *N*-hydroxyimide salts with F_2NOSO_2F produced new difluoraminoxy derivatives [86, 87].

Khisamutdinov et al. showed that the difluoramino group in various (difluoramino)dinitromethane derivatives could be displaced by azide anion to make azidodinitromethyl derivatives [88, 89]. In contrast, a *nitro* group in (difluoramino)trinitromethane was displaced by nucleophiles such as fluoride and nitrite [90, 91]. The nitrite by-product keeps yields low by competitive reaction with the starting material, resulting in (difluoramino)dinitromethane anion side product. With bromide nucleophile, the course of reaction was unexpectedly complex, as the initial bromo(difluoramino)dinitromethane un-

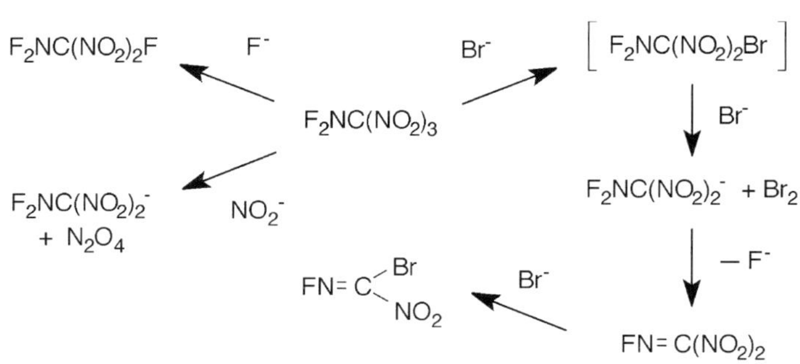

Scheme 9 Nucleophilic displacement reactions in (difluoramino)trinitromethane

derwent further reactions (bromide additions and fluoride elimination) to produce bromo(*N*-fluoro)nitromethanimine (Scheme 9).

3.2.2 Thermochemical and Other Properties

Pepekin's first report [74, 75] of products in this class gave experimental values of thermodynamic properties such as heats of formation. Also prior to the synthetic methodology descriptions [80, 81], Litvinov et al. [76, 77] reported various physical properties (melting point and/or boiling point, density) of several members of this class, including a particularly dense and energetic one: bis[2-(difluoramino)-2,2-dinitroethyl]nitramine, $[NF_2C(NO_2)_2CH_2]_2N{-}NO_2$ (**13**). This compound had an attractive density of 2.045 g cm^{-3}, a melting point of 103 °C, and thermal stability better than bis(2,2,2-trinitroethyl)nitramine. However, the authors made an assessment of these members of this new class: "Nevertheless, it should be noted that the considered compounds exhibit a rather high sensitivity to the action of external factors, and thus have no practical application" [76, 77]. Indeed, Shchetinin [92, 93] later reported that bis[2-(difluoramino)-2,2-dinitroethyl]nitramine (**13**) had a sensitivity comparable to the primary explosive lead azide: the critical pressure of explosion excitation measured by the disintegrating shell method was $\bar{p}_{cr}^{DS} = 0.42 \pm 0.02$ for $[NF_2C(NO_2)_2CH_2]_2N{-}NO_2$ (**13**) compared to $\bar{p}_{cr}^{DS} = 0.38 \pm 0.03$ for lead azide; $\bar{p}_{cr}^{DS} = 0.93 \pm 0.03$ for PETN; and $\bar{p}_{cr}^{DS} = 1.15 \pm 0.03$ for RDX.

Nazin and coworkers measured thermal decomposition of α-(difluoramino)polynitroalkanes and concluded that (difluoramino)dinitroalkanes differ little in their stability from their nitro analogs, trinitroalkanes [94, 95]. Bis[2-(difluoramino)-2,2-dinitroethyl]nitramine (**13**) was not one of the examples in this study, however.

Pepekin [96] measured and reported experimental detonation parameters and the metal acceleration ability of bis[2-(difluoramino)-2,2-dinitroethyl]-nitramine (**13**), as well as compositions of it with HMX and with aluminum, and compared them to these parameters for HMX as well as formulations of bis(2,2,2-trinitroethyl)nitramine. The highest metal acceleration ability was offered by a 60 : 40 composition of **13** with HMX, which performed at 110.8% of HMX. Pure **13** also exhibited a higher detonation velocity than HMX. Literature involving this class of compound has mostly subsided in recent years, though there is another report by Pepekin and Gubin [97, 98] that **13** exhibits propellant performance that is 105.2% of that of HMX, measured by the "M-40 technique" (acceleration of steel plates 40 mm in diameter and 4 mm thick accelerated from the end of a cylindrical 40 × 40 mm charge placed in a thick-walled steel shell). Hexanitrobenzene is reported as having a performance 106% of that of HMX.

3.3 Other Organic Difluoramines

The compounds mentioned in this subsection are examples of specific classes not otherwise reviewed above, and specifically examples that have a structural feature not commonly encountered in the conventional chemistry of difluoramines or compounds that show prospective new applicability.

3.3.1 Polymers

Although its difluoramination chemistry is straightforward, a relatively recent preparation of an NF_2-substituted polymer (a binder ingredient candidate) was reported by Jiang et al. (Hubei Redstar Chemical Institute, China) [99]. 4,4-Bis(difluoramino)pentyl glycidyl ether polymer was prepared by reaction of 4,4-bis(difluoramino)pentanol with epichlorohydrin followed by homopolymerization of the resulting oxirane ether monomer (Scheme 10) or copolymerization with THF.

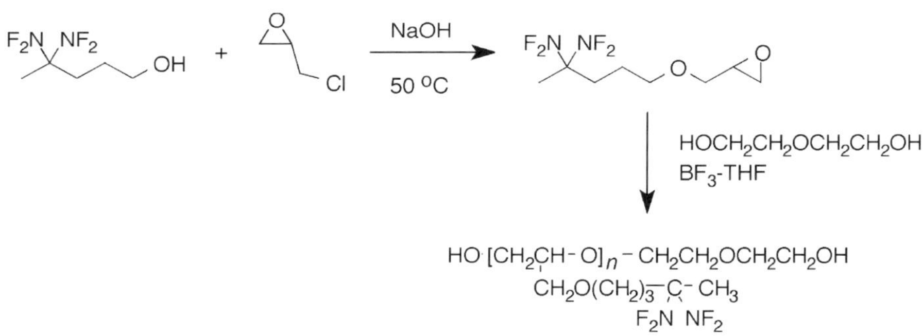

Scheme 10 Preparations of a *gem*-bis(difluoramino)-substituted monomer and polymer [99]

Archibald, Manser, and coworkers developed a series of analogs of nitratoalkyl polymers (e.g., poly-NMMO and poly-BNMO). 3-(Difluoraminomethyl)-3-methyloxetane (**14**) and 3,3-bis(difluoraminomethyl)oxetane (**15**) [100] monomers were prepared by direct fluorination of corresponding carbamate ester derivatives [101, 102]. The monomers were polymerized (Scheme 11) to form homopolymers and copolymers with load-bearing polyether backbones as well as energetic NF_2 groups.

Solomun et al. [103] have demonstrated that it is feasible to produce large concentrations of NF_2 constituency embedded in a polyamide matrix by brief exposure to elemental fluorine (1 mbar of 1–10% F_2 in N_2) of bulk and thin-film samples of polycaprolactam (PA6) and polylauryllactam (PA12). The

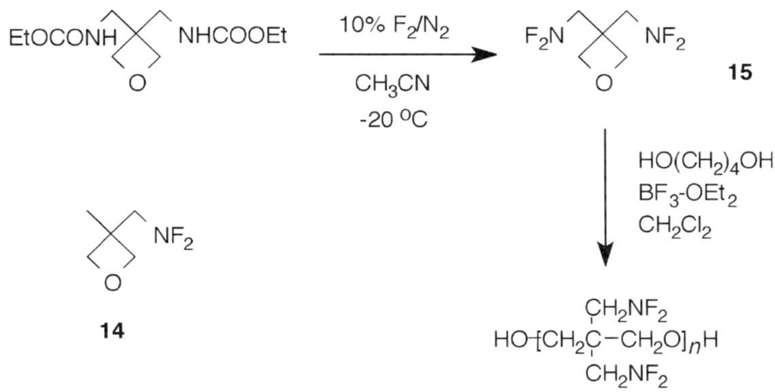

Scheme 11 Preparations of a difluoramino-substituted oxetane monomer and polymer

resulting fluorinated films were characterized by reflection–absorption FTIR, X-ray photoelectron, and NMR spectroscopies. The process utilizes the same transformation that small-molecule amides undergo in producing (difluoramino)alkanes by direct fluorination [104].

3.3.2 Plasticizers

Archibald and Manser prepared NF_2-substituted analogs of pentaerythrityl esters (such as PETN). Specifically, 3-(difluoraminomethyl)-3-methyloxetane (**14**) and 3,3-bis(difluoraminomethyl)oxetane (**15**) [101] were subjected to ring-opening nitration by nitric acid (Scheme 12), giving corresponding 1,3-propanediyl dinitrate esters in quantitative yield [105, 106]. As neopentyl difluoramines, the products had superior stability compared to many other mono(difluoramino)alkyl-substituted compounds. 2,2-Bis(difluoraminomethyl)-1,3-propanediyl dinitrate (**16**) showed a melting point of 58 °C and DSC onset/peak of 177/188 °C.

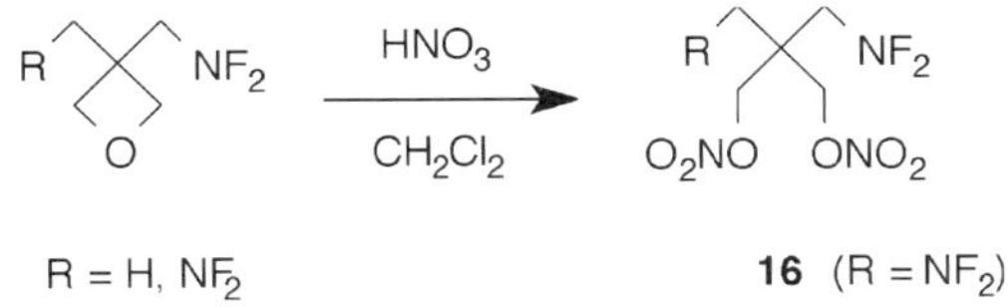

Scheme 12 Preparations of difluoramino-substituted nitrate plasticizers

Rice and Russell (US Naval Research Laboratory) conducted a high-pressure study of the thermal interaction between elemental boron and Aerojet's 2,2-bis(difluoraminomethyl)-1,3-propanediyl dinitrate (**16**), as well as with PETN [107]. The mixture compressed in a cubic zirconia anvil cell was heated by absorption of a single laser pulse (Nd : YAG frequency-doubled laser at 532 nm), followed by recooling of the sample. Products observed at 1.7 GPa and 50 K included O=B–F, identified by infrared absorptions.

3,3-Bis(difluoramino)-1,5-dinitratopentane (**17**), as a potentially useful plasticizer, was prepared by Adolph and Trivedi [108] by conventional difluoramination of 1,5-bis(trifluoroacetoxy)-3-pentanone followed by transesterification with methanol, and then nitration with mixed acid (Scheme 13).

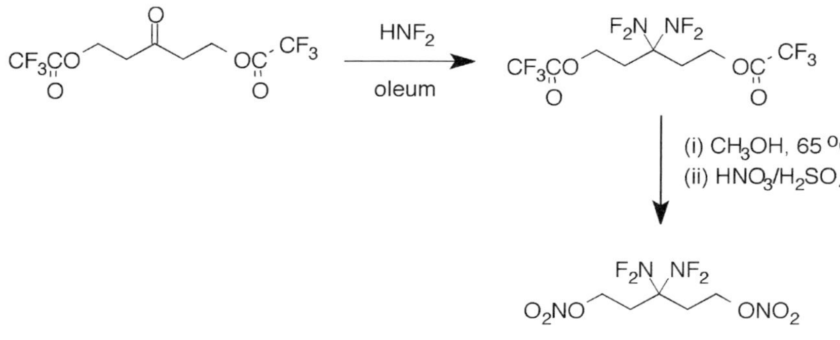

17

Scheme 13 Preparation of a *gem*-bis(difluoramino)-substituted nitrate plasticizer [108]

3.3.3 Miscellaneous Organic Derivatives

Synthesis

Fokin et al. directly fluorinated several aminofurazans (Scheme 14) to prepare 3-(difluoramino)furazans [109, 110]. Similarly, certain *N*-alkyl-5-aminotetrazoles were amenable to direct fluorination to 5-(difluoramino)tetrazole derivatives [111, 112]. NF_2-containing 1,2,4-oxadiazoles were prepared by Shreeve and coworkers by reactions of sodium 5-[(difluoraminomethyl)difluoromethyl]tetrazolate with oxalyl chloride [113].

1,1,2,2-Tetrakis[*N*-(difluoraminomethyl)nitramino]ethane (**18**) was first reported by Zheng et al. in a 1985 conference [114], and its chemistry (Scheme 15) was described further in later publications [28–30].

An interesting example of a unique product of difluoramination of a heterocyclic imide carbonyl group appeared in the context of a report by Gilardi and coworkers of structural features among a group of *gem*-bis(difluoramino)-substituted organics [115]: 1-{3-[5,5-bis(difluoramino)-2-oxopyrrolidinyl]-2-oxopropyl}pyrrolidine-2,5-dione (**19**). Imide carbonyls are usually not subject to difluoramination under typical reaction conditions.

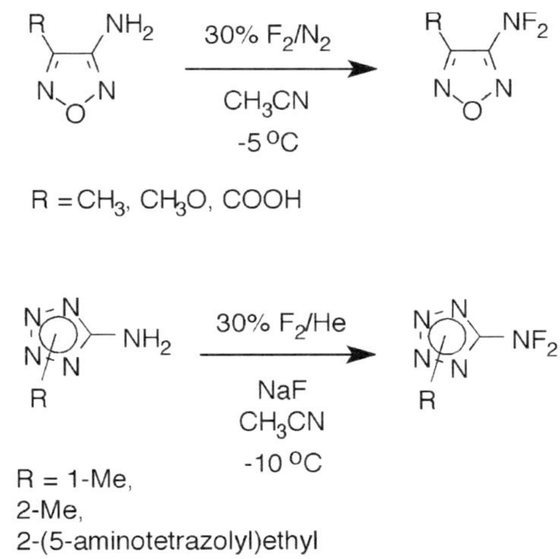

Scheme 14 Direct fluorination of various aminoazoles to (difluoramino)azoles

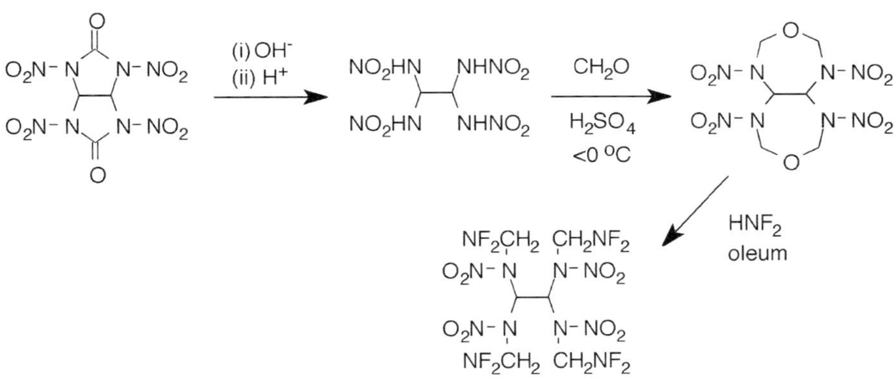

18

Scheme 15 Preparation of 1,1,2,2-tetrakis[*N*-(difluoraminomethyl)nitramino]ethane (**18**)

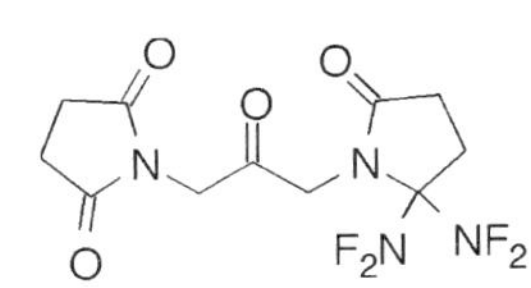

19

Other examples of typical reaction products, *N*-[*gem*-bis(difluoramino)-alkyl]imides, from ketone-substituted imides are provided in the same paper as well as in an earlier work by Fokin et al. [116, 117].

Many examples of addition of tetrafluorohydrazine (N_2F_4) to unsaturated organics had been discovered in the course of early work on difluoramines [20, 21]. A new example (Scheme 16) of this general transformation was the addition of N_2F_4 to 1,3-dehydroadamantane to prepare 1,3-bis(difluoramino)adamantane [118].

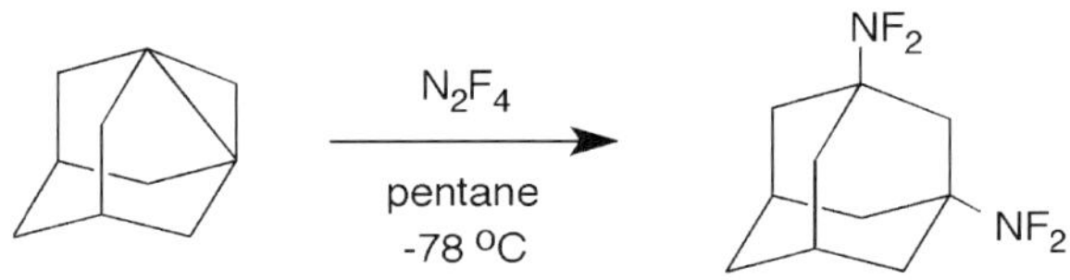

Scheme 16 Addition of tetrafluorohydrazine to 1,3-dehydroadamantane [118]

Though more accurately an organic silazane than a simple "organic difluoramine", the first stable trialkyl(difluoramino)silane was prepared by Shreeve, Williams, and coworkers by two stepwise fluorinations (Scheme 17) of tris(*tert*-butyl)silazane by Selectfluor (**20**) fluorinating agent [119].

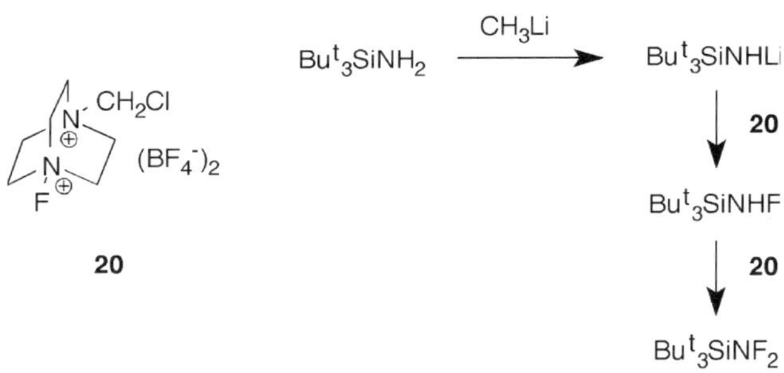

Scheme 17 Preparation of tris(*tert*-butyl)(difluoramino)silane using Selectfluor (**20**)

Thermochemistry

Thermochemical characterizations of several organic difluoramines prepared during the Project Principia era were reported throughout the 1960s and 1970s and will not be reviewed here. Studies of the specific class of products of electrophilic difluoramination were reviewed above (Sect. 3.2.2).

More recently, two significant papers by Fokin, Grebennikov, Manelis, and Nazin [120–123] reported kinetic studies of thermal decomposition of a good number of organic difluoramine derivatives. The structures were chosen in order to ascertain inductive and steric effects on thermal decomposition, as well as effects of conjugation within the reactant and the presence of an α-hydrogen (relative to NF_2). The authors report that internal alkyl-substituted difluoramines (i.e., those lacking α-hydrogen) are more stable than analogous nitro compounds, owing to higher activation energies and somewhat smaller preexponential factors. The presence of α-hydrogen (e.g., in benzyldifluoramine) fundamentally changes the decomposition rate, and decomposition probably occurs by cleavage of HF. Otherwise, the decomposition rate does not depend on electronic effects of substituents but is decreased by steric shielding of the NF_2 group. (These results were also recently reviewed by Manelis et al. [124].) One of the specific systems studied, 2,2-bis(difluoramino)propane, was recently reinvestigated by Lin and coworkers [125]. In agreement with Fokin et al. [120, 121], it was concluded that the primary fragmentation process in this system, similarly to other internal *gem*-bis(difluoramino)alkanes, is homolytic cleavage of a C–NF_2 bond.

Pepekin's study of the thermodynamic properties of difluoramino and nitro compounds [74, 75] included many organic difluoramines besides the products of electrophilic difluoramination cited above. Properties reported include heats of combustion, formation, and atomization, Clausius–Clapeyron equation parameters, and the enthalpies and entropies of evaporation and sublimation. This collection of properties allowed estimation of group additivity parameters for general calculations of thermodynamic properties of organic difluoramines, which were compared to those of corresponding nitro groups.

Organic difluoramines were among the many classes of high-energy compounds analyzed by Nazin et al. in assessing the influence of molecular structure on their thermal stability [126, 127]. The sites of primary breakdown were discussed, as was the mutual influence of different functional groups.

Several difluoramines (along with nitro compounds) were also among the collection of energetic materials in a study by Matveev, Nazin, and coworkers [128, 129] of polyfunctional compounds with several of the same type of reactive substituent, to determine the stepwise character of successive thermal degradation steps. It was found that the degradation of most of the studied classes occurred stepwise via long-lived intermediate products.

4 New Synthetic Methodology

The technique of electrophilic difluoramination [80, 81], reviewed above, has produced the greatest variety of new products in the class of organic difluoramines. Otherwise, most recent results have employed the conventional methodology that was devised in the era of Project Principia, as reviewed elsewhere [16–25]. Process improvements in conventional difluoramination of ketones by HNF_2 [35, 37] have made the transformation more scalable, particularly by employing an inert cosolvent to alleviate hazards of difluoramine, but employ the same fundamental chemical reactions as in the 1960s. However, the recognition of a requisite nitrogen-protection strategy made the transformation applicable to heterocycles with proximate nitrogens.

The feasibility of isolating relatively stable salts of difluorosulfamic acid and their utility in performing difluoraminations of model ketones was demonstrated by Chapman et al. [130, 131], offering the prospect of alleviating many of the problems inherent in the use of hazardous difluoramine (HNF_2) in conventional difluoraminations. Sodium difluorosulfamate was prepared by direct fluorination of aqueous sodium sulfamate; it could be isolated by separation from sodium fluoride by-product and water after neutralization; and it could be safely stored cold. Difluoramination of 1,3-dibromoacetone by sodium difluorosulfamate in sulfuric acid (1% SO_3) produced 1,3-dibromo-2,2-bis(difluoramino)propane in 38% crude yield, comparable to conventional difluoramination. Later, Christe and coworkers [132] better characterized sodium difluorosulfamate (Raman and IR spectra) and other difluorosulfamate salts (X-ray crystallography).

A direct fluorination of nucleophilic primary (as well as secondary) amines with Selectfluor fluorinating agent (**20**), rather than elemental fluorine, has been reported by Singh and Shreeve [133], making small-scale preparations of (non-geminal) organic difluoramines more convenient.

The utility of trityldifluoramine (Ph_3CNF_2), a known source of difluoramine via acidic hydrolysis [134], for effecting in-situ difluoramination of ketones [22, 23, 135, 136] has been recently rediscovered [137]. The latter system invokes the mechanism of difluoramine alkylation by carbenium ions generated in strong acids, as traditionally employed in conventional difluoraminations of ketones with added difluoramine [138, 139].

A fundamentally new preparation of internal and terminal *gem*-bis(difluoramino)alkanes was demonstrated by Chapman et al. [140] via the direct fluorination of the corresponding *gem*-bisacetamides, with 1,1-bisacetamidocyclohexane and 1,1-bisacetamidopropane used as examples, leading to 1,1-bis(difluoramino)cyclohexane and 1,1-bis(difluoramino)propane, respectively (Scheme 18). 1,1-Bis(difluoramino)cyclohexane behaved less than ideally as a model system: its yield by conventional difluoramination is atypically

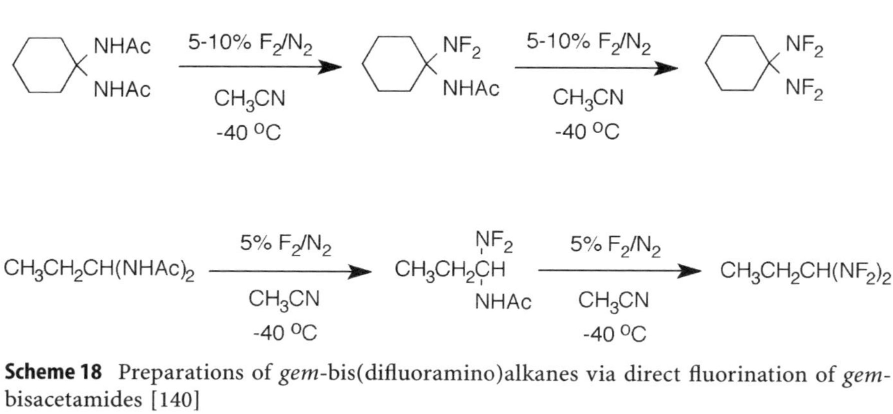

Scheme 18 Preparations of *gem*-bis(difluoramino)alkanes via direct fluorination of *gem*-bisacetamides [140]

low (31%) for ketone difluoramination, and its yield by the new transformation maximized at about 10%. 1,1-Bis(difluoramino)propane's yield (~ 40%) was more typical of conventional difluoraminations.

5 Theoretical Treatments

An early report on computational estimation of the thermodynamic properties of (difluoramino)alkanes, outside of the context of the American and Russian research programs, was by İnel (Boğaziçi University, Turkey), who estimated heats of formation, entropies, and bond dissociation energies of several simple difluoramines [141, 142], derived from published experimental data on perfluorinated amines and thermochemical relationships.

Heats of formation of known and hypothetical difluoramino (and nitro) compounds were estimated by Leroy, Bourasseau, and coworkers [143], initially by a semiempirical computational procedure (of Pople et al.). Calculation was based on the concept of isodesmic reactions, which conserves, as far as possible, the number and nature of chemical bonds. A reaction energy change, calculated at the SCF level, approximates the experimental heat of reaction by incorporating experimental heats of formation of reference compounds. The new compounds had been studied at ab initio level using a 6-31G basis set for geometry optimization and total energy determination. These authors also developed group additivity parameters for calculating heats of formation of new energetic compounds (nitro and difluoramino) by Benson's method [144].

The most significant contributions to computational studies of organic difluoramines, which have also directed synthetic research efforts, have been by Politzer and coworkers. (Difluoramino)aromatics were among the first systems studied by calculations of average local ionization energies [145] and electrostatic potentials [146] for predicting chemical reactivity and substituent effects. Politzer et al. also developed the application of density functional theory (DFT) and electrostatic potentials to estimation of thermodynamic properties (especially dissociation energies and solid-state heats of formation) of many new hypothetical NF_2-containing structures [147–150], including predictions of thermochemical processes undergone by difluoramine derivatives [151]. This approach led to early estimations [9, 26] of important thermodynamic properties of the difluoramino-nitramines that were being synthetically pursued. Politzer's computational methodology has been presented in recent overviews [152, 153].

Thermochemical properties of difluoramino-substituted cage hydrocarbons, specifically, poly(difluoramino)cubanes [154] and poly(difluoramino)-adamantanes [155], have been estimated by Ju et al. in order to compare results from a variety of computational methods: density functional theory, Hartree–Fock, and MP2 methods (6-31G* and 6-311G** basis sets), as well as semiempirical MO methods MNDO, AM1, and PM3 for the former chemical system. The semiempirical methods did not produce accurate and reliable results for heats of formation. Interactions between difluoramino substituents were determined. Ju and coworkers also calculated (using DFT and semiempirical methods) heats of formation of a variety of difluoramino pentaerythrityl derivatives [156], including some that had been considered by Archibald and Manser [106] as prospective plasticizers, for example, a tetrakis(difluoramino) analog of PETN, $C(CH_2NF_2)_4$, as a theoretical structure of interest.

References

1. Furin GG, Fainzilberg AA (1999) Russ Chem Rev 68:653
2. Taylor SD, Kotoris CC, Hum G (1999) Tetrahedron 55:12431
3. Li Y (2001) Huaxue Shiji [Chemical Reagents] 23:143
4. Colburn CB (1965) Endeavour 24:138
5. Fokin AV, Kosyrev YuM (1966) Russ Chem Rev 35:791
6. Freeman JP (1967) Inorg Chim Acta Rev 1:65
7. Ruff JK (1967) Chem Rev 67:665
8. Klapötke TM (2006) J Fluorine Chem 127:679
9. Politzer P, Lane P (1997) Adv Mol Struct Res 3:269
10. Brown RC, Kolb CE, Yetter RA, Dryer FL, Rabitz H (1995) Combust Flame 101:221
11. Yetter RA, Dryer FL, Rabitz H, Brown RC, Kolb CE (1998) Combust Flame 112:387
12. Ammon HL (2001) Struct Chem 12:205
13. Ruff O, Giese M (1936) Ber Dtsch Chem Ges B 69B:598
14. Davenas A (2003) J Propul Power 19:1108
15. Holzmann RT (ed) (1966) Advanced propellant chemistry. American Chemical Society, Washington, DC
16. Fokin AV, Kosyrev YuM (1970) Zh Vses Ova im D I Mendeleeva 15:81
17. Fokin AV, Kosyrev YuM (1970) Mendeleev Chem J 15:109
18. Baum K (1971) Intra-Sci Chem Rep 5:69
19. Fokin AV, Studnev YuN, Kuznetsova LD (1976) Reakts Metody Issled Org Soedin 24:7
20. Fokin AV, Studnev YuN (1982) Izv Akad Nauk SSSR Ser Khim 1812
21. Fokin AV, Studnev YuN (1982) Bull Acad Sci USSR Div Chem Sci 31:1609
22. Fokin AV, Kosyrev YuM, Shevchenko VI (1982) Izv Akad Nauk SSSR Ser Khim 1831
23. Fokin AV, Kosyrev YuM, Shevchenko VI (1982) Bull Acad Sci USSR Div Chem Sci 31:1626
24. Urbański T (1984) Chemistry and technology of explosives, vol 4, chap 9. Pergamon, Oxford
25. Hanefeld W (1990) *N,N*-Dihalogen-amine. In: Klamann D (ed) Methoden der organischen chemie (Houben-Weyl), vol E16a, organische Stickstoff-Verbindungen I, Part 2. Georg Thieme, Stuttgart, p 893
26. Miller RS (1996) Mater Res Soc Symp Proc 418:3
27. Chapman RD, Archibald TG, Baum K (1989) Research in energetic compounds. Report ONR-7-1 (interim). Fluorochem, Azusa, CA; NTIS accession number

ADA214106, available from http://www.ntis.gov/help/ordermethods.asp, last visited: 23 April 2007
28. Zheng Y, Huang T, Zhang M, Wang X (1987) A study on the syntheses and detonation properties of difluoroamino compounds. In: Ding J (ed) Proceedings of the international symposium on pyrotechnics and explosives. China Academic, Beijing, p 234
29. Zheng Y, Zhou J, Zhou D, Zhang M (1988) Binggong Xuebao [Acta Armamentarii] 59
30. Zheng Y, Zhou J, Zhou D, Zhang M (1988) Chem Abstr 109:189782
31. Sayles DC (1972) US Patent 3 636 154
32. Tyler WE, Lovett JR (1972) US Patent 3 687 954
33. Baum K, Trivedi NJ, Lovato JM, Iyer VK (1993) Novel approaches to the synthesis of fluorodinitromethane and fluorodinitroethanol. Report NRO-1-1 (final). Fluorochem, Azusa, CA; available at http://stinet.dtic.mil/cgi-bin/GetTRDoc?AD=A269158&Location=U2&doc=GetTRDoc.pdf, last visited: 26 April 2007
34. Fluorochem (1991) Research in energetic compounds. Report ONR-7-1 (final) to the Office of Naval Research (Arlington, VA) on Contract N00014-88-C-0536. Fluorochem, Azusa, CA. Described in: Baum K, Trivedi NJ, Lovato JM, Iyer VK (1993) Novel approaches to the synthesis of fluorodinitromethane and fluorodinitroethanol. Report NRO-1-1 (final). Fluorochem, Azusa, CA; NTIS accession no. ADA269158, available at http://stinet.dtic.mil/cgi-bin/GetTRDoc?AD=A269158, last visited: 26 April 2007
35. Chapman RD, Welker MF, Kreutzberger CB (1998) J Org Chem 63:1566
36. Chapman RD, Gilardi RD, Welker MF, Kreutzberger CB (1999) J Org Chem 64:960
37. Chapman RD, Groshens TJ (2004) US Patent Application 11/010059
38. Adolph HG, Stern AG (2006) US Patent 7 145 003
39. Ammon HL, Holden JR, Du Z (2002) Structure and density predictions for energetic materials. In: Energetic materials design MURI kick-off meeting [extended abstracts]. Aberdeen Proving Ground, MD. http://www.chem.missouri.edu/Thompson/MURI02/extended/Ammon_MURI_extended_abstract_4.pdf, last visited: 20 April 2007
40. Oyumi Y, Brill TB, Rheingold AL (1986) J Phys Chem 90:2526
41. Achuthan CP, Jose CI (1990) Propellants Explos Pyrotech 15:271
42. Ryzhkov LR, McBride JM (1997) J Am Chem Soc 119:4826
43. Peiris SM (2002) Experimental and computational studies of molecular and lattice symmetries of energetic materials at high pressure. DTIC, available at http://stinet.dtic.mil/str/quick-tr.html, last visited: 23 April 2007
44. Gobin C, Orwig J, Nicol M (2006) American Physical Society March Meeting, Baltimore, MD, abstract Q1.150. http://meetings.aps.org/link/BAPS.2006.MAR.Q1.150, last visited: 20 April 2007
45. Nicol M, Gobin C, Kim E (2007) American Physical Society March Meeting, Denver, CO, abstract V23.6. http://meetings.aps.org/link/BAPS.2007.MAR.V23.6, last visited: 20 April 2007
46. Degirmenbasi N, Peralta-Inga Z, Olgun U, Gocmez H, Kalyon DM (2006) J Energ Mater 24:103
47. Peralta-Inga Z, Degirmenbasi N, Olgun U, Gocmez H, Kalyon DM (2006) J Energ Mater 24:69
48. Chew JW, Black SN, Chow PS, Tan RBH, Carpenter KJ (2007) Cryst Eng Comm 9:128
49. HNFX Rev. 1.3 (2005) Material safety data sheet (MSDS). Naval Air Weapons Station, China Lake, CA

50. Cady HH, Smith LC (1962) Studies on the polymorphs of HMX. Report LAMS-2652. Los Alamos Scientific Laboratory, Los Alamos, NM
51. Licht HH (1971) In: Hansson J (ed) Proceedings of symposium on chemical problems connected with the stability of explosives. Sektionen för Detonik och Förbränning, Sundbyberg, Sweden, p 168
52. McCrone WC (2001) Microscope 49:47
53. Zhang J, Oxley J, Smith J, Bedford C, Chapman R (2000) J Mass Spec 35:841
54. Oxley JC, Smith JL, Zhang J, Bedford C (2001) J Phys Chem A 105:579
55. Kuo KK, Young G (2002) Proc Combust Inst 29:2947
56. Chapman RD, Nguyen BV (2001) US Patent 6 310 204
57. Emmons WD, Freeman JP (1955) J Am Chem Soc 77:6061
58. Luk'yanov OA, Ternikova TV (1983) Izv Akad Nauk SSSR Ser Khim 667
59. Luk'yanov OA, Ternikova TV (1983) Bull Acad Sci USSR Div Chem Sci 32:605
60. Garcia J, González J, Segura R, Urpí F, Vilarrasa J (1984) J Org Chem 49:3322
61. Axenrod T, Guan X-P, Sun J, Qi L, Chapman RD, Gilardi RD (2001) Tetrahedron Lett 42:2621
62. Park Y-D, Cho S-D, Kim J-J, Kim H-K, Kweon D-H, Lee S-G, Yoon Y-J (2006) J Heterocycl Chem 43:519
63. Gilardi R (2004) Polytypical polymorphs occurring in an energetic material. In: Virtual course on polymorphism: diversity amidst similarity, International School of Crystallography – 35th course, Erice, Sicily, Italy; http://erice2004.docking.org/vcourse/polymorph/15tue/1830-Gilardi/Gilardi.ppt, last visited: 20 April 2007
64. Dunn BP (2001) Rocket engine specific impulse program. http://dunnspace.com/isp.htm, last visited: 20 April 2007
65. Hong W-L, Tian D-Y, Liu J-H, Wang F (2001) Guti Huojian Jishu [J Solid Rocket Technol] 24:41
66. Hong W-L, Tian D-Y, Liu J-H, Wang F (2001) Chem Abstr 135:275005
67. Politzer P, Murray JS, Grice ME, Sjoberg P (1991) Computer-aided design of monopropellants. In: Olah GA, Squire DR (eds) Chemistry of energetic materials, chap 4. Academic, San Diego, CA
68. Feng Z (2000) Huaxue Jinzhan [Prog Chem] 12:171
69. Feng Z (2000) Chem Abstr 134:88422
70. Tan G-H (2000) 火炸药科技动态 2 [Trends in Explosives]
71. Li S-W, Zhao F-Q, Yuan C, Luo Y, Gao Y (2002) Guti Huojian Jishu [J Solid Rocket Technol] 25:36
72. Li S-W, Zhao F-Q, Yuan C, Luo Y, Gao Y (2002) Chem Abstr 138:6183
73. Kim YG (2004) BK21 Program in Chemical Engineering, Seoul National University, Newsletter 5(1):22, http://web.archive.org/web/20041112214526/http://bk21-chem-eng.snu.ac.kr /news/newsletter/2004_spring/5-1.pdf, last visited: 23 April 2007
74. Pepekin VI (1994) Khim Fiz 13:42
75. Pepekin VI (1994) Chem Phys Rep 13:67
76. Litvinov BV, Fainzil'berg AA, Pepekin VI, Smirnov SP, Loboiko BG, Shevelev SA, Nazin GM (1994) Dokl Akad Nauk 336:67
77. Litvinov BV, Fainzil'berg AA, Pepekin VI, Smirnov SP, Loboiko BG, Shevelev SA, Nazin GM (1994) Dokl Chem 336:86
78. Dalinger IL, Vinogradov VM, Shevelev SA, Kuz'min VS (1996) Mendeleev Commun 6:13
79. Dalinger IL, Vinogradov VM, Shevelev SA, Kuz'min VS, Arnautova EA, Pivina TS (1998) Propellants Explos Pyrotech 23:212

80. Fokin AV, Studnev YuN, Kuznetsova LD (1996) Dokl Akad Nauk 346:358
81. Fokin AV, Studnev YuN, Kuznetsova LD (1996) Dokl Chem 346:22
82. Fokin AV, Studnev YuN, Rapkin AI, Kuznetsova LD (1996) Izv Akad Nauk Ser Khim 2689
83. Fokin AV, Studnev YuN, Rapkin AI, Kuznetsova LD (1996) Russ Chem Bull 45:2547
84. Fokin AV, Studnev YuN, Kuznetsova LD (1996) Izv Akad Nauk Ser Khim 2056
85. Fokin AV, Studnev YuN, Kuznetsova LD (1996) Russ Chem Bull 45:1952
86. Fokin AV, Studnev YuN, Stolyarov VP, Valiev RSh (1999) Izv Akad Nauk Ser Khim 130
87. Fokin AV, Studnev YuN, Stolyarov VP, Valiev RSh (1999) Russ Chem Bull 48:131
88. Khisamutdinov GKh, Slovetsky VI, Golub YuM, Shevelev SA, Fainzil'berg AA (1997) Izv Akad Nauk Ser Khim 338
89. Khisamutdinov GKh, Slovetsky VI, Golub YuM, Shevelev SA, Fainzil'berg AA (1997) Russ Chem Bull 46:324
90. Khisamutdinov GKh, Shevelev SA (2001) Izv Akad Nauk Ser Khim 706
91. Khisamutdinov GKh, Shevelev SA (2001) Russ Chem Bull 50:736
92. Shchetinin VG (1999) Fiz Goreniya Vzryva 35:116
93. Shchetinin VG (1999) Combust Explos Shock Waves (Engl Transl) 35:570
94. Grebennikov VN, Nazin GM, Manelis GB (1995) Izv Akad Nauk Ser Khim 649
95. Grebennikov VN, Nazin GM, Manelis GB (1995) Russ Chem Bull 44:628
96. Pepekin V (1996) Development of high-efficiency energetic explosives. 27th international annual conference of ICT, 25–28 June 1996, Karlsruhe, Germany,
97. Pepekin VI, Gubin SA (2007) Fiz Goreniya Vzryva 43:99
98. Pepekin VI, Gubin SA (2007) Combust Explos Shock Waves (Engl Transl) 43:84
99. Jiang Z, Xu G, Chen Y (1990) 21st International Annual Conference of ICT, Karlsruhe, Germany
100. Gilardi R, Evans RN, Manser GE (2003) Acta Crystallogr Sect E Struct Rep Online E59:o2032
101. Archibald TG, Manser GE, Immoos JE (1993) US Patent 5 272 249
102. Archibald TG, Manser GE, Immoos JE (1995) US Patent 5 420 311
103. Solomun T, Schimanski A, Sturm H, Illenberger E (2005) Macromolecules 38:4231
104. Grakauskas V, Baum K (1970) J Org Chem 35:1545
105. Archibald TG, Manser GE (1994) PCT Patent WO 94/05 643
106. Archibald TG, Manser GE (1998) US Patent 5 789 617
107. Rice JK, Russell TP (1996) Mater Res Soc Symp Proc 418:373
108. Adolph HG, Trivedi NJ (2001) US Patent 6 325 876
109. Fokin AV, Tselinskii IV, Mel'nikova SF, Vergizov SN, Studnev YuN, Stolyarov VP, Il'in SS (1986) Izv Akad Nauk SSSR Ser Khim 2086
110. Fokin AV, Tselinskii IV, Mel'nikova SF, Vergizov SN, Studnev YuN, Stolyarov VP, Il'in SS (1986) Bull Acad Sci USSR Div Chem Sci 35:1901
111. Fokin AV, Studnev YuN, Stolyarov VP, Mel'nikov AA (2000) Izv Akad Nauk Ser Khim 950
112. Fokin AV, Studnev YuN, Stolyarov VP, Mel'nikov AA (2000) Russ Chem Bull 49:949
113. John EO, Kirchmeier RL, Shreeve JM (1990) J Fluorine Chem 47:333
114. Zheng YY, Zhou JZ, Huang TJ, Zhou DL, Zhang MN (1985) J Fluorine Chem 29:216
115. Butcher RJ, Gilardi R, Baum K, Trivedi NJ (2002) Thermochim Acta 384:219
116. Fokin AV, Voronkov AN, Timofeenko IA (1978) Izv Akad Nauk SSSR Ser Khim 2644
117. Fokin AV, Voronkov AN, Timofeenko IA (1978) Bull Acad Sci USSR Div Chem Sci 27:2366

118. Prakash GKS, Bae C, Kroll M, Olah GA (2002) J Fluorine Chem 117:103
119. Majumder U, Armantrout JR, Williams RV, Shreeve JM (2002) J Org Chem 67:8435
120. Fokin AV, Grebennikov VN, Manelis GB, Nazin GM (1993) Dokl Akad Nauk 332: 735
121. Fokin AV, Grebennikov VN, Manelis GB, Nazin GM (1993) Dokl Chem 332:247
122. Grebennikov VN, Manelis GB, Nazin GM, Fokin AV (1994) Izv Akad Nauk Ser Khim 336
123. Grebennikov VN, Manelis GB, Nazin GM, Fokin AV (1994) Russ Chem Bull 43:315
124. Manelis GB, Nazin GM, Rubtsov YuI, Strunin VA (2003) Thermal decomposition and combustion of explosives and propellants. Taylor & Francis, London, chap 7
125. Park J, Chakraborty D, Jamindar S, Xia WS, Lin MC, Bedford C (2002) Thermochim Acta 384:101
126. Nazin GM, Prokudin VG, Manelis GB (2000) Izv Akad Nauk Ser Khim 231
127. Nazin GM, Prokudin VG, Manelis GB (2000) Russ Chem Bull 49:234
128. Dubikhin VV, Karnauh GE, Lagodzinskaya GV, Matveev VG, Nazin GM, Prokudin GV (2003) 34th International Annual Conference of ICT, 24–27 June 2003, Karlsruhe, Germany
129. Matveev VG, Nazin GM (2003) Kinet Catal 44:735
130. Chapman RD, Kreutzberger CB, Welker MF (1995) Difluoramino energetic materials. JANNAF novel ingredients for liquid and solid propellants specialist session, 4 April 1995, Pasadena, CA; http://stinet.dtic.mil/cgi-bin/GetTRDoc?AD=A451187, last visited: 20 April 2007
131. Chapman RD, Yee RY, Gilardi RD, Pinkerton AA (1996) Chemistry of difluoramine derivatives. 15th annual working group institute on synthesis of high energy density materials, 12–13 June 1996,Picatinny Arsenal, NJ; http://stinet.dtic.mil/cgi-bin/GetTRDoc?AD=A451459, last visited: 20 April 2007
132. Haiges R, Wagner R, Boatz JA, Yousufuddin M, Etzkorn M, Prakash GKS, Christe KO, Chapman RD, Welker MF, Kreutzberger CB (2006) Angew Chem Int Ed Engl 45:5179
133. Singh RP, Shreeve JM (2001) Chem Commun, p 1196
134. Graham WH, Parker CO (1963) J Org Chem 28:850
135. Fokin AV, Kosyrev YuM, Shevchenko VI, Potarina TM (1971) USSR Patent 299 505
136. Fokin AV, Kosyrev YuM, Shevchenko VI, Potarina TM (1971) Chem Abstr 75:76265
137. Prakash GKS, Etzkorn M, Olah GA, Christe KO, Schneider S, Vij A (2002) Chem Commun, p 1712
138. Fokin AV, Kosyrev YuM, Makarov VA, Novoselov NP (1969) Dokl Akad Nauk SSSR 186:112
139. Fokin AV, Kosyrev YuM, Makarov VA, Novoselov NP (1969) Proc Acad Sci USSR Chem Sect 186:350
140. Chapman RD, Davis MC, Gilardi R (2003) Synth Commun 33:4173
141. Inel Y (1976) Bogazici Univ Derg Temel Bilimler Kim 4/5:9
142. Inel Y (1977) Chem Abstr 92:170261
143. Leroy G, Sana M, Wilante C, Peeters D, Bourasseau S (1989) J Mol Struct: THEOCHEM 187:251
144. Sana M, Leroy G, Wilante C, Peeters D, Bourasseau S (1991) J Energ Mater 9:137
145. Sjoberg P, Murray JS, Brinck T, Politzer P (1990) Can J Chem 68:1440
146. Haeberlein M, Murray JS, Brinck T, Politzer P (1992) Can J Chem 70:2209
147. Politzer P, Lane P, Grice ME, Concha MC, Redfern PC (1995) J Mol Struct: THEOCHEM 338:249
148. Politzer P, Lane P, Sjoberg P, Grice ME, Shechter H (1995) Struct Chem 6:217

149. Politzer P, Murray JS, Grice ME (1996) Mater Res Soc Symp Proc 418:55
150. Politzer P, Lane P (1996) J Mol Struct: THEOCHEM 388:51
151. Politzer P, Lane P, Grice ME (1996) J Mol Struct: THEOCHEM 365:89
152. Politzer P, Murray JS, Seminario JM, Lane P, Grice ME, Concha MC (2001) J Mol Struct: THEOCHEM 573:1
153. Politzer P, Lane P, Concha MC (2003) Computational approaches to heats of formation. In: Politzer P, Murray JS (eds) Energetic materials, part 1. Decomposition, crystal and molecular properties, chap 9. Elsevier, Amsterdam
154. Ju X-H, Li Y-M, Xiao H-M (2005) J Phys Chem A 109:934
155. Ju X-H, Ji G-F, Xiao H-M (2006) Chem Phys 326:395
156. Fan X-W, Ju X-H, Xiao H-M, Qiu L (2006) J Mol Struct: THEOCHEM 801:55

Struct Bond (2007) 125: 153–194
DOI 10.1007/430_2006_053

Published online: 18 January 2007

Computational Aspects of Nitrogen-Rich HEDMs

Betsy M. Rice (✉) · Edward F. C. Byrd · William D. Mattson

U.S. Army Research Laboratory, AMSRD-ARL-WM-BD, Aberdeen Proving Ground, Aberdeen, MD 21005-5069, USA
betsyr@arl.army.mil

Abstract A variety of computational procedures used to predict properties of energetic materials is presented. These procedures, based on standard atomistic simulation methods, demonstrate the ability to predict key properties of these materials related to performance or hazard. Several applications of the various methods for nitrogen-rich materials are provided to illustrate capabilities. Also, an overview of theoretical efforts in computational design of novel all-nitrogen materials is given.

Keywords Computational chemistry · Crystal density · Energetic materials · Heats of formation · Quantum mechanics

Abbreviations

A7	*Strukturbericht* designation for α-arsenic crystal lattice structure
B3LYP	Becke 3-parameter hybrid density functional using non-local correlation provided by Lee, Yang and Parr
BP	Black phosphorus crystal lattice structure. *Strukturbericht* designation is A17
CCSD(T)	Coupled cluster with single, double and perturbative triple excitations
CG	Cubic gauche
CHNO	Carbon-hydrogen-nitrogen-oxygenal

CSD	Cambridge Structural Database
DFT	Density functional theory
EM	Energetic material
ESP	Electrostatic potential
G2	Gaussian-2 theory for calculating total molecular energies
G3	Gaussian-3 theory for calculating total molecular energies
G3(MP2)	Variant of Gaussian-3 theory for calculating total molecular energies in which the basis set extensions are obtained at the second-order Møller–Plesset level
G3(MP2)//B3LYP	Variant of Gaussian-3 theory for calculating total molecular energies in which the geometries and zero-point energies are obtained from B3LYP density functional theory instead of those used in conventional G3 calculations
GGA	Generalized gradient approximation
GIPF	General interaction properties function
Gx	Variant of the G2 or G3 method
HEDM	High energy density materials
HF	Hartree–Fock
HPC	Hexagonally packed chain
MC	Monte Carlo
MD	Molecular dynamics
MM	Molecular modeling
MOLPAK	MOLecular PAcKing crystal structure prediction software
MP	Molecular packing
MP2	Second order Møller–Plesset theory
NVE	Microcanonical
NVT	Canonical
PES	Potential energy surface
QCISD(T)	Quadratic configuration interaction with single double and perturbative triple excitations
QM	Quantum mechanics
QMD	Quantum molecular dynamics
QSAR	Quantitative structure activity relationship
QSPR	Quantitative structure property relationship
RMS	Root mean square
WMIN	Lattice energy minimization software

1 Introduction

The development of accurate models and simulations of energetic materials (EM) has been aggressively pursued within the EM research community since the advent of computational capabilities. Numerous analytic and computational models that predict the performance of an EM in a variety of applications exist, but most have a significant dependence on empirical data that previously could be obtained only through measurement. Due to the cost and time associated with collection of such empirical data (i.e., synthesis and testing), the EM community has recognized that for purposes of screening new materials, its dependence on such information must be reduced. Therefore, great attention has been given towards developing physics-based atomistic models for use in EM research and has resulted in a dramatic evolution of methods and applications of these to EM.

A variety of models based on atomistic simulation methods have been developed to predict key properties of an EM that are used to assess potential performance in a weapon or its sensitivity to impact [1, 2]. While showing a measure of success, these models still include some empiricism and have been developed for conventional explosives, most of which are composed of carbon, hydrogen, nitrogen, and oxygen (CHNO). Therefore, one cannot immediately conclude that the same models and methods will be applicable to the new class of high-nitrogen molecular and ionic crystals that are showing great promise as high energy density materials (HEDMs), and thus are limited in their use as screening tools. Also, concern for potential environmental hazards associated with the use and synthesis of EMs calls for the development of methods that will predict the environmental impact of any new material being considered for synthesis. Unfortunately, there are only a few methods that exist for such a purpose, both for CHNO and high-nitrogen materials, and significant efforts in development and assessment of such methods are still required. Efforts are being made to address all of these limitations, and will be discussed herein.

Atomistic methods in EM research are not limited to use as tools for screening purposes only. There are several examples where quantum mechanical characterizations of isolated molecules or elementary reactions are used to augment experimental information of newly synthesized high-nitrogen condensed phase materials [3–13]. Quantum mechanical approaches are also used to identify novel forms of nitrogen in which structural energy is stored, such as exotic all-nitrogen molecular species and high pressure polymorphic phases of solid nitrogen. Several examples will be discussed in this chapter. Atomistic simulation methods can also be used to predict probable reaction mechanisms and for exploring the dynamic response of a material to an initiating event. Simulations for the latter process using molecular dynamics (MD) methods, however, have been limited to conventional CHNO explosives and to our knowledge, only two have been performed for triazolium-based compounds, with a focus on predicting physical properties [14, 15]. Since there are no widespread applications of this methodology to high-nitrogen systems, these types of simulations will not be discussed in this chapter.

Section 2 will describe the various computational approaches used to develop predictive procedures for high-nitrogen solids. Section 3 will describe applications of these procedures to existing high-nitrogen HEDMs (both molecular and ionic crystals). Section 4 will be devoted to a discussion of predictions of novel all- or high-nitrogen species, followed by Sect. 5, in which predicted non-molecular high pressure phases of solid nitrogen are presented. Section 6 will contain concluding remarks, including identification of needs for further advancement of predictive capabilities for this emerging class of EMs.

2
Computational Methods

Atomistic simulation approaches encompass a wide variety of methods that can be loosely categorized into three areas: methods based on 1) classical physics, 2) quantum physics, and 3) empirical models that use atomistic simulation results. The methods based on classical physics used in EM research include molecular modeling/molecular packing (MM/MP), molecular dynamics (MD), and Monte Carlo (MC). MM/MP applications in EM research are used mainly for ab initio crystal prediction, a method in which the most probable packing of a molecule in the crystalline state is predicted using information about a single molecule. The ab initio crystal prediction of densities of EM will be discussed in the next section. MC methods are used to predict thermodynamic properties, while MD simulations predict the time-dependent behavior of a material (although the results can also be used to predict thermochemical properties). The accuracy of results using any of these methods is almost completely dependent on the quality of the description of interatomic interactions within a system. While some promising interaction potentials have been developed and successfully used in MC or MD simulations of conventional CHNO EMs [16–19], we are unaware of similar simulations for high-nitrogen HEDMs. Therefore, our discussion of simulation methods based on classical physics will be limited in this chapter.

Quantum mechanical methods are widely used in research of high-nitrogen energetic materials; applications include the evaluation of elementary reactions, establishing stability rankings among possible conformers or the generation of molecular properties, such as equilibrium structures, vibrational spectra, electrostatic potentials (ESP), electron densities, and thermodynamic information. Various QM theories ranging from the highly accurate CCSD(T) method to more computationally tractable approaches [such as density functional theory (DFT) or second-order perturbation theory (MP2)] have been applied to prediction of various molecular properties, many of which will be discussed in the next section. Most calculations using the extremely accurate quantum mechanical methods are performed to characterize all nitrogen molecules (Sect. 4), determine stability, and, in the case of species that are not stable, computationally design derivatives that will generate a stable species. Many of these studies have been limited to a small number of atoms, due to prohibitive computational requirements associated with high levels of quantum mechanical theory. Less accurate approaches (i.e., DFT using modest basis sets) have been used for predicting molecular properties associated with screening since these can be calculated for large molecules rapidly yet provide sufficiently accurate results. Atomistic simulation of novel high pressure phases of nitrogen has been dominated by solid state DFT treatments, the only computationally tractable quantum mechanical method for the condensed phase. These calculations will be reviewed in Sect. 5. Although there are numerous classical molecular simulations of highly compressed nitrogen, none of the interaction potentials used in those simulations correctly model phase transitions occurring in the nitrogen system and will not be discussed herein.

QM methods are also used in the third category of atomistic simulation methods we will include in this discussion. We focus on quantitative structure activity relationship/quantitative structure property relationship (QSAR/QSPR), a computational chemistry methodology that is extremely popular within the pharmaceutical community for new drug design. This approach establishes mathematical correlations between molecular descriptors (an inherent property of the chemical system) and various physico-chemical properties and behavior for various classes of compounds. An extensive compilation and description of over two thousand molecular descriptors are provided by Todeschini and Consonni [20]. While many of the descriptors are empirically derived or obtained from experimental information, many of them can be generated using quantum mechanical methods [21]. QSPR/QSAR methods using conventional and quantum-mechanical molecular descriptors have been used successfully to predict certain key properties of EMs for conventional CHNO materials [1] and applications will be described in the next section. Drawbacks to this method are its empiricism and reliance on experimental information; predictive capability of this method is not ensured for systems that are outside of the data set to which the QSPRs were parameterized. As with all empirical models, there is a possibility that a QSPR will perform poorly for dissimilar chemical systems. To our knowledge, this approach has not been widely applied to high-nitrogen materials.

The theoretical methods that will be discussed in this chapter represent only a subset of the various atomistic simulation methods used in computational materials research, and we refer the interested reader to the various comprehensive reviews on each method [22–30]. For the purposes of this chapter, however, we will highlight important points associated with various theories in application to the high nitrogen materials.

3
Prediction of Properties of High-Nitrogen Solids (Neutrals, Ionics)

The majority of efforts in developing predictive capabilities have focused on calculating properties that are indicative of the performance of a material. In recent years, however, vulnerability requirements and environmental restrictions have demanded that these factors be given equal or greater weight in the development and design of new materials, thus facilitating development of predictive models to address these aspects. In this section, we will describe the various models and methods used to predict these properties, and various applications to high-nitrogen HEDMs.

3.1 Performance Properties

Special attention has been given to the prediction of two properties that are used to provide an initial assessment of the potential performance of a material in a gun or warhead: the heat of formation and the density of the material. A variety of computational chemistry methods to accurately predict such performance properties [1, 31] exists, but applications to EM have been limited almost exclusively to conventional CHNO explosives. Unfortunately, there has been no similar wide-scale application of these computational methods to high nitrogen compounds. Additionally, there have not been extensive applications to ionic crystals (either high-nitrogen or CHNO salts), since the majority of EMs are neutral molecular organic crystals. Previous tools that have been developed to predict either of these properties for CHNO energetic neutral crystals are not suited for ionic materials, as will be described hereafter. Further, applications of existing tools that work well for CHNO neutral molecular crystals have shown inconsistent (and sometimes poor) behavior when applied to high nitrogen materials. Thus, the existing methods must be reassessed and modified to accommodate the high nitrogen systems, and alternative methods should also be explored. In this section, we will present some of these alternatives.

3.1.1 Crystal Densities: Neutral Molecular Crystals

We will begin this section with a discussion of the prediction of crystal densities for neutral molecular crystals. Several approaches have been used to predict crystal densities without a priori knowledge of the system. The first, and most sophisticated, is that of ab initio crystal prediction, in which the crystal microstructure is determined using information about a single, isolated molecule. While a number of procedures have been developed and assessed [32], all methods generally follow a three-step computational approach. The first step corresponds to generating a three-dimensional model of the molecule (the packing moiety) that will be used to construct candidate crystals of different symmetries. In the next step hypothetical crystal structures using the molecular models are created; the contents of the structures are dependent upon the crystal symmetry being explored and the orientation of the packing moiety. In the third step the energy of each hypothetical crystal is minimized with respect to lattice and molecular orientation parameters. After generating a large series of candidate crystal structures, the crystals can then be ranked (usually by energy or density). Although it has been shown to be a very useful predictive methodology, it suffers from certain limitations. The majority of computational methods assume that the crystal structure with the lowest lattice energy corresponds to the thermodynamically favored structure rather than the structure with the lowest free energy. This assumption effectively ignores entropic and vibrational enthalpic contributions to the free energy and does not consider kinetic factors associated with crystal growth, such as solvent effects and crystallization conditions. Another of the major limitations of this method of crystal density prediction is its reliance on a description of interatomic interactions. As in any atomistic modeling procedure, the quality of the result is dependent on the accuracy of the description of the interatomic forces. Finally, current methods require significant improvement in their search methods used to generate the candidate crystals, particularly for systems in which the asymmetric unit of a crystal (Z') is greater than one [32]. However, the utility of the procedure in EM research has been demonstrated, in spite of the aforementioned assumptions and limitations [33, 34]. A survey study [34] was performed in which 174 CHNO molecular crystals whose molecules contained functional groups common to CHNO explosives (i.e., nitramines, nitroaliphatics, nitrate esters, and nitroaromatics) were subjected to the method of ab initio crystal prediction using a transferable CHNO interaction potential [19] and the method developed by Ammon and co-workers(MOLPAK/WMIN) [33]. The systems chosen were restricted to crystalline space groups and systems that could be treated by MOLPAK/WMIN (i.e., $Z' \leq 1$). The study showed that for 85% of the chemical systems simulated, the method and model produced a crystal structure whose lattice parameters and contents of the crystal (i.e., molecular positions and orientations) matched the experimental crystal. Additionally, approximately 75% of these were the low energy structures of all possible candidates generated in the crystal prediction process. Predicted densities (calculated at 0 K) had a root-mean-square (rms) deviation from experiment of 3%. Inclusion of thermal effects is expected to bring these results more in line with the experimental values. This study showed that the method of ab initio crystal prediction is suitable to predict crystal densities of a new EM (provided the description of the interatomic interactions are reasonably accurate). However, because it utilizes an empirically derived interatomic model that was parameterized to CHNO systems, it cannot be assumed that the model will adequately predict structures of systems that are dissimilar from the original set to which it was parameterized. We demonstrate this in an application of the same interaction potential [19] and the method of ab initio crystal prediction to six high-nitrogen neutral molecular crystals synthesized by Klapötke [35]. The results were unsatisfactory; in three cases the experimental crystal was not found among the various polymorphs that were generated in the calculations. For the remaining three, two of the structures identified as matching the experimental crystals were significantly higher in energy than those of other polymorphs generated in the calculations. Furthermore, the contents of the unit cells and the cell dimensions had unacceptably large deviations from the measured values. This exercise clearly demonstrated a need for an interaction potential to treat such high-nitrogen compounds. Ammon has recently refined the default interatomic interaction potential used in the MOLPAK/WMIN suite of programs to describe high-nitrogen systems [36]. Application of the MOLPAK/WMIN methods and this newly refined interatomic potential produces densities of eight high nitrogen crystals within a few percent. These results are given in Table 1. Further assessments of this model and method for high-nitrogen systems are being performed as experimental information is being collected.

Ab initio crystal prediction provides a density value and important information about the positions of the atoms in the unit cell, invaluable information that can be used to construct molecular models for use in other simulations or might be useful in analysis of experimental results of dynamic response (e.g., directional shock sensitivity). Unfortunately, the methods are computationally intensive and analysis of the results is not trivial. For assessment of potential performance, often the user only needs the value of the crystal density at room conditions. For such calculations, QSAR/QPSR approaches are extremely attractive due to their ease of use and speed. Several QSPR-type approaches have been used to predict condensed phase densities of molecular organic systems [37–45]; some of which are included in EM research. In some of the QSPR-type approaches, a property called "molecular

Table 1 Crystal densities predicted using molecular volumes within the 0.001 a.u. isosurface of electron density

System	Expt	MOLPAK/ WMIN [36]	6-31G**		6-311+G(2df,2p)	
			0 K	Temperature corrected	0 K	Temperature corrected
g088 [35]	1.473	1.44	1.462	1.436	1.423	1.398
gn085 [35]	1.381	1.33	1.335	1.313	1.290	1.269
Mincob [224]	1.515	1.484	1.524	1.500	1.473	1.449
weigla [35]	1.444	1.491	1.488	1.453	1.440	1.406
g098 [35]	1.515	1.479	1.489	1.462	1.441	1.415
g096 [35]	1.522	1.495	1.555	1.527	1.504	1.476
Jahxog [225]	1.719	1.766	1.686	1.646	1.635	1.596
Iceduq [226]	1.384	1.423	1.382	1.349	1.345	1.314
Average % deviation		- 0.4	- 0.3	- 2.2	- 3.3	- 5.2
Rms % deviation		2.7	2.0	2.9	3.9	5.6

volume" is used in the development of the QSPR. The concept of molecular volume is ambiguous and can be defined in a variety of ways. For example, the experimental molecular volume in the Cambridge Structural Database (CSD) [46] is defined as the ratio of the volume of a unit cell to the number of molecules within the cell. Molecular volume can also be approximated assuming group or atom additivity,

$$V_{\mathrm{VA}} = \sum n_i V_i \,, \qquad (1)$$

where V_{VA} is the molecular volume assuming additivity, V_i is the volume of the ith constituent atom or functional group, and n_i is the number of the ith atoms or functional groups contained within the molecule. The atomic or group volumes can be defined by parameterizing to a large set of experimental crystal data [41, 43, 47, 48], or can be derived using accepted standard values, such as van der Waals radii [45]. Density predictions [39, 40, 44] have also been made using the general interaction properties function (GIPF) methodology [49–51], developed by Politzer for quantum-mechanical-based QSPR-like applications to predict macroscopic properties of a variety of materials. In this methodology, a GIPF is a function that uses statistical descriptors of the electrostatic potential (ESP) mapped onto a molecular surface to describe some macroscopic property. The parameters of the GIPF are determined by fitting to experimental information. In all of these calculations, the molecular surface is defined to be the 0.001 electrons/bohr^3 isosurface of electron density [52]. In addition to the statistical descriptors of features of the surface ESP, Politzer frequently uses the area of the molecular surface in his GIPFs [49–51]. Politzer and co-workers developed GIPFs to predict liquid and crystal densities for numerous compounds [44]; the crystal density GIPF was a function of the area of the molecular surface and the variance of the surface ESP, where the electron density ESPs were calculated at the HF/STO-3G level. These authors indicated that this GIPF could be modified to use molecular volume rather than surface area, but that the resulting GIPF was sufficiently accurate to render such a calculation unnecessary. Subsequent studies using the GIPF approach for crystal density prediction [39, 40, 45] showed improvement upon replacing the surface area term with molecular volume, defined as either the volume contained within the molecular surface used in the Politzer approach [39, 40] or the van der Waals volume [45]. Pan and Lee showed that the original Politzer GIPF for crystal density prediction was not transferable to cyclic and caged compounds (many of which included EMs), and required reparameterization to produce acceptable results [40]. Additionally, Bouhmaida and Ghermani, who used experimental electron densities in generating the surface ESPs, found that the volume within the molecular surface was adequate for predicting crystal density, and that the surface variance was a poor descriptor in establishing this correlation [39]. They concluded that the differences between the results generated by Politzer [44] and their application might be due to alterations of the ESPs of the molecules when in the solid state.

We have undertaken a study in which molecular volume is used to predict crystal densities for 181 CHNO species for which experimental crystallographic information exists. In this study, we defined our molecular volume to be that contained within the 0.001 a.u. isosurface of electron density of a molecule calculated at the B3LYP/6-31G** level. This level of QM theory is somewhat of an improvement over that used by Politzer and Murray [44] and Pan and Lee [40]. Two sets of molecular volumes were generated: one set was composed of volumes corresponding to the molecules in configurations consistent with those of the experimental crystals (denoted V_{Expt}), and the other (V_{opt}) was composed of molecules in equilibrium gas phase configurations (optimized at the B3LYP/6-31G** level). Since the geometry optimizations produce a 0 K result, we imposed a thermal correction to each system. This correction assumes isotropic expansion and has the form

$$V_{\mathrm{opt,Corrected}} = V_{\mathrm{opt}}(1 + \alpha T) , \tag{2}$$

where T denotes the temperature at which the crystal structure was measured and α was determined by fitting the right-hand side of Eq. 2 to the experimental molecular volumes (as defined earlier in [46]). For the set of molecules whose configurations correspond to the experimental structure, thermal and crystal field effects are already included; therefore, no thermal correction was imposed on molecular volumes or densities calculated using these structures.

V_{Expt} and V_{opt} were then compared with experimental values; molecular volumes calculated using experimental structures had average and rms deviations from experiment of – 4.2 and 5.8%, respectively. This corresponds to 6.3% rms deviation in crystal density. Average and rms deviation of V_{opt} from experiment are – 0.9 and 3.7%, respectively. Application of the thermal correction to V_{opt} produced average and rms deviations from experiment of 1.3 and 3.8%, indicating thermal effects are minimal. These V_{opt} correspond to a 3.6% rms deviation from experiment in crystal density, indicating far better agreement with experiment than results using the molecular configuration corresponding to the measured crystal structure. The maximum deviations of molecular volume and density from experimental results from this set are 42.8 Å^3 and 0.166 g/cc, respectively. We note that this method of crystal density prediction produced statistical agreement with experiment as good as those generated in the study of 174 CHNO systems using ab initio crystal prediction [34], at a substantially reduced computational cost. However, we emphasize that this method does not provide any information about the arrangement of the atoms within the unit cell.

We next calculated molecular volumes for eight high-nitrogen neutrals using two different basis sets [6-31G** and 6-311+G(2df,2p)] and the B3LYP density functional to explore the effect of basis set size on the molecular vol-

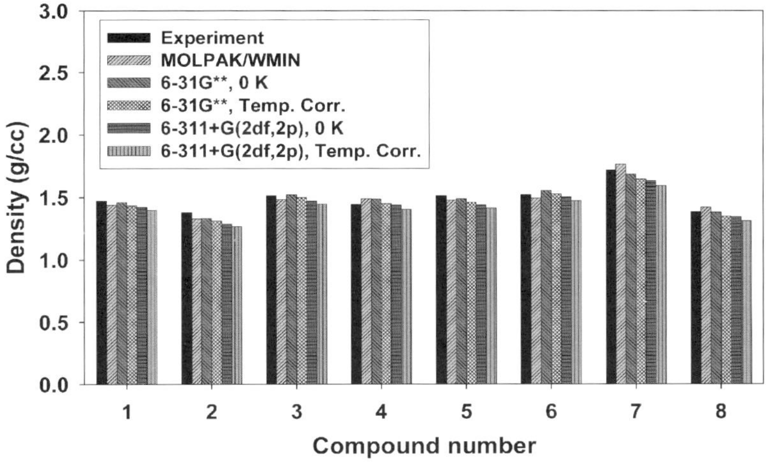

Fig. 1 Comparison of densities of high-nitrogen EM crystals (g/cc) using experimental crystal densities, MOLPAK/WMIN predictions, direct evaluation of 0 K and temperature-corrected molecular volumes using the 6-31G** basis set, direct evaluation of 0 K and temperature corrected molecular volumes using the 6-311+G(2df,2p) basis set

ume. In this study, geometries were optimized at the B3LYP/6-31G** level. Densities calculated using molecular volumes within the 0.001 a.u. isosurface of electron density with and without the temperature correction described above are listed in Table 1; a visual comparison with experimental values is given in Fig. 1. Additionally, results generated by Ammon using the recently modified force field (described earlier) and the MOLPAK/WMIN procedure [36] are also shown in Fig. 1. Temperature-corrected and 0 K crystal densities calculated using the larger basis set both have larger percent deviations from experiment than the other theoretical treatments. The MOLPAK/WMIN predictions (temperature corrected using factor recommended by Hoffman [41]) and the temperature-corrected 6-31G** calculations have approximately the same rms % deviation from experiment, while the uncorrected 6-31G** results have the smallest rms % deviation from experiment of all theoretical treatments.

3.1.2
Crystal Densities: Ionic Molecular Crystals

In principle, the various methods used for crystal density prediction of neutral molecular crystals should be applicable to ionic systems. Indeed, group additivity schemes have been developed for ionic crystals [53, 54]. Unfortunately, the ionic partners that make up the formula unit in the ionic crystal introduce complications in implementing some of the other methods. The main challenge is properly defining the arrangement of the ionic partners relative to one another. For several implementations of ab initio crystal prediction, the search algorithms cannot independently position or orient multiple ions that make up the formula unit during the generation of candidate crystals [32]. However, efforts are being made to extend this capability and initial results are promising [32, 55].

Attempting to energetically rank various arrangements of isolated ionic partners using quantum mechanical calculations might not be useful in all cases, since the most probable relative positions and orientations corresponding to the solid phase will be strongly influenced by the crystalline field, a property that cannot captured in these calculations. Further, quantum mechanical geometry optimizations of isolated ionic partners can result in spontaneous reactions, e.g., proton transfer [56, 57]; the preservation of charge separation between the ionic partners in such a calculation is difficult to maintain.

However, quantum-mechanically calculated ionic volumes can be used to predict crystal densities of ionic crystals, where the ionic volume is defined to be that contained within a selected isosurface of electron density. Following Jenkins et al. [58], the volume of the formula unit M_pX_q of an ionic crystal is simply the sum of the volumes of the ions contained in the formula unit:

$$\text{Volume} = pV_+ + qV_- , \tag{3}$$

where M denotes the cation and X denotes the anion. This method of formula unit volume prediction was applied to 34 high-nitrogen ionic salts provided by Klapötke [35]. In this application, the ionic volume is assumed to be that contained within the 0.001 a.u. isosurface of electron density of the ion calculated at the B3LYP/6-31G** level. Two sets of ionic volumes were generated. The first set were calculated using molecular structures consistent with the experimental crystal and the second used equilibrium geometries resulting from a B3LYP/6-31G** geometry optimization. Unlike what we observed in our earlier application of this method to neutral CHNO molecular crystals, the formula unit volumes calculated using the experimental structures were in closer agreement with experimental values than those using the optimized structures. Upon examination of the optimized geometries of the cationic and anionic components of the ionic volumes, we found the following: for those ions that had no hydrogen atoms, the differences between the volumes of ions assuming the experimental structures with those using the optimized structures were very small (fractions of $Å^3$). The largest differences were for ions that contained hydrogen atoms. Also, the magnitude of the difference in volume is directly proportional to the number of hydrogen atoms in the ion. This difference appears to be due to differences in the X – H (X = C, N) bond distances in the optimized and experimental structures. The B3LYP/6-31G** X – H bond distances are ~ 0.1 Å larger than those reported for the experimental structures.

In order to establish a correction factor to be used for calculations using optimized geometries of hydrogen-containing high-nitrogen ions, we first averaged experimental ionic volumes for each ion that is present in different ionic crystals. This averaged experimental ionic volume is then compared with the volume of the optimized ion. The difference between the averaged experimental and optimized ionic volumes scales almost linearly with the number of hydrogen atoms within the ion. This relationship allowed us to generate a correction factor for the high-nitrogen ionic crystals.

To calculate a "corrected" ionic volume using structures optimized at the B3LYP/6-31G** level, the formula is:

$$V(\text{corrected}) = V(\text{uncorrected}) - 0.976(\text{No. of Hydrogen Atoms in the ion}). \quad (4)$$

We did not attempt to define a thermal correction for these crystals. Calculated formula unit volumes using the experimental structures from the 34 high-nitrogen ionic crystals synthesized by *T.* Klapötke [35] had a rms deviation from experimental values of 4.6%. Formula unit volumes calculated using optimized geometries and corrected for the number of hydrogens had a 4.2% rms deviation from experimental values, whereas the uncorrected values had a 6.7% rms deviation from experiment. The rms deviation

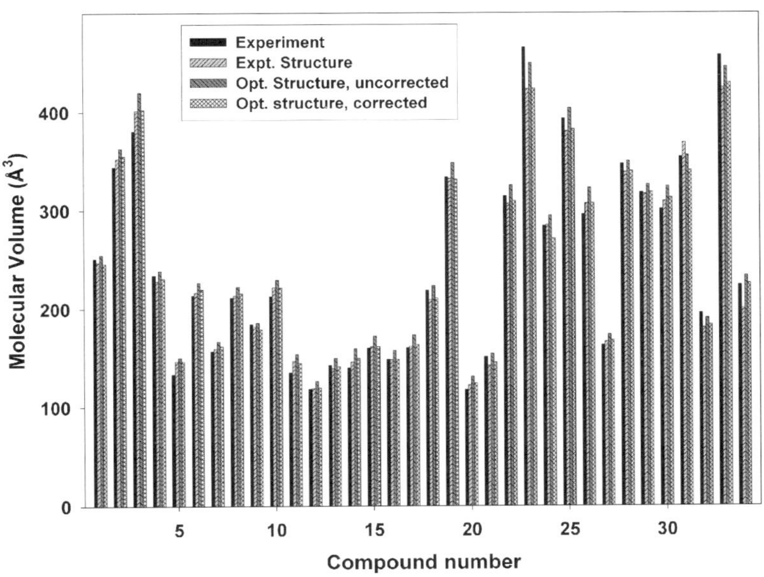

Fig. 2 Comparison of formula unit volumes of 34 high-nitrogen ionic crystals (g/cc) using experimental volume, predicted volumes using the experimental structures, predicted uncorrected volumes using optimized structures, and predicted volumes using optimized structures, corrected for the number of hydrogens (see text)

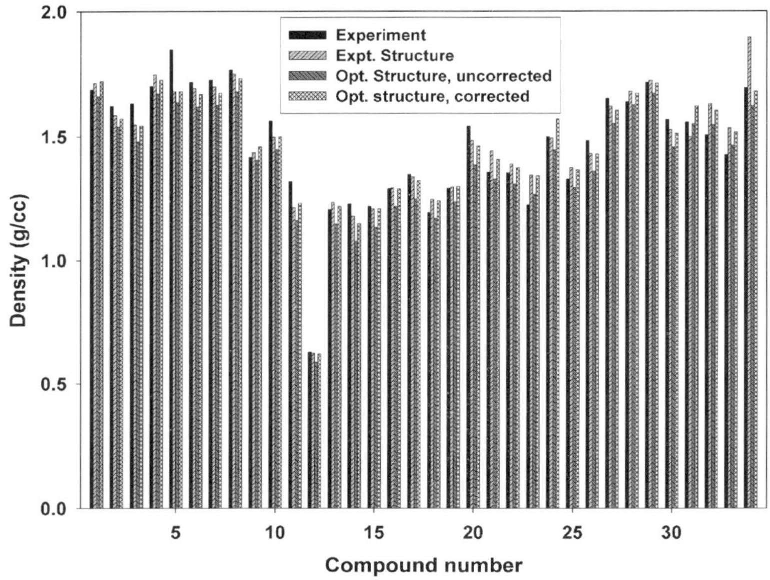

Fig. 3 Same as Fig. 2, except this provides a comparison of crystal densities

from experimental crystal densities using the volumes corresponding to the experimental structures is 4.7%, whereas the corresponding values using uncorrected and corrected formula unit volumes generated from optimized structures are 6.1 and 4.2%, respectively. Visual comparisons of the predicted and experimental molecules volumes and crystal densities are given in Figs. 2 and 3.

3.1.3
Solid Phase Heats of Formation: Neutrals

It can be argued that the solid phase heats of formation of a crystal can be calculated using highly accurate solid state quantum mechanical methods. However, computational obstacles rule out this approach. First, the only computationally feasible quantum mechanical approach to calculate the lattice energy of molecular crystals at this time is density functional theory. DFT does not adequately describe dispersion interactions, which are the main components of the binding energy in such crystals at ambient conditions [59, 60]. Secondly, this approach requires knowledge of the crystalline environment, and thus requires empirical information, which works against the overall goal to reduce dependence on such information through modeling. Solid phase heats of formation can be predicted using QSPR approaches [61–64], but their predictive capability might be limited to the chemical systems to which they were parameterized. To our knowledge, there are no QSPR methods for predicting the solid phase heats of formation for high-nitrogen compounds.

The solid phase heats of formation of weakly bound molecular crystals can also be obtained using predictions of gas phase heats of formation and heats of sublimation through the following relation [65]:

$$\Delta H^{\circ}_{f(s)} = \Delta H^{\circ}_{f(g)} - \Delta H_{sub} \,. \tag{5}$$

There are a variety of quantum-mechanically based methods to predict the gas phase heats of formation for neutral species. Politzer et al. have provided a detailed review of the various methods and their use in EM research [1], and we refer the interested reader to these. However, we will point out salient points from this review to illustrate applications of some of these methods to high-nitrogen compounds.

The numerous quantum-mechanically based schemes that exist for predicting the gas phase heats of formation include semi-empirical approaches that use atom or group equivalents. In this method, the gas phase heat of the formation is the difference in the energy of the molecule and empirically corrected energies of the molecule's component atoms or functional groups. The corrected energies, denoted as atom equivalent or group equivalent energies, are determined by fitting to experimentally measured values of heats of formation for representative systems. Several groups, including ours, have used this approach to develop a computational procedure for predicting the gas phase heats of formation of CHNO species [66–71]. Other more general methodologies that have a lesser degree of empiricism exist (e.g., the Gx methods [72]), although they require substantially more computational resources. As described by Politzer et al. [1], many of these methods predict gas phase heats of formation with a remarkable degree of accuracy. Calculations of heats of sublimation are almost exclusively performed using QSPR-like methods [66–68, 73–76]. The GIPF approach developed by Politzer for predicting this quantity has been particularly successful in applications to molecular crystals [66–71, 73–76]. Standard QSPRs using conventional molecular descriptors contained within commercial QSPR software [67] have also been developed for CHNO systems; however, these did not perform as well as the GIPF-based QSPRs when applied to a test set of molecules that were not included in the parameterization.

It cannot be assumed that any of the aforementioned computational methods that were parameterized using experimental information for CHNO systems would be transferable to high-nitrogen systems. In fact, application of the Byrd and Rice method [67] to a newly synthesized high-nitrogen compound (1-Methyl-5-(methylnitramino)-1*H*-tetrazole [4]) produced a solid phase heat of formation of 70.1 kcal/mol, far larger than the measured value (2.8 kcal/mol). Such a large discrepancy suggests that the method might not transferable, and since there are so few (if any) data for gas phase heats of formation or heats of sublimation for high-nitrogen molecular crystals, reparameterization of the equations used in Eq. 5 would be difficult. Thus, alternative approaches must be explored to treat high-nitrogen compounds.

The values for heats of sublimation of the CHNO crystals we surveyed typically ranged between 10–30 kcal/mol, with very few systems falling outside this range. Additionally, the heat of sublimation must always be a positive number. Conversely, the experimental gas phase heats of formation for the CHNO systems we have surveyed fall within a much larger range (from − 70 to + 100 kcal/mol [67]). In considering the disagreement of our approach to predicting solid phase heats of formation with the measured results for 1-Methyl-5-(methylnitramino)-1*H*-tetrazole, we concluded that the major source of the difference in our calculations and the measured value might be in the value of the gas phase heat of formation. We therefore undertook other quantum-mechanically based methods for predictions of the gas phase heat of formation of high-nitrogen compounds. In all calculations reported hereafter, the heats of sublimation used in Eq. 5 were calculated using the GIPF-methodology reported in [67].

The alternative methods used for predicting gas phase heats of formation are variants of the popular G3 methods [77] specifically the G3(MP2) [78] and G3(MP2)//B3LYP [79] approaches. The overall procedure involves multiple calculations at the HF/B3LYP, MP2, and QCISD(T) levels. The G3 method also attempts to correct for basis set size effects, zero point energy, and includes a general empirical correction. Therefore, like the atom/group equivalent approaches described above, this method includes some empiricism. However, G3 parameters are fitted to widely different chemical systems and a larger number of thermodynamic properties than those in our atom/group equivalent method [67, 68]. Additionally, the empiricism is not dependent on atom or group type, as are the atom/group-equivalent methods [1]. Further the G3 and atom/group-equivalent approaches use completely different methodologies for computing the energies. While the G3 and atom/group-equivalent approaches for predicting as phase heats of formation rely on empirical fits, more accurate quantum mechanical approaches (such as CCSD(T) using a complete basis set) cannot be applied to any of the systems under consideration in this project (i.e., EMs) due to computational constraints. In fact, the G3(MP2) and G3(MP2)//B3LYP methods were developed to reduce computational cost compared to the original G3 [80] method at a small cost in accuracy. However, we found even these modified G3 calculations to be computationally infeasible for high-nitrogen systems larger than $C_{12}H_{14}N_8$. For this system, the G3(MP2) and G3(MP2)//B3LYP methods require excessive computational resources are not readily available outside of supercomputer clusters at this time.

The G3(MP2), G3(MP2)//B3LYP and our atom/group equivalent method were used to calculate the gas phase heats of formation of a set of neutral high nitrogen crystals; the values were added to the predicted heats of sublimation (using the GIPF method as described in Ref. [67]) to produce solid phase heats of formation. The results and experimental values for the resulting solid phase heats of formation are shown in Fig. 4.

As evident in Fig. 4, the theoretical results are in reasonable agreement with each other for all compounds. Additionally, all calculated values are in agreement with the experiment except for Compound 2 (1-Methyl-5-(methylnitramino)-1*H*-tetrazole [4]). This suggests that either re-measurement of the heat of formation of Compound 2 should be performed or that there is some aspect of this system that is not captured by the quantum mechanical calculations. We emphasize that the two quantum mechanical approaches (atom/group equivalents versus G3 methods) for prediction of the gas phase heats of formation are sufficiently dissimilar that it is extremely unlikely that the error is within this portion of Eq. 1. There is a possibility, of course, that the heat of sublimation is poorly predicted for this system. That would not, however, explain the large discrepancy between the theoretical values and the measured result for Compound 2. Average and rms percent deviations were calculated using the seven other high-nitrogen compounds. Average percent deviations of the predictions from experiment are approximately the same for the group-equivalent and G3(MP2) approaches (these overestimate, on average, the experimental value by 5.3 and 5.6%, respectively). The average percent deviation of the G3(MP2)//B3LYP approach is

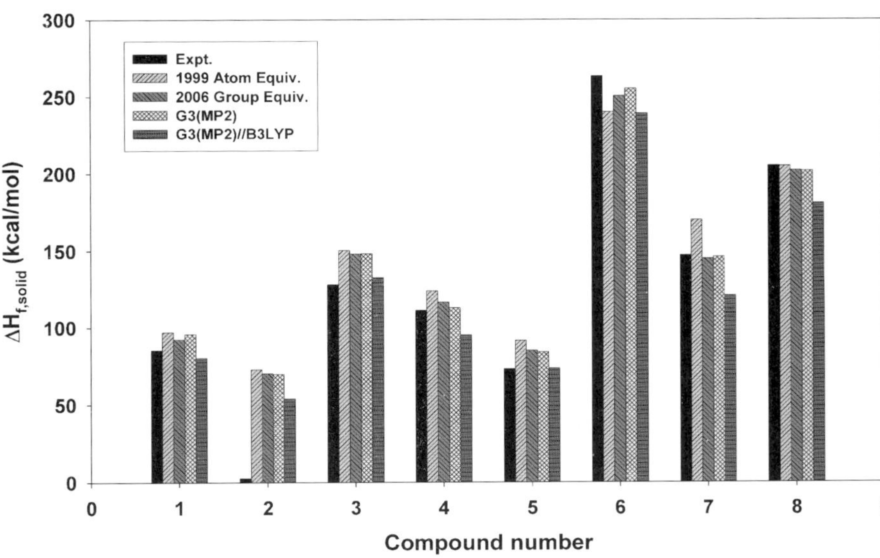

Fig. 4 Solid phase heats of formation for high-nitrogen crystals. Experimental values are taken from [4] and [223]. "1999 Atom Equiv" denotes calculations using the atom equivalent method described in [68]. "2006 Group Equiv" denotes calculations using the group equivalent method described in [67]. G3(MP2) and G3(MP2)//B3LYP denote calculations using variants of the G3 method [78, 79], respectively

larger (– 7.9 kcal/mol) and underestimates the experimental value. The rms percent deviations from experiment are comparable for the group equivalent and G3 approaches, with the group equivalent and G3(MP2) having the best overall agreement with experiment (rms % deviation is $\sim$ 9.5%).

Of particular interest to us are the consistently close agreement between the group-equivalent method [67] and the G3(MP2) values. We remind the reader that the group-equivalent method was fitted to a smaller and more chemically specific set of experimental data (CHNO compounds with functional groups common to explosives) whereas the G3 parameters were fitted to a more general and substantially larger set of chemical compounds. Additionally, the group-equivalent approach requires considerably less computational resources. Since this good agreement between the two approaches has only been demonstrated for eight compounds, further investigation is required to determine whether this trend holds for a larger number of high-nitrogen compounds.

3.1.4
Solid Phase Heats of Formation: Ionic Crystals

As for the neutral crystals, we are not aware of any QSPRs developed to predict the solid phase heats of formation of high nitrogen salts, probably due to the relative scarcity of such experimental data required for establishing the correlations.

Since the crystalline binding energy in an ionic crystal is dominated by electrostatic interactions that are orders of magnitude larger than those in a neutral molecular crystal, the same scheme used for predicting the solid phase heat of formation for a neutral molecular crystal (Eq. 5) cannot be applied to ionics. Rather, the heat of formation of ionic compounds is determined using Born–Haber cycles [65] in which a series of reactions involving the ionic components are employed to produce the overall final result. Unfortunately, error is always introduced when quantum mechanical calculations are used in evaluating the reactions. Therefore, care must be taken in the choice of reactions used in the cycle.

The simplest of cycles might consist of three steps in which formation energies for each of the components can be calculated.

$$M_{(g)} + X_{(g)} \rightarrow M^+_{(g)} + X^-_{(g)} \rightarrow M^+X^-_{(solid)} \,. \tag{6}$$

In this simple scheme, the first step requires the evaluation of the heats of formation of the neutral form of the charged components. The formation energies for the ionic components are then determined by evaluating the electron affinity or ionization energy and adding these to the heats of formation of the neutral moieties. The third step of the cycle requires the evaluation of the lattice enthalpy. Chemical reactions that generate the gas phase ionic components in Step 2 can also be used. Also, an alternative procedure for generating the gas phase formation energies of the ions directly is presented by Beaucamp et al. [81]. In this work, gas phase heats of formation for the ions were determined using the method of atom equivalents. The atom equivalents were derived for a series of neutral compounds [82] and applied to five ammonium salts. These values and cohesive energies for these systems were used to generate solid phase heats of formation. The results are in reasonable agreement with the experiment, and further study on a larger number of high-nitrogen salts should be performed to determine the suitability of this method.

Unfortunately, there are limited (if any) experimental data that can be used to assess the quality of the calculations associated with each individual step in the aforementioned cycle. We are limited to comparing the overall final result with the experimental value of the heat of formation of the ionic crystal. This makes establishing the sources of errors in this cycle complicated.

It is possible to introduce significant error in the first step of this cycle. The magnitude of the error introduced in a quantum mechanical prediction of any reaction (including atomization) is dependent on the choice of reaction (i.e., the types and number of bonds that are broken or formed). The most accurate quantum mechanical approach for predicting the gas phase heat of formation of a new compound is to use isodesmic reactions [83]. This approach works very well if there are reliable heat of formation values for the products of the isodesmic reaction. If there is more than one possible reaction that will lead to the gas-phase ions in the cycle, it is possible that the quantum mechanical predictions will produce different results that are dependent on which reaction is chosen [84]. Also, generation of the gas-phase ions in Step 2 might require a sequence of reactions rather than a single reaction; this would introduce further error. Also, while there might be many different possible reactions leading to the charged moieties of interest, corresponding experimental data of the reaction products must be available. In lieu of that, Gx procedures [72] (or atom/group equivalent methods, if applicable to the system) can be used to predict the heats of formation of the various reaction products, but again, error is introduced.

To avoid the complications associated with identifying reaction sequences that can be used to produce gas phase formation energies of the ionic species, we will illustrate predictions of the solid phase heats of formation of a few high-nitrogen salts using the Born–Haber cycle given in Eq. 6. In this, gas phase heats of formation of the neutral forms of the charged species are first calculated (Step 1), and are followed by calculations of the electron affinity or ionization energy (Step 2) to generate the heats of formation of the charged moieties in the gas phase. The main drawback to this approach is the requirement that the neutral form of the charged species must be a minimum on the potential energy surface (PES). If it is not, then the Born–Haber cycle would require inclusion of reactions leading to the ionic moieties in Step 2.

We were encouraged by the good agreement between experimental and predicted values (using the G3 method) of gas phase heats of formation for molecules contained in the eight high-nitrogen neutral crystals as described in the previous section. Therefore, we had a measure of confidence in using the G3 methods to predict gas phase formation energies of the neutral form of the charged species (Step 1). Unfortunately, there were many cases in which the optimized structure that is generated in the first step of the G3 calculations did not have the same chemical connectivity as that of the charged moiety in Step 2, indicating that the neutral form of the ion is not a minimum on the PES. The subsequent steps in the G3 calculations are all dependent on the molecular structure resulting from this optimization; therefore, it is essential that the molecular structure have the same chemical connectivity as that of the charged moiety. For the 26 ions corresponding to the various high-nitrogen salts provided to us by Klapötke [35], the most common result of a problematic geometry optimization was the dissociation of a hydrogen atom from the molecule. Since the geometry optimization performed in the first step of both of the two G3 approaches use a small basis set (6-31G*) and either the Hartree–Fock (HF) or B3LYP approach, it is possible that the problems are due to the application of an inadequate level of theory to treat these species. Tests of this hypothesis were performed by simply increasing the basis set size in the geometry optimization to include additional polarization and a diffuse function (6-31+G**). For all but four compounds a stable structure was found using B3LYP (we did not test HF). Applying the remainder of the G3(MP2) or G3(MP2)//B3LYP methodologies using these structures and the corresponding B3LYP/6-31+G** vibrational frequencies (scaled appropriately) produced gas phase formation energies. These revised G3 methods are denoted hereafter as modG3. The G3(MP2) and G3(MP2)//B3LYP empirical corrections will be applied accordingly, as modG3 has not been fitted for the correct empirical factor. Performances of the modG3 approaches were then evaluated through application to the original 8 neutral high-nitrogen compounds described in the previous section. The modG3 values track closely the G3 method from which the empirical correction is taken. For the G3(MP2) method, the average and maximum differences between the original and modified methods are 1.4 and 2.2 kcal/mol, respectively. For the G3(MP2)//B3LYP method, the average and maximum differences between the original and modified methods are 0.0 and 0.4 kcal/mol, respectively.

When determining the heats of formation of the ionic components beginning with the neutral forms of the species (Step 2), it is incorrect to use the vertical ionization energy, i.e., the energy required to add or subtract an electron without allowing structural relaxation upon ionization. In many of the cases we have examined, the neutral form of the ionic component has a substantially different structural conformation than that of the charged species; in such cases, the ionization energies or electron affinities must include the effects due to structural relaxation upon ionization. The energies resulting from such calculations will be referred to as "relaxed ionization energies". Figure 5 provides a visual example of the structural relaxation upon ioniza-

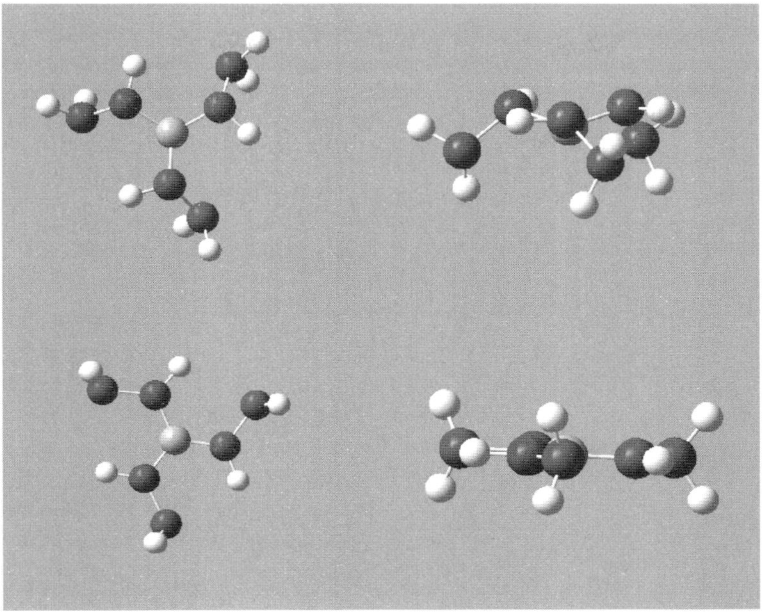

Fig. 5 Molecular structures of CH_9N_6 before (upper half of the figure) and after (lower half of the figure) ionization. Structures depicted in the left-most portion of the figure are top-down views of the moieties; structures in the right-most portions are side views of the moieties

tion for the CH_9N_6 moiety calculated at the B3LYP/6-31+G** level; it is clear that in some cases, structural relaxation would be a significant contributor to the final value of the ionization energy.

To determine the effect of structural relaxation on the ionization energy, we have calculated the relaxed and vertical ionization energies for the 20 high-nitrogen cations and 6 anions contained within various high-nitrogen salts synthesized by Klapötke [35]. The results are shown in Fig. 6. We found the effect to be particularly pronounced for the ionization energies; half of the compounds had structural relaxation energies on the order of 50 kcal/mol.

The final step in the evaluation of the Born–Haber cycle given in Eq. 6 is the determination of the lattice enthalpy, a measure of the energy required to dissociate the ionic crystal into its gaseous ions. The magnitude of this energy is quite large compared to the weaker intermolecular cohesive energies associated with organic molecular crystals (which are due mainly to van der Waals interactions), since it is a result of numerous long-range electrostatic interactions of ionic partners within the crystal. This value can be directly calculated by adding up all interatomic interactions within the crystal lattice if the positions of all atoms in the crystal are known and a reasonable description

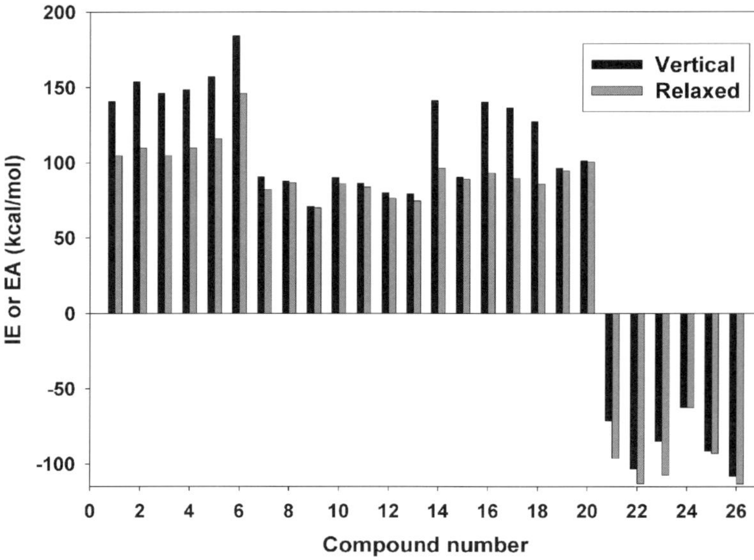

Fig. 6 Vertical and relaxed ionization energies for 20 cations and 6 anions used in high-nitrogen ionic crystals synthesized by Klapötke [35]

of the interatomic interactions is available (see, for example [58, 81, 85, 86]). Unfortunately, these two conditions cannot always be met. Therefore, numerous empirical schemes have been introduced to predict this quantity without knowledge of the crystal structure. One of the most widely used methods was developed by Kapustinskii, in which the lattice energy of a salt is estimated using values of the ionic radii [87]. Extensions to this approach have been made to better predict lattice energies of complex ionic crystals [88, 89]. Another method uses lattice energies calculated using point charges and experimentally crystallographic structures for a larger number of systems to generate a QSPR for cohesion energies in salts [85]. Politzer and Murray applied the GIPF approach to generate GIPF relations between lattice energies of ammonium, sodium and potassium cations with various anions [90]. A different QSPR-type method developed by Jenkins et al. has been applied to several high-nitrogen ionic crystals [58, 91]. This approach correlates the inverse cube root of the volume of the formula unit of an ionic crystal M_pX_q with its lattice potential energy $U_{pot}(M_pX_q)$. The lattice potential energy is related to the lattice enthalpy as follows:

$$\Delta H_L = U_{pot}(M_pX_q) + [p(n_M/2 - 2) + q(n_X/2 - 2)]RT \,, \tag{7}$$

where the values of n_M and n_X are dependent on whether the ions are monatomic, linear polyatomic, or non-linear polyatomic. Jenkins et al. [58, 91] parameterized the functional form for U_{pot} using lattice potential energies from [92]. These data consist of values derived from thermochemical cycles using experimental information and those from full scale calculations of the lattice potential energy of a crystal. In these calculations, crystal structural data (atomic positions) are specified, and an interatomic interaction potential (van der Waals terms plus Coulombic interactions, with partial charges assigned to nuclei) are used to produce the lattice energies. The information used in the parameterization was limited to salts containing alkali metal and alkaline earth cations (Li, Na, K, Rb, Cs, Mg, Ca, Sr, Ba). Goldschmidt radii of the cations (r_+) were used to define the cation volumes ($V_+ = 4/3\pi r_+^3$). Anion volumes were determined first by calculating the molecular volume of the alkaline earth/alkali metal salts (using experimental values, $V_{\text{unit cell}}/N_{\text{unit}}$) and subtracting the volumes of the cations.

A possible source of error associated with this approach might be due to the variable reliability of the fitting data [92]. Another source of error associated with this approach when applied to high-nitrogen salts is the types of salts used in the parameterization. In this parameterization of the QSPR, the cations of the salts were limited to alkaline metal or alkaline earth atoms (small and spherical volumes). The anion volumes were derived from experimental information about the salts and the values assigned to the alkali metal or alkaline earth cations. It is not known whether this method is transferable to salts that have significantly different chemical compositions, and for which the cations are substantially different in shape and size (such as some of the new high-nitrogen salts).

A rough assessment of the Jenkins approach for predicting lattice energy for high-nitrogen salts can be performed through direct calculations assuming the experimental structure and reasonable interaction potentials (such as that developed by Ammon, described earlier). At this time, we have not attempted nor are we aware of such an assessment. We will, however, use the Jenkins approach for purposes of illustrating prediction of the heat of formation of ionic crystals, specifically, three recently synthesized high-nitrogen salts [10]. These are presented and compared with experimental and other theoretical predictions in Table 2. The theoretical result generated by Gálvez-Ruiz et al. [10] invoked a Born–Haber cycle using reactions along with proton affinities, whereas we utilize the procedure outlined above. Additionally, the

Table 2 Heats of formation of tetrazolium salts

Species	ΔH_f, MX(s) (kcal/mol) Expt. [10]	G2 [10]	modG3(MP2)
1-5-Diamino-4-methyl-tetrazolium nitrate	8.4	41.7	47.5
1-5-Diamino-4-methyl-tetrazolium azide	138.1	161.6	165.4
1-5-Diamino-4-methyl-tetrazolium dinitramide	45.3	92.1	94.9

Gálvez-Ruiz et al. calculations used the G2 method [93], which has different empirical corrections and approaches than the G3 methods we used. Note that the two theoretical approaches, which would have distinctly different sources of error, are very close in value for the three compounds. They both, however, overshoot the experimental values by 30–40 kcal/mol. Since the common element between the two theoretical methods is the Jenkins approach [58, 91] for predicting lattice energies, it is possible that the major discrepancy between experiment and theory is within this step. However, before this conclusion can be made, the Jenkins method should be evaluated as suggested above.

3.2 Prediction of Vulnerability and Environmental Hazard

The performance potential of a new EM is not the only factor considered when determining whether to pursue synthesis or full-scale development. Of equal concern are the environmental impact of the material and its vulnerability to accidental initiation. Unfortunately, both categories of potential hazard encompass wide ranging phenomena. Environmental hazards include effects of the material on aquatic and mammalian life, ground water and atmospheric fate and transport, human toxicity or carcinogenicity. Vulnerability of an EM refers to different ways in which accidental initiation can occur, including electrostatic discharge, friction, shock, or impact, the results of which differ depending on the type of initiation. Both types of hazard represent diverse sets of extremely complex and interrelated physical-chemical properties and phenomena that have not been extricated or identified. Consequently, little progress has been made in developing a fundamental, detailed characterization of either type of hazard. Rather, most efforts have been directed to developing QSAR/QSPRs to predict potential hazards. This approach has been used to a small extent for predicting environmental hazards of conventional EM due to limited empirical information [94–101]. To our knowledge, there have been no published reports of predictions of environmental hazards of high-nitrogen EMs.

There exists a large body of empirical vulnerability data for conventional EM for which numerous researchers have correlated molecular or material properties [102–156]. Of these, a large number of the molecular properties used in the correlations were predicted using semi-empirical or quantum mechanical methods. While many of these are quite useful in identifying potential vulnerability of an EM, they should not be used to justify mechanistic arguments [157, 158]. Additionally, as with all QSPR approaches, the predictive capability is strongly dependent on the quality of the empirical information used in the parameterization. Unreliable empirical information used in the parameterization could result in a highly inaccurate tool. For vulnerability, the majority of the empirical data consists of results of dropweight impact tests, a crude and rapid method used to qualitatively assess the sensitivity of a material to impact. This test is notoriously inaccurate and its results are strongly dependent on the conditions under which the experiments are performed. Therefore, correlations made with the results could incorporate flawed measurements. Further, it is well established that the sensitivity of a material to accidental initiation is influenced by material morphology, something that cannot be captured with such simple correlations. Finally, there is never a guarantee that a QSPR developed for one series of compounds is transferable to a different chemical family. Since several correlations developed for conventional CHNO explosives are not transferable across chemical families, it is not expected that such correlations would be maintained for high-nitrogen HEDMs. We note, however, that two studies [5, 6] applied a QSPR-type method to assess impact sensitivity using quantum mechanically calculated electrostatic potentials [102, 159] to several high-nitrogen solids and showed that the correlations were maintained. However, since these methods were developed for CHNO explosives, similar applications to a larger number of high-nitrogen systems must be performed before it is concluded that such a correlation is indicative of sensitivity to impact for these systems. Clearly, advances in development of predictive methodologies in this area are needed.

4 Novel Polynitrogen Species

The earlier portions of this chapter have focused on computational methods of high-nitrogen compounds that are produced using conventional concepts of inorganic and organic chemical synthesis and known functionality [160]. This portion of the chapter will examine recent efforts to predict exotic forms of all-nitrogen molecules using quantum mechanical theories. Earlier predictions have been described in previous reviews of such compounds [161–163] and will not be repeated here. Since then, many larger all-nitrogen compounds in a variety of cyclic, acyclic or caged conformations have been explored using theoretical chemistry. The species investigated include ionic clusters [164–166], cylinders [167], cages [168–176], nanoneedles and nanotubes [177] and helices [178]. Isomers of smaller systems (N_7 [179], N_{10} [180] and N_{12} [181]) have also recently been reported. While the majority of these studies have focused on the evaluation of these novel forms of molecular nitrogen, at least one study uses theory to provide guidance in developing synthesis routes for an all-nitrogen system [182].

As shown in the earlier studies on smaller systems [161–163], stability of these compounds is perhaps the overriding issue that must be addressed. Thus, all of these calculations have several common goals: To establish a) whether the structure is a local minimum on the potential energy surface

(PES); b) its degree of stability with respect to unimolecular decomposition; and c) the structural features that stabilize the species. A successful demonstration of the ability of theory to predict stability of such compounds is given by Dixon et al. [183], in which high-level quantum mechanical calculations [CCSD(T)] indicated that $N_5^+N_3^-$ and $N_5^+N_5^-$ salts are not stable; these results were subsequently confirmed by experiment. Theory has also been used to computationally engineer structures to enhance stability through the addition of non-nitrogen atoms [184–191]. All of these studies have the potential to aide in the selection of the most promising all nitrogen molecular system for synthesis. The only caveat is that in most of these studies, only modest levels of quantum mechanical theory (mainly DFT and MP2) have been applied due to the system sizes, thus precluding the application of more accurate quantum mechanical methods. DFT and MP2 methods are adequate to identify local minima on the PES and provide estimates of stability to unimolecular decomposition. However, the extreme complexity of the electronic structure of these exotic forms of nitrogen requires higher levels of QM theory in order to quantify the degree of stability. This is illustrated by recent conflicting theoretical calculations on N_7 clusters using DFT and MP2 and a variety of basis sets [179, 192]. Using the G3 method, Wang et al. [192] predicted that the most stable of N_7 conformers is an N_5 ring with an N_2 side chain, but Zhao et al. found that the stability of the conformer was dependent on level of theory and basis set [179]. For such a case, only a higher level quantum mechanical treatment could resolve the discrepancies. It is unfortunate that at this time, the sizes of systems being explored exceed the computational bounds required for the necessary high-level treatments. But as history has shown, advances in computational power will allow for increasingly larger systems to be investigated using more accurate quantum mechanical treatments.

5
Novel High Pressure Phases of Nitrogen

Interest in high and all-nitrogen materials is not exclusive to the energetic materials community; the high-pressure physics community has been searching for novel, high-pressure forms of nitrogen for years [193–203]. As for all molecular systems, it is presumed that sufficient application of pressure will lead to the destruction of covalent bonds. For nitrogen, whose triple bond introduces extra complexity over singly bonded molecular solids, it is expected that any pressure-induced transformation will proceed first through formation of an intermediate polymeric network of singly or doubly bonded atoms, similar to that of other group V elements (phosphorus and arsenic) before losing covalency completely at higher pressures. Thus, experimental verification of such high-pressure non-molecular phases of nitrogen has long been sought, but it has only been recently that such was achieved [199–203]. In addition to these non-molecular phases of nitrogen, several molecular phases over wide temperature and pressure ranges have been experimentally determined [193–198, 204, 205].

Theoretical studies, on the other hand, have predicted a variety of non-molecular structures at pressures for which only molecular phases have been observed. These include a monatomic simple-cubic phase [206–208], a semi-metallic arsenic structure A7 [206], a metallic, simple-tetragonal phase [206, 209], and the cubic gauche (CG) structure [205, 207, 210–213], for which recent experimental evidence has been given [203]. Results from theoretical calculations also predict other polymeric forms of nitrogen, as will be described hereafter.

The disparity between the theoretical predictions and experimental verifications can be attributed to several factors, including large energy barriers that might inhibit any transition leading to the polymeric forms upon compression of the molecular crystals. These can be explored using theoretical methods, as will be shown here. This section will describe detailed, ab initio descriptions of the electronic structure of high-pressure phases of nitrogen, and ab initio quantum molecular dynamics (QMD) calculations, in which integration of the classical equations of motion using quantum mechanical forces produce time-dependent atomic positions and velocities. The latter calculations have been used to identify stable structures through quenching of materials simulated at high temperatures and pressures. QMD calculations are also used to establish ranges of metastability. This section will be limited to discussions of quantum mechanical calculations, with a focus on density functional theory treatments of these systems, and will not include discussions of classical molecular simulations of high-pressure phases of molecular nitrogen.

It was generally assumed that pressure-induced transformation of molecular nitrogen would progress first to a chain-like structure, then to a layered structure and finally to an bulk, or extended solid, structure. Early theoretical work, therefore, focused on establishing the simple chain, layered, and bulk structures known from other Group V elements, including A7 [206], and black phosphorus (BP) [207]. Early calculations of the various forms of non-molecular nitrogen predicted that the layered A7 and BP structures are lower in energy than the chain and bulk structures [206]. Later calculations predicted that the transition from molecular nitrogen would progress directly to the bulk CG structure, which is unique to nitrogen, bypassing formation of layered and chain structures [207, 210–213]. The CG form of nitrogen is a fully coordinated three-dimension structure, and is an analog to the diamond structure of carbon. Calculations predict it to be substantially lower in energy than the other non-molecular phases A7 and BP. Another chain structure was recently predicted that is energetically competitive with CG and will be detailed hereafter.

The majority of the early calculations in which new structures of nitrogen were identified resulted from deformations of a simple cubic lattice of nitrogen. One of us pursued a series of DFT studies to identify structures resulting from compression of the highest pressure known phase of nitrogen, the non-cubic molecular ε phase [214]. These results will be presented here. Some of these calculations were performed for systems that had been previously identified through other theoretical calculations using a lower level of quantum mechanical theory [207]. These calculations were redone at the higher level [using the generalized gradient approximation (GGA)] to verify relative energies.

We first confirmed the CG form (Fig. 7), and also found that CG can exist in either a left or right-handed form. These two forms are degenerate in energy and suggest that equilibration to one form or the other might be difficult. Instead, these features on the potential energy surface suggest that an amorphous mix of these two forms might result upon compression.

Although Barbee [211] has shown the CG structure to be stable at low temperature and zero pressure, we performed QMD simulations for various temperatures and pressures ranging from 0 and 50 GPa to test for instability. NVE simulations with an initial temperature of 300 K and a volume consistent with the low-pressure form of CG resulted in the dissociation of the structure into molecular nitrogen and a large energy release. The same result was observed for an NVE simulation with an initial temperature of 400 K and a volume corresponding to 50 GPa at 0 K. When these structures destabilized, the transition was extremely rapid and the temperature increased to over 10 000 Kelvin in a few time steps.

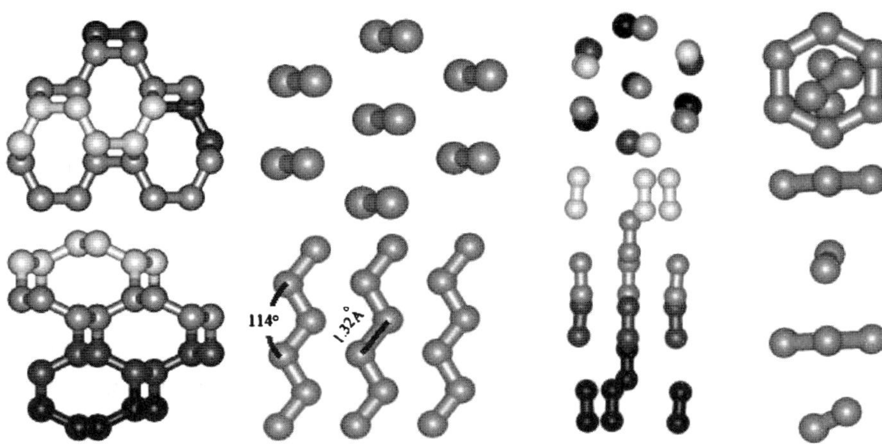

Fig. 7 Diagrams of various forms of compressed nitrogen. The structures are, from *left* to *right*, cubic gauche (CG), the hexagonally packed chain (HPC), ε-nitrogen, and Hexagonal N6 rings

Calculations performed by us and others produced an assortment of chain structures [207, 210, 212, 213], all but one of which are not energetically competitive with CG. The only energetically competitive chain structure was found in DFT optimizations of a forty-eight atom cell initially arranged in the high pressure molecular ε phase of nitrogen. Minimization of this cell at 220 GPa produced a series of hexagonally packed chains (HPC). At pressures above 400 GPa, this structure transforms into a fully connected structure with each atom being bonded to three others. With only a small increase in energy, this chain structure can be idealized as a deformed simple hexagonal lattice and a two-atom basis. The chains zig-zags, with the major axis of the chain parallel with the c-axis. The minor axes of all the chains are parallel to each other and perpendicular to one of the hexagonal lattice vectors. The idealized unit cell with a simple hexagonal lattice is given by the lattice vectors: $(a, 0, 0)$, $(a/2, \sqrt{3}a/2, 0)$, $(0, 0, c)$; the Cartesian atomic coordinates: $\pm(0, b, c/4)$, where $b = \sqrt{d^2 - c^2/4}$, and d is the bond length of 1.32 Å, and the a/c ratio is 0.9 with a lattice constant of 2.555 Å. The chains have bond angles of about 114° at low pressure. The HPC structure is shown in Fig. 7.

Two tests of stability were performed for this phase at pressures near ambient. For the first test, each atom in a sixty-four atom cell was randomly displaced by a maximum of 1% of the bond length in each Cartesian direction. A subsequent geometry optimization was performed, with a target pressure near that of the initial structure ($\sim$ 100 bar) in which atoms were not displaced. The system converged to the undisturbed state. This test was repeated, except all of the atoms were displaced by 5% of the bond lengths; the results did not change. For the second test, atoms were displaced along the line connecting atom pairs within the chain. This test was designed to facilitate transformation to a molecular nitrogen phase. Alternating bonds in the chain were contracted by 2% (resulting in an approximate two-percent expansion of the remaining bonds), after which a geometry optimization was performed. The system again converged to the undisturbed state. The same test was repeated except the bonds were contracted or expanded by 10%. The energies of the states in which the atoms are randomly displaced or displaced along bonds were higher than the HPC state by as much as 0.6 eV/atom. This indicates that there are substantial barriers to transition along these two paths. The HPC chain structure is distinguished from the other chain structures we investigated since it is energetically competitive with the CG phase, whereas the other chain structures were substantially higher in energy. The rapid increase in energy upon perturbation of atoms within the HPC from the equilibrium position and the higher energy of very similar chains suggest that the specific packing in the HPC phase is the basis for its low energy and stability. It is interesting that theoretical predictions of all-nitrogen molecules include molecular structures that resemble the hexagonally packed chains [163]. In these structures, the intramolecular bond lengths and angles are very similar to those of the chain. It is conceivable that such molecules could be packed into the crystalline state, perhaps at near ambient conditions.

Figure 8 shows the energy curves for the HPC and CG structures. The energy for the chain structure is slightly lower than CG at pressures well above the transition pressure (i.e., $V < 5$ Å^3), but is approximately the same or slightly higher than CG for the pressure range near the transition pressure. The difference in the energies of these structures is within the error of the GGA approximations, making the HPC structure practically degenerate in energy with the CG structure between approximately 5 and 6.5 Å^3. At pressures less than the transition pressure the HPC is lower in energy than CG.

In the HPC phase, each atom has valence electrons symmetrically bonded to two other atoms. For such a case, it is expected that the bands at the Fermi energy would be π-bands, similar to those in graphite. However, the nearly one-dimensional nature of the HPC chain structure should lead to a large, metallic, nearly flat Fermi surface unlike the semi-metallic Fermi surface in graphite. Indeed, the chain structure is metallic in nature as evidenced by the band structure in Fig. 9. The band structure of an identical single isolated chain is provided for comparison. Only the bands along the major axis of the chain are shown, as directions normal to the major axis are uninteresting, as expected for chains. We explored the possibility of a Peierls distortion leading to an insulator [215], but this is not found in the tests on distorted structures

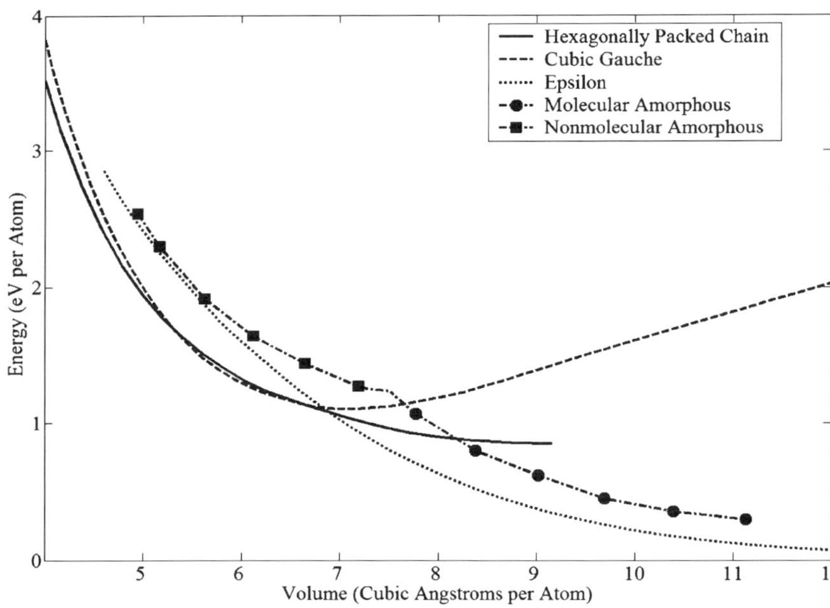

Fig. 8 Energy versus volume curves for the HPC, CG, ε, amorphous molecular and amorphous nonmolecular phases of nitrogen

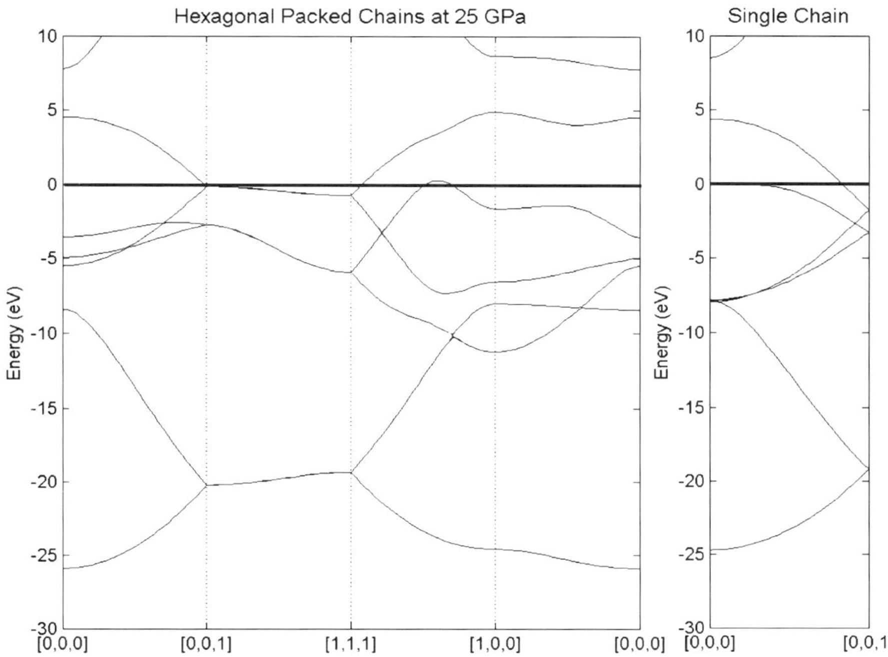

Fig. 9 Band structure of the HPC structure (*left-hand side*) and a single isolated chain (*right-hand side*). The Fermi energy is represented by the *solid dark line*

reported above. Presumably the stability of the metallic nature of the HPC phase is due to the fact that the system is not truly one-dimensional.

A metallic form of nitrogen is of great interest because of the possibility of superconductivity at a high transition temperature in such a phase. This possibility is suggested both by the small mass and by the large valence of nitrogen. Indeed, one the highest temperature elemental superconductor known is sulfur at high pressure [216–218], first predicted by theory [219]. The small mass of nitrogen implies a large Debye temperature, and the large number of valence electrons of nitrogen implies a large carrier density in a metallic phase. Both of these properties will tend to enhance the transition temperature. The electron–phonon interactions should be large for this element, as for other first row elements. On the one hand, the electron–phonon interaction is responsible for the instability of structures like the simple cubic; on the other hand, if the stable structure is metallic, then the electron-phonon interaction could lead to a high superconducting transition temperature T_c. Thus nitrogen can be an example of the competition between high T_c and structural instabilities [206]. Such a question invites further study.

An interesting molecular phase that was identified during the exploration of the conformational space of nitrogen was that of hexagonal N_6, shown on the right side of Fig. 7. While this is a molecular phase, it is not a diatomic molecular phase, and thus has the potential for large energy release upon transition to the lower-energy diatomic molecular phase. When the ε structure is minimized at ~ 60 GPa, this non-diatomic molecular phase appears. The ε structure transitions into a hexagonal-close-packed structure that consists of a unit cell containing one N_6 and one N_2 molecule, with the N_2 molecule centered coaxially between the N_6 molecules on alternating levels. The transition to this phase creates hexagonal N_6 from the disk-like molecules forming columns. The sphere-like molecules remain molecular and are centered between two coaxial N_6 molecules. These columns of N_6 are packed hexagonally, with adjacent columns offset along the axis of the column by a third the distance between N_6 in the column. This offset alternates direction going clockwise for each of the six neighboring columns. This structure has an energy that is equivalent to the molecular ε structure at the same volume and is a local energy minimum between 18 and 145 GPa. Since this is a relatively high-energy structure, no further stability testing was performed. At the lower end of the pressure range, the hexagonal structure transforms back into the ε molecular structure. Interestingly, as the pressure on this structure is increased beyond 145 GPa, the N_6 molecules begin to bond to each other at adjacent corners of the hexagons, forming a three dimensional structure not entirely dissimilar to the cubic gauche structure while maintaining the encapsulated N_2 molecules. While the energy of this structure is clearly above the other structures shown, it may still be physically observable if the transition barriers are large enough. It may be possible to obtain this phase by heating then cooling the ε phase at pressures below 100 GPa. Thus, this structure may be a candidate for the θ phase [202]. In addition, N_6 rings have been observed by Vogler et al. [220].

A second series of calculations were performed that were different in spirit than those described heretofore, in which the ε-form of molecular nitrogen was compressed to generate new structures. In this series, atomic nitrogen gas at high temperature and pressure was rapidly cooled at constant pressure. These calculations used QMD to produce the quench. The quenches of the hot atomic nitrogen gas produced, as expected, amorphous final states, but the final pressure at the end of the quench determined whether the final state would be molecular or a connected network of nitrogen. The molecular phase was only found below 100 GPa. At 100 GPa and above, the quenched configurations were interconnected networks, some of these would have a few diatomic nitrogen molecules, but all other nitrogen atoms are connected by a path of bonds to the other atom in the structure. The transition pressure of 100 GPa agrees well with that obtained from analysis of experimental results [200] (~ 100 GPa). In addition Raman studies of nitrogen have found evidence for the nitrogen diatomic bond destabilization at about 100 GPa [197, 220].

One of our major findings is that quenches to higher pressures yield amorphous networks of singly, doubly and triply connected atoms. The number of singly connected atoms is small, but the ratio of doubly to triply connected atoms is close to one. This may be interpreted in one of two ways, one being that the doubly connected atoms are either stacking faults or an artifact of the unphysical nature of the quench, but not energetically favored. Alternatively, and in light of the predictions of the HPC chain, these doubly connected atoms may actually be small examples of such chains and thus energetically competitive with the triply bonded form.

A typical amorphous structure is shown in Fig. 10. Bonds that cross the periodic boundary are not shown, thus giving the appearance or more singly and doubly connected atoms than are in the structure. For the structure represented in Fig. 10, there are eight singly connected atoms, twenty-eight doubly connected atoms, and twenty-eight triply connected atoms. All of the nitrogen networks produced in our simulations had several common characteristics. First, all atoms are bonded to one, two, or three other atoms. The number of singly bonded atoms ranged from four to ten, but the number of atoms bonded to two atoms or to three atoms was the same in each of the networks tested ($\sim$ 27–30). Secondly, the dihedral angles for pairs of triply connected atoms in the amorphous structure were similar to those in the CG structure. Thirdly, the amorphous networks transition to a liquid with increasing temperature. The atoms do not move freely, but bonded pairs will swap neighbors.

To test the stability of the amorphous networks, the pressure on the simulation cell was slowly released until the structure began to break apart. Even

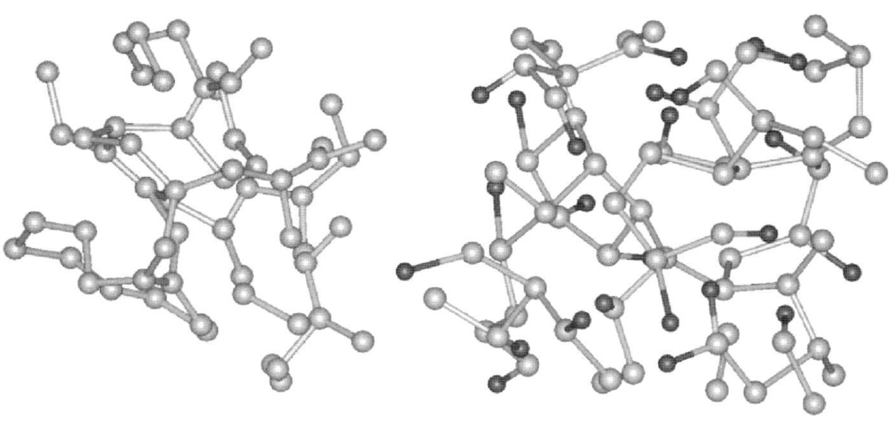

Fig. 10 *Left-hand side*: A typical amorphous structure, quenched to 200 GPa. The bonds that cross the periodic boundaries are not shown. *Right-hand side*: Passivated amorphous nitrogen structure. The hydrogen atoms are shown in *dark gray* and the nitrogen in *light gray*

at low temperature the amorphous structure became unstable between 80 and 100 GPa, the same pressure range where the quenches of the hot reactive nitrogen switched from producing molecular to non-molecular structures. For the amorphous configurations, the transition from the non-molecular to molecular phase begins with a dangling nitrogen atom breaking away with its single nearest neighbor forming a diatomic molecule. In all cases we observed, this dangling nitrogen was bound to only one other atom. We concluded that these singly bonded, dangling nitrogen atoms are the seeds of instability in the amorphous structure. The formation of the diatomic molecule completely destabilizes the system and the transformation from the polymeric phase proceeds rapidly. At higher temperatures the transformation occurs in the same pressure range, indicating that it is not a temperature-driven transition. The transition pressure of the quenches and the back transition pressure agree with the experimentally predicted equilibrium transition pressure of 100 GPa [201].

We were able to stabilize the amorphous network at pressures greater than 80 GPa by passivating all singly connected atoms. A passivated structure (Fig. 10) was constructed from an amorphous structure (used in the stability tests) by adding hydrogen atoms in the following manner. For a singly connected nitrogen (denoted N′), two hydrogen atoms were added such that all $X-N'-X$ angles are 120°, (X = N, H). The $X-N'$ bond lengths were set to 1 Å. Doubly connected nitrogen atoms had a single hydrogen atom added so that both $H-N-N$ angles are the same and are obtuse. Therefore, the passivated structure is one in which all nitrogen atoms are bonded to three other atoms. The coordination of this structure is consistent with the experimental material, which estimates a coordination of 2.5 at pressures of 180 GPa [199]. This passivated amorphous nitrogen structure was minimized with respect to energy at constant pressure. The stability of the passivated amorphous structure was examined in the same way as the amorphous structure. The passivated structure became unstable between 60 and 80 GPa irrespective of the temperature of the simulation. These theoretical studies clearly demonstrate that passivating the amorphous structure with hydrogen has a stabilizing effect; experiments are currently being performed to explore these ideas [221].

Given the number of energetically competing structures and the experimentally implied large barrier to transition, formation of an amorphous structure upon compression of nitrogen is not surprising. However, new crystallographic data [222] show that there might be some regularity to the experimental structure. Upon examination of the amorphous structures produced by the quench, we find evidence of both the cubic gauche and chain structures. The triply connected atoms have dihedral angles that are consistent with cubic gauche structure. The doubly connected atoms are chain-like. The fact that they occur with roughly equal probability in the amorphous structures implies that neither is strongly favored over the other and thus, this amorphous phase might contain an equimolar mixture of such doubly and triply connected atoms. NVT QMD simulations of the amorphous liquid did not produce a change in the number of bonds an individual atom has but shows that bonded atoms swap neighbors. This suggests that transforming between doubly and triply connected bonding may be energetically improbable. These theoretical results suggest that the observed experimental structure could be composed of an amorphous collection of small clusters of the different competing structures.

Although the results described in this section have not completely elucidated all of the complexities associated with pressure-induced structural phase transitions of nitrogen, the calculations have demonstrated how carefully designed theoretical simulations can be used to predict novel and interesting high pressure phases of nitrogen.

6
Concluding Remarks

The design and synthesis of novel high-energy, high-nitrogen materials is of great importance in advancing the field of energetic materials. Developments in computational chemistry and physics-based modeling such as those described herein are crucial in producing rapid breakthroughs in the development of these new materials. Computational exploration of notional materials allows for screening among potential candidates, assessing possible synthesis routes, or identifying conditions under which the materials exist. This, in turn, leads to a substantial reduction of time, money and waste streams associated with synthesis and testing of inferior materials or exploring materials under inappropriate experimental conditions. While progress is being made to develop accurate computational tools that are applicable to this emerging class of EMs, there are still significant challenges to overcome, particularly in the development of methods to assess environmental and vulnerability hazard. However, the potentially large cost and time savings compel further investment in developing theoretical procedures to complement experimentation leading to the synthesis and design of these new materials.

References

1. Politzer P, Murray JS (eds) (2003) Energetic Materials: Part 1. Decomposition, Crystal and Molecular Properties (Theoretical and Computational Chemistry). Elsevier Science, Amsterdam
2. Politzer P, Murray JS (eds) (2003) Energetic Materials: Part 2. Detonation, Combustion (Theoretical and Computational Chemistry). Elsevier Science, Amsterdam
3. Drake GW, Hawkins TW, Boatz J, Hall L, Ashwani V (2005) Propellants, Explos, Pyrotech 30:156

4. Karaghiosoff K, Klapötke TM, Mayer P, Piotrowski H, Polborn K, Willer RL, Weigand JJ (2006) J Org Chem 71:1295
5. Klapötke TM, Mayer P, Schulz A, Weigand JJ (2004) Propellants, Explos, Pyrotech 29:325
6. Hammerl A, Klapötke TM, Mayer P, Weigand JJ, Holl G (2005) Propellants, Explos, Pyrotech 30:17
7. Klapötke TM, Karaghiosoff K, Mayer P, Penger A, Welch JM (2006) Propellants, Explos, Pyrotech 31:188
8. Klapötke TM, Mayer P, Schulz A, Weigand JJ (2005) J Am Chem Soc 127:2032
9. Hammerl A, Klapötke TM, Nöth H, Warchhold M (2001) Inorg Chem 40:3570
10. Gálvez-Ruiz JC, Holl G, Karaghiosoff K, Klapötke TM, Löhnwitz K, Mayer P, Nöth H, Polborn K, Rohbogner CJ, Suter M, Weigand JJ (2005) Inorg Chem 44:4237
11. Geith J, Klapötke TM, Weigand J, Holl G (2004) Propellants, Explos, Pyrotech 29:3
12. Hammerl A, Klapötke TM, Nöth H, Warchhold M, Holl G (2003) Propellants, Explos, Pyrotech 28:165
13. Hammerl A, Klapötke TM (2002) Inorg Chem 41:906
14. Alavi S, Thompson DL (2005) J Phys Chem B 109:18127
15. Cadena C, Maginn EJ (2006) J Phys Chem B 110:18026
16. Strachan A, van Duin ACT, Chakraborty D, Dasgupta S, Goddard III WA (2003) Phys Rev Lett 91:098301
17. Boyd S, Gravelle M, Politzer P (2006) J Chem Phys 124:104508
18. Smith GD, Bharadwaj RK (1999) J Phys Chem B 103:3570
19. Sorescu DC, Rice BM, Thompson DL (1997) J Phys Chem B 101:798
20. Todeschini R, Consonni V (2000) Handbook of Molecular Descriptors. In: Mannhold R, Kubinyi H, Timmerman H (eds) Series of Methods and Principles of Medicinal Chemistry, Vol. 11. Wiley, New York, p 667
21. Karelson M, Lobanov V, Katritzky AR (1996) Chem Rev 96:1027
22. Bartlett RJ (2003) Electron Correlation from Molecules to Materials. In: Gonis T, Kioussis N, Ciftan M (eds) Electron Correlations and Materials Properties 2. Kluwer Academic/Plenum Publishers, New York, p 219
23. Helgaker T, Jorgensen P, Olsen J (2000) Molecular Electronic-Structure Theory. Wiley, Berlin
24. Frenkel D, Smit B (2002) Understanding Molecular Simulation: From Algorithms to Applications. Elsevier (USA), San Diego
25. Perkins R, Fang H, Tong W, Welsh WJ (2003) Environ Toxicol Chem 22:1666
26. Roy K (2004) Mol Diversity 8:321
27. Hafner J, Wolverton C, Ceder G (2006) MRS Res Bull 31:659
28. Martin RM (2004) Electronic Structure: Basic Theory and Practical Methods. Cambridge University Press, Cambridge
29. Gavezzotti A (2002) Cryst Eng Comm 4:343
30. Dunitz JD, Scheraga HA (2004) Proc Natl Acad Sci USA 101:14309
31. Rice BM (2003) Applications of theoretical chemistry in assessing energetic materials for performance or sensitivity. In: Thompson DL, Brill TB, Shaw RW (eds) Overviews of Recent Research in Energetic Materials. World Scientific, Singapore, London, Hong Kong, New Jersey
32. Day GM, Motherwell WDS, Ammon HL, Boerrigter SXM, Della Valle RG, Venuti E, Dzyabchenko A, Dunitz JD, Scheweizer B, van Eijck BP, Erk P, Facelli JC, Bazterra VE, Ferraro MB, Hoffman DWM, Leusen FJJ, Liang C, Pantelides CC, Karamertzanis L, Price SL, Lewis TC, Nowell H, Torrisi A, Scheraga HA, Arnautova YA, Schmidt MU, aVerwer P (2005) Acta Crystallogr, Sect B: Struct Sci 61:511

33. Holden JR, Du Z, Ammon HL (1993) J Comput Chem 14:422
34. Rice BM, Sorescu DC (2006) J Phys Chem B 108:17730
35. Klapötke TM, University of Munich, private communication
36. Ammon HL, Maryland U, private communication
37. Karelson M, Perkson A (1999) Comput Chem (Oxford) 23:49
38. Dunitz JD, Filippini G, Gavezotti A (2000) Tetrahedron 56:6595
39. Bouhmaida N, Ghermani NE (2005) J Chem Phys 122:114101
40. Pan J-F, Lee Y-W (2004) Phys Chem Chem Phys 6:471
41. Hofmann DWM (2002) Acta Crystallogr, Sect B: Struct Sci 57:489
42. Cho SG, Goh EM, Kim JK (2001) Bull Korean Chem Soc 22:775
43. Immirzi A, Perini B (1977) Acta Crystal A 33:216
44. Murray JS, Brinck T, Politzer P (1996) Chem Phys 204:289
45. Jalbout AF, Zouh ZY, Li X, Solimannedjad M, Ma Y (2003) J Mol Struct (THEOCHEM) 15:664–665
46. Allen FH (2002) Acta Crystallogr, Sect B: Struct Sci 58:380
47. Ammon HL, Mitchell S (1998) Propellants, Explos, Pyrotech 23:260
48. Ammon HL (2001) Struct Chem 12:205
49. Politzer P, Murray JS, Brinck T, Lane P (1994) Immunoanalysis of Agrochemicals. In: Nelson JO, Karu AE, Wong RB (eds) ACS Symposium Series 586. American Chemical Society, Washington DC, p 109
50. Murray JS, Politzer P (1994) Quantitative Treatment of Solute/Solvent Interactions. In: Politzer P, Murray JS (eds) Theoretical and Computational Chemistry. Elsevier Scientific, Amsterdam, p 243
51. Murray JS, Brinck T, Lane P, Paulsen K, Politzer P (1994) J Mol Struct (THEOCHEM) 307:55
52. Bader RWF, Carroll MT, Cheeseman JR, Change C (1987) J Am Chem Soc 109:7968
53. Beaucamp S, Marchet N, Mathieu D, Agafonov V (2003) Acta Crystallogr, Sect B: Struct Sci 59:498
54. Trohalaki S, Pachter R, Drake GW, Hawkins T (2005) Energy Fuel 19:279
55. van Eijck BP, Kroon J (2000) Acta Crystallogr, Sect B: Struct Sci 56:535, (erratum 56:745)
56. Schmidt MW, Gordon MS, Boatz JA (2005) J Phys Chem A 109:7285
57. Zorn DD, Boatz JA, Gordon MS (2006) J Phys Chem B 110:11110
58. Jenkins HDB, Roobottom HK, Passmore J, Glasser L (1999) Inorg Chem 38:3609
59. Wu X, Vargas MC, Nayak S, Lotrich V, Scoles G (2001) J Chem Phys 115:8748
60. Reimers JR, Cai Z-L, Bilic A, Hush NS (2003) Ann NY Acad Sci 1006:235
61. Sukhachev DV, Pivina TS, Volk FS (1994) Propellants, Explos, Pyrotech 19:159
62. Muthurajan H, Sivabalan R, Talawar MB, Anniyappan M, Venugopalan S (2006) J Hazard Mater 133:30
63. Keshavarz MH (2005) Thermochim Acta 428:95
64. Keshavarz MH (2005) J Hazard Mater 119:25
65. Atkins PW (1982) Physical Chemistry. Oxford University Press, Oxford
66. Politzer P, Lane P, Concha MC (2003) Computational Approaches to Heats of Formation In: Politzer PA, Murray JS (eds) Energetic Materials: Part 1. Decomposition, Crystal and Molecular Properties (Theoretical and Computational Chemistry). Elsevier Science, Amsterdam, p 247
67. Byrd EFC, Rice BM (2006) J Phys Chem A 110:1005
68. Rice BM, Pai SV, Hare J (1999) Combust Flame 118:445
69. Habibollahzadeh D, Grice ME, Concha MC, Murray JS, Politzer P (1995) J Comput Chem 16:654

70. Politzer P, Lane P, Concha MC (2004) Struct Chem 15:469
71. Mathieu D, Simonetti P (2002) Thermochim Acta 384:369
72. Raghavachari K, Curtiss LA (2005) G2, G3 and Associated Quantum Chemical Models for Accurate Theoretical Thermochemistry. In: Dykstra CE (ed) Theory and Applications of Computational Chemistry: The First Forty Years. Elsevier BV, Amstersdam, p 785
73. Kim CK, Lee KA, Hyun KH, Park HJ, Kwack IY, Kim CK, Lee HW, Lee B-S (2004) J Comput Chem 25:2073
74. Politzer P, Murray JS, Grice ME, Desalvo M, Miller E (1997) Mol Phys 91:923
75. Mathieu D, Bougrat P (1999) Chem Phys Lett 303:601
76. Politzer P, Ma Y, Lane P, Concha MC (2005) Int J Quantum Chem 105:341
77. Curtiss LA, Raghavachari K (2002) Theor Chem Acc 108:61
78. Curtiss LA, Redfern PC, Raghavachari K, Rassolov V, Pople JA (1999) J Chem Phys 110:4703
79. Baboul AG, Curtiss LA, Redfern PC, Raghavachari K (1999) J Chem Phys 110:7650
80. Curtiss LA, Raghavachari K, Redfern PC, Rassolov V, Pople JA (1998) J Chem Phys 109:7764
81. Beaucamp S, Bernand-Mantel A, Mathieu D, Agafonov V (2004) Mol Phys 102:253
82. Rousseau E, Mathieu D (2000) J Comp Chem 21:367
83. Hehre WJ, Ditchfield R, Radom L, Pople JA (1970) J Am Chem Soc 92:4796
84. Raghavachari K, Stefanov BB, Curtiss LA (1979) J Chem Phys 106:6764
85. Audoux J, Beaucamp S, Mathieu D, Poullain D, Agafonov V, Allouchi H (2004) Proceedings of the 35th International Annual Conference of ICT, p 33/1
86. Beaucamp S, Mathieu D, Agafonov V (2005) J Phys Chem B 109:16469
87. Kapustinskii AF (1956) Q Rev, Chem Soc 10:283
88. Liu D, Zhang S, Wu Z (2003) Inorg Chem 42:2465
89. Glasser L (1995) Inorg Chem 34:4935
90. Politzer P, Murray JS (1998) J Phys Chem A 102:1018
91. Jenkins HDB, Tudela D, Aglasser L (2002) Inorg Chem 41:2364
92. Jenkins HDB (1998) In: Lide DR (ed) Handbook of Chemistry and Physics, 79th ed. CRC Press, Boca Raton, p 1222
93. Curtiss LA, Raghavachari K, Trucks GW, Pople JA (1991) J Chem Phys 94:7221
94. Mathur KC, Chauhan UK, Shrivastava R, Khadikar PV (2001) Orient J Chem 17:253
95. Rendic S, Jurisic B, Medic-Saric MM (1995) Biological and environmental properties of nitro-, nitramine- and nitrate compounds: Explosives and drugs application of QSPR and QSAR studies in environmental toxicology assessment. In: Richardson M (ed) Environmental Toxicology Assessment. Taylor & Francis, London, p 303
96. Nikolic S, Medic-Saric M, Rendic S, Trinajstic N (1994) Drug Metabol Rev 26:717
97. Karmarkar S, Saxena A, Verma RG, Ramendra G, Karmarkar S, Mathur K, Mathur S, Singh S, Khadikar P (2000) Pollution Res 19:337
98. Khadikar PV, Lukovits I, Agrawal VK, Shrivastava S, Jaiswal M, Gutman I, Karmarkar S, Shrivastava A (2003) Indian J Chem, Sect A: Inorg, Bio-inorg, Phys, Theor 42:1436
99. Mathur KC, Khadikar PV, Chauhan UK, Shrivastava R (2001) Res J Chem Environ 5:68
100. Cenas N, Nemeikaite-Ceniene A, Sergediene E, Nivinskas H, Anusevicius Z, Sarlauskas J (2001) Biochim Biophys Acta, Gen Subj 1528:31
101. Purohit V, Basu AK (2000) Chem Res Toxicol 13:673
102. Rice BM, Hare JJ (2002) J Phys Chem A 106:1770
103. Kamlet MJ (1976) Proceedings 6th Syposium (International) on Detonation, San Diego, California, p 312

104. Kamlet MJ, Adolph HG (1979) Propellants Explos 4:30
105. Delpuech A, Cherville J (1978) Propellants Explos 3:169
106. Delpuech A, Cherville J (1979) Propellants Explos 4:121
107. Delpuech A, Cherville J (1979) Propellants Explos 4:61
108. Adolph HG, Holden JR, Cichra DA (1981) Naval Surface Weapons Center Technical Report NSWC TR 80–495
109. Fukuyama I, Ogawa T, Miyake A (1986) Propellants, Explos, Pyrotech 11:140
110. Mullay J (1987) Propellants, Explos, Pyrotech 12:60
111. Mullay J (1987) Propellants, Explos, Pyrotech 12:121
112. Jain SR (1987) Propellants, Explos, Pyrotech 12:188
113. Fried LE, Ruggiero AJ (1994) J Phys Chem 98:9786
114. McNesby KL, Coffey CS (1997) J Phys Chem B 101:3097
115. Wu CJ, Fried LE (2000) Proceedings 6th Syposium (International) on Detonation, Snowmass, Colorado, p 490
116. Vaullerin M, Espagnacq A, Morin-Allory L (1998) Propellants, Explos, Pyrotech 23:237
117. Xiao H-M, Fan J-F, Gu Z-M, Dong H-S (1998) Chem Phys 226:15
118. Fan J, Gu Z, Xiao H, Dong H (1998) J Phys Org Chem 11:177
119. Fan J, Xiao H (1996) J Mol Struct (THEOCHEM) 365:225
120. Zhao XC, Heming X, Shulin Y (1999) Chem Phys 250:243
121. Belik AV, Potemkin VA, Sluka SN (1999) Combust Explos Shock Waves 35:562
122. Zeman S (2000) Propellants Explos Pyrotech 25:66
123. Owens FJ, Jayasuriya K, Abrahmsen L, Politzer P (1985) Chem Phys Lett 116:434
124. Murray JS, Lane P, Politzer P, Bolduc PR (1990) Chem Phys Lett 168:135
125. Politzer P, Murray JS, Lane P, Sjoberg P, Adolph HG (1991) Chem Phys Lett 181:78
126. Murray JS, Lane P, Politzer P (1995) Mol Phys 85:1
127. Murray JS, Lane P, Politzer P (1998) Mol Phys 93:187
128. Politzer P, Alper HE (1999) Comput Chem (Singapore) 4:271
129. Owens FJ (1996) J Mol Struct (THEOCHEM) 370:11
130. Politzer P, Murray JS (1996) J Mol Struct 376:419
131. Politzer P, Murray JS (1995) Mol Phys 86:251
132. Politzer P, Murray JS (1996) In: Marinkas PL (ed) Organic Energetic Compounds. Nova Science Publishers, Inc, New York
133. Sundararajan R, Jain SR (1983) Ind J Technol 21:474
134. Afanas'ev GT, Pivina TS, Sukhachev DV (1993) Propellants Explos Pyrotech 18:309
135. Nefati H, Diawara B, Legendre JJ (1993) SAR QSAR Environ Res 1:131
136. Litvinov BV, Solkan VN, Fainzilberg AA, Garmasheva NV, Loboyko BG (1995) Dokl Akad Nauk 341:487
137. Lin Q, Xia X, Yang G, Zhang G, Liao H (1999) Gaoya Wuli Xuebao 3:64
138. McCormack J, Norton ES, Scullion HJ (1969) Explosivstoffe 17:225
139. Zanirato P (2006) Chim Ind (Milan, Italy) 88:130
140. Badders NR, Wei C, Aldeeb AA, Rogers WJ, Mannan MS (2006) J Energ Mater 24:17
141. Shu Y, Zhang C, Wang X, Zhao X (2005) Theoretical studies on the properties of some energetic materials. In: Vagenknecht J (ed) New Trends in Research of Energetic Materials, Proceedings of the Seminar, 8th, Pardubice, Czech Republic, Apr 19–21, 2005. University of Pardubice, Pardubice, p 779
142. Zhang C, Shu Y, Huang Y, Wang X (2005) J Energ Mater 23:107
143. Zhang C, Shu Y, Huang Y, Zhao X, Dong H (2005) J Phys Chem B 109:8978
144. Koch E-C (2005) Propellants, Explos, Pyrotech 30:5
145. Zeman S (2003) Propellants, Explos, Pyrotech 28:308

146. Zeman S, Krupka M (2003) Propellants, Explos, Pyrotech 28:301
147. Zeman S, Friedl Z, Huczala R (2002) Study of the impact reactivity of polynitro compounds part III. Relationship between electronic charges at nitrogen atoms of primarily split off nitro groups and impact sensitivity of some polynitro arenes. In: Vagenknecht J (ed) New Trends in Research of Energetic Materials, Proceedings of the Seminar, 5th, Pardubice, Czech Republic, Apr. 24–25, 2002. University of Pardubice, Pardubice, p 426
148. Zeman S, Krupka M (2002) Study of the impact reactivity of polynitro compounds part II. Impact sensitivity as a function of the intermolecular interactions. In: Vagenknecht J (ed) New Trends in Research of Energetic Materials, Proceedings of the Seminar, 5th, Pardubice, Czech Republic, Apr 24–25, 2002. University of Pardubice, Pardubice, p 415
149. Koshi M, Widijaja J, Kuroki Y (2001) Kayaku Gakkaishi 62:283
150. Lee HW, Chung K-H, Cho SG, Lee B-S, Kim CY (2006) International Annual Conference of ICT (2006), 37th (Energetic Materials), p 177/1
151. Zhang C (2006) Chem Phys 324:547
152. Zhao Q, Zhang S, Li QS (2005) Chem Phys Lett 407:105
153. Pacheco-Londono LC, De la Torre-Quintana LF, Primera-Pedrozo OM, Herrera GM, Ballesteros LM, Hernandez-Rivera SP (2004) Proceedings of SPIE-The International Society for Optical Engineering 5403:269
154. Saraf SR, Rogers WJ, Mannan MS (2003) J Hazard Mater 98:15
155. Keshavarz MH, Jaafari M (2006) Propellants, Explos, Pyrotech 31:216
156. Keshavarz MH, Pouretedal HR (2005) J Hazard Mater 124:27
157. Brill TB, James KJ (1993) J Phys Chem 97:8752
158. Brill TB (2005) Connecting molecular properties to decomposition, combustion, and explosion trends. In: Thompson DL, Brill TB, Shaw RW (eds) Overviews of Recent Research in Energetic Materials. World Scientific, Singapore, London, Hong Kong, New Jersey, p 1
159. Politzer P, Murray JS (2003) Sensitivity Correlations. In: Politzer PA, Murray JS (eds) Energetic Materials, Part 2: Detonation, Combustion, Theoretical and Computational Chemistry. Elsevier Science, Amsterdam, p 5
160. Singh RP, Verma RD, Meshri DT, Shreeve JM (2006) Angew Chem Int Ed 45:3584
161. Kwon O, McKee ML (2003) Polynitrogens as Promising High-Energy Density Materials: Computational Design. In: Politzer PA, Murray JS (eds) Energetic Materials: Part 1. Decomposition, Crystal and Molecular Properties (Theoretical and Computational Chemistry). Elsevier Science, Amsterdam, p 405
162. Nguyen MT (2003) Coord Chem Rev 244:93
163. Bartlett RJ, Fau S, Tobita M, Wilson K, Perera A (to be published)
164. Li PC, Guan J, Li S, Qian SL, Wen GX (2003) Phys Chem Chem Phys 5:1116
165. Cheng L, Li Q Xu W, Zhan S (2003) J Molec Model 9:99
166. Liu YD, Yiu PG, Guan J, Li QS (2002) J Mol Struct (THEOCHEM) 588:37
167. Zhou H, Wong N-B, Zhou G, Tian A (2006) J Phys Chem A 110:7441
168. Zhou H, Wong N-B, Zhou G, Tian A (2006) J Phys Chem A 110:3845
169. Zhou G, Pu X-M, Wong N-B, Tian A, Zhou H (2006) J Phys Chem A 110:4107
170. Wang LJ, Zgierski MZ (2003) Chem Phys Lett 376:698
171. Strout DL (2005) J Phys Chem A 109:1478
172. Bruney LY, Bledson TM, Strout DL (2003) Inorg Chem 42:8117
173. Strout DL (2004) J Phys Chem A 108:10911
174. Sturdivant SE, Nelson FA, Strout DL (2004) J Phys Chem A 108:7087
175. Strout DL (2004) J Phys Chem A 108:2555

176. Strout DL (2006) J Phys Chem A 110:7228
177. Wang JL, Lushington GH, Mezey PG (2006) J Chem Inf Model 46:1965
178. Wang L, Mezey PG (2005) J Phys Chem A 109:3241
179. Zhao JF, Li QS (2003) Chem Phys Lett 368:12
180. Wang LJ, Mezey PG, Zgierski MZ (2004) Chem Phys Lett 391:338
181. Li QS, Zhao JF (2002) J Phys Chem A 106:5367
182. Wang LJ, Warburton P, Mezey PJ (2002) J Phys Chem A 106:2748
183. Dixon DA, Feller D, Christe KO, Wilson WW, Vij A, Vij V, Jenkins HDB, Olson RM, Gordon MS (2004) J Am Chem Soc 126:834
184. Wilson KJ, Perera SA, Bartlett RJ, Watts JD (2001) J Phys Chem A 105:7693
185. Cottrell R, Jones J, Gilchrist A, Shields D, Strout DL (2006) J Phys Chem A 110:9011
186. Strout DL (2006) J Phys Chem A 110:4089
187. Colvin KD, Cottrell R, Strout DL (2006) J Chem Theory Comput 2:25
188. Colvin KD, Strout DL (2005) J Phys Chem A 109:8011
189. Strout DL (2005) J Chem Theory Comput 1:561
190. Bruney LY, Strout DL (2003) J Phys Chem A 107:5840
191. Strout DL (2003) J Phys Chem A 107:1647
192. Wang X, Tian A-M, Wong NB, Law C-K, Li W-K (2001) Chem Phys Lett 338:367
193. Reichlin R, Schiferl D, Martin S, Vanderborgh C, Mills RL (1985) Phys Rev Lett 55:1464
194. Mills RL, Olinger B, Cromer DT (1986) J Chem Phys 84:2837
195. Schuch AF, Mills RL (1970) J Chem Phys 52:6000
196. Bini R, Ulivi L, Kreutz J, Jodl HJ (2000) J Chem Phys 112:8522
197. Olijnyk H, Jephcoat AP (1999) Phys Rev Lett 83:332
198. Olijnyk H (1990) J Chem Phys 93:8968
199. Goncharov AF, Gregoryanz EA, Mao HK, Liu Z, Hemley RJ (2000) Phys Rev Lett 85:1262
200. Eremets MI, Hemley RJ, Mao HK, Gregoryanz E (2001) Nature 411:170
201. Gregoryanz E, Goncharov AF, Hemley RJ, Mao H (2001) Phys Rev B 64:52103
202. Gregoryanz E, Goncharov AF, Hemley RJ, Mao H, Somayazulu M, Shen G (2002) Phys Rev B 66:224108
203. Eremets MI, Gavriliuk AG, Trojan IA, Dzivenko DA, Boehler R (2004) Nat Mater 3:558
204. LeSar R (1984) J Chem Phys 81:5104
205. Mitas L, Martin RM (1994) Phys Rev Lett 72:2438
206. Lewis SP, Cohen ML (1992) Phys Rev B 46:11117
207. Mailhiot C, Yang LH, McMahan AK (1992) Phys Rev B 46:14419
208. McMahan AK, LeSar R (1985) Phys Rev Lett 54:1929
209. Martin RM, Needs RJ (1986) Phys Rev B 34:5082
210. Mattson WD, Sanchez-Portal D, Chiesa S, Martin RM (2004) Phys Rev Lett 93: 125501
211. Barbee III TW (1994) In: Schmidt SC et al. (ed) In High-Pressure Science and Technology – 1993, AIP Conf Proc No 309. AIP, New York, p 163
212. Yu HL, Yang GW, Yan XH, Xiao Y, Mao YL, Yang YR, Cheng NX (2006) Phys Rev B 73:012101
213. Alemany MMG, Martins JL (2003) Phys Rev B 68:024110
214. Mattson WD (2003) The Complex Behavior of Nitrogen Under Pressure: Ab Initio Simulation of the Properties of Structure and Shock Waves, Thesis, Univ Illinois at Urbana-Champaign, Urbana-Champaign
215. Kittel C (1996) Introduction to Solid State Physics. Wiley, New York

216. Kometani S, Eremets M, Shimizu K, Kobayashi M, Amaya K (1997) J Phys Soc Jpn 66:2564
217. Struzhkin VV, Hemley RJ, Mao HK, Timofeev YA (1997) Nature 390:382
218. Gregoryanz E, Struzhkin VV, Hemley RJ, Eremets MI, Mao H-K, Timofeev YA (2002) Phys Rev B 65:064504
219. Zakharov O, Cohen ML (1995) Phys Rev B 52:12 572
220. Vogler A, Wright RE, Kunkely H (1980) Angew Chem Int Ed 19:717
221. Ciezak JA (2006) US Army Research Laboratory, private communication
222. Hemley RJ (2002) The Carnegie Institution Geophysical Laboratory, private communication
223. Huynh MHV, Hiskey MA, Chavez DE, Naud DL, Gilardi RD (2005) J Am Chem Soc 127:12537
224. Hammerl A, Holl G, Klapötke TM, Mayer P, Noth H, Piotrowski H, Warchhold M (2002) Eur J Inorg Chem 4:834
225. Hang M, Huynh V, Hiskey MA, Archuleta JG, Roemer EL, Gilardi R (2004) Angew Chem Int Ed 43:5658
226. Patra S, Sarkar B, Ghumaan S, Fiedler J, Kaim W, Lahiri GK (2004) Inorg Chem 43:6108

Struct Bond (2007) 125: 195–271
DOI 10.1007/430_2006_052

Published online: 18 January 2007

Sensitivities of High Energy Compounds

Svatopluk Zeman

Institute of Energetic Materials,
University of Pardubice, 532 10 Pardubice, Czech Republic
svatopluk.zeman@upce.cz

Abstract The article presents a survey of the development trends in studies of sensitivity (initiation reactivity) of energetic materials (EMs) over the last nine years, focusing mainly on impact and shock sensitivities. Attention is given to the initiation by heat, laser, electrostatic discharge, impact and shock, including the influence of hydrostatic compression, crystal defects, molecular structure and desensitizing admixtures on the initiation reactivity. Problems of the initiation of nitromethane are examined with a special accent. It is stated that one of the best-developed theories for such studies is Dlott's Model of the Multiphonon Up-Pumping. Also significant is the model based on Non-Equilibrium Zeldovich–von Neuman–Döring theory. Very important are approaches devised by Politzer and Murray, updated by Price et al. as a hybrid model of prediction of the impact sensitivity of CHNO explosives. The physical organic chemistry approach to the sensitivity problem (POC model) is discussed with special emphasis. In this way it has been found that the electron structure and close neighborhood of the primarily leaving nitro group are dominant factors in the initiation by shock, electric spark and heat of polynitro compounds.

Keywords Detonation · Energetic materials · Electric spark · Impact · Shock · Sensitivity · Thermal decomposition

Abbreviations

1,3-DNB	1,3-dinitrobenzene
1,5-DNN	1,5-dinitronaphthalene
1,8-DNN	1,8-dinitronaphthalene
AcAn	1,9-diacetoxy-2,4,6,8-tetranitro-2,4,6,8-tetraazanonane
BITNT	3,3′-dimethyl-2,2′,4,4′,6,6′-hexanitro-1,1′-biphenyl
BTX	5,7-dinitro-1-(2,4,6-trinitrophenyl)-1H-1,2,3-benzotriazole
CPX	1,3-dinitroimidazolidine
CTB	2-chloro-1,3,5-trinitrobenzene
DADN	1,5-diacetyl-3,7-dinitro-1,3,5,7-tetrazocane
DATB	2,4-diamino-1,3,5-trinitrobenzene
DCTB	2,4-dichloro-1,3,5-trinitrobenzene

Decagen	1,3,5,7,9-pentanitro-1,3,5,7,9-pentazecane[1]
Digen	1-nitro-1-azaethylene[2]
DIPAM	3,3′-diamino-2,2′,4,4′,6,6′-hexanitro-1,1′-biphenyl
DIPS	1,3,5-trinitro-2-[(2,4,6-trinitrophenyl)thio]benzene
DIPSO	1,3,5-trinitro-2-[(2,4,6-trinitrophenyl)sulfonyl]benzene
DMDIPS	2,4,6-trinitro-3-[(3-methyl-2,4,6-trinitrophenyl)thio]toluene
DMEDNA	2,5-dinitro-2,5-diazahexane
DMNA	2-nitro-2-azapropane
DMNO	2,5-dinitro-2,5-diazahexane-3,4-dione
DNAt	4,6-dinitro-2,1-benzoisoxazole (4,6-dinitroanthranil)
DNBF	4,6-dinitro-2,1,3-benyoxadiayol-1-ium-1-olate (4,6-dinitrobezofuroxane)
DNDAH	3,5-dinitro-3,5-diazaheptane
DNDC	1,4-dinitropiperazine
DODECA	2,2′,2″,2‴,4,4′,4″,4‴,6,6′,6″,6‴-dodecanitro-[1,3′,1′,1″]quaterphenyl
DPE	1,3,5-trinitro-2-[2-(2,4,6-trinitrophenyl)ethyl]benzene
DPM	1,3,5-trinitro-2-(2,4,6-trinitrobenzyl)benzene
DPT	3,7-dinitro-1,3,5,7-tetraazabicyclo[3.3.1]nonane
HMX	1,3,5,7-tetranitro-1,3,5,7-tetrazocane (octogen)
HNB	2,2′,4,4′,6,6′-hexanitro-1,1′-biphenyl
HNIW	2,4,6,8,10,12-hexanitro-2,4,6,8,10,12-hexaazaisowurtzitane
HNO	*N*,*N*′- bis(2,4,6-trinitrophenyl)ethanediamide
HNS	1,3,5-trinitro-2-[(*E*)-2-(2,4,6-trinitrophenyl)vinyl]benzene
HOMO	1,3,5-trinitro-1,3,5-triazepane
MADNBF	7-amino-4,6-dinitro-2,1,3-benzoxadiazol-1-ium-1-olate
MDN	1,1-dinitro-1-azaethane
NONA	2,2′,2″,4,4′,4″,6,6′,6″-nonanitro[1,1′,3′,1″]-terpheny
NTFA	*N*,*N*-bis(2,4-dinitrophenyl)-2,4,6-trinitroaniline
OHMX	2,4,6,8-tetranitro-2,4,6,8-tetraazanonane
ONT	2,4,6,4′,6′,2″,4″,6″-octanitro[1,1′,3′,1″]-terphenyl
ORDX	2,4,6-trinitro-2,4,6-triazaheptane
PA	2,4,6-trinitrophenol
PAM	2,4,6-trinitroaniline
PDNBTO	4,6-dinitro-2-(2,4,6-trinitrophenyl)-2H-1,2,3-benzotriazol-1-ium-1-olate
PETN	penterythritol tetranitrate
PYX	*N*,*N*′-bis(2,4,6-trinitrophenyl)-3,5-dinitropyridine-2,6-diamine
RDX	1,3,5-trinitro-1,3,5-triazinane (hexogen)
TACOT-Z	1,3,7,9-tetranitro[1,2,3]benzotriazolo[2,1-*a*]-[1,2,3]benzotriazole-5,11-diium-6,12-diide
TATB	2,4,6-triamino-1,3,5-trinitrobenzene
TCTB	2,4,6-trichloro-1,3,5-trinitrobenzene
TENN	1,4,5,8-tetranitronaphthalene
TENPO	1,3,7,9-tetranitrophenoxazine
Tetrogen	1,3-dinitro-1,3-diazetidine[3]
TETRYL	*N*-methyl-*N*,2,4,6-tetranitroaniline
TEX	4,10-dinitro-2,6,8,12-tetraoxa-4,10-diazaisowurtzitane
TMPM	*N*,*N*′,*N*″-tris(3-methyl-2,4,6-trinitrophenyl)-[1,3,5]triazine-2,4,6-triamine

[1] It has not been synthesized yet, for a prediction of some its properties see [16, 98, 148, 176].

[2] It has not been synthesized yet, for a prediction of some of its properties see [16, 98, 148, 176].

[3] It has not been synthesized yet, for a prediction of some of its properties see [16, 98, 148, 176].

TNA	2-methoxy-1,3,5-trinitrobenzene
TNAD	*trans*-1,4,5,8-tetranitrodecahydro-pyrazino[2,3-*b*]pyrazine
TNADEC	2,4,7,9-tetranitro-2,4,7,9-tetraazadecane
TNAZ	1,3,3-trinitroazetidine
TNB	1,3,5-trinitrobenzene
TNBA	2,4,6-trinitrobenzoic acid
TNCr	3-hydroxi-2,4,6-trinitrotoluene
TNMs	2,4,6-trimethyl-1,3,5-trinitrobenzene
TNN	1,4,5-trinitronaphthalene
TNPMO	1,3,7,9-tetranitrophenothiazine-5-oxide
TNPTD	1,3,7,9-tetranitrophenothiazine-5,5-dioxide
TNR	2,4-dihydroxi-1,3,5-trinitrobenzene
TNT	2,4,6-trinitrotoluene
TNX	2,4-dimethyl-1,3,5-trinitrobenzene
TPM	*N*,*N*′,*N*″-tris(2,4,6-trinitrophenyl)-[1,3,5]triazine-2,4,6-triamine
TPT	2,4,6-tris(2,4,6-trinitrophenyl)-[1,3,5]triazine

1
Introduction

Sensitivity of high energy materials (EMs) is primarily due to the chemical character of the materials; this means it is possible to use the term "initiation reactivity of EMs" in this case. However, the means of transfer of the initiation impulse to the reaction centre of the EM molecule or the molecule of the most reactive component of the explosive mixture is also of great importance. Therefore, according to Dlott a complex solution to the problem of initiation must involve the areas of continuum mechanics, chemistry and quantum mechanics (quantum chemistry) [1]. The main interest has been focused on studies of shock and impact sensitivities of EMs. In the last 16 years the preferred tools for the solution of these sensitivities have involved quantum chemistry [1–5]. The application of chemistry to these problems is relatively reluctant and mostly without any broader contexts. Nevertheless, the approach of physical organic chemistry has been applied not only to studies of impact and shock reactivity [6, 7], but also sensitivity to electric spark [6, 8], and in part to thermal reactivity of EMs [7] as well. This survey presents development trends of studies of initiation reactivity of EMs over the last nine years with emphasis on the contribution of physical organic chemistry to these studies. Research results presented at conferences and seminars are quoted here only as the exception.

2
Primary Splitting of EM Molecules

Study of EM sensitivity has become an implicit aspect of understanding the primary chemical processes of their initiation. In this respect, initiation by heat (low-temperature thermal decomposition) is the most frequently studied to date. With regard to the above-mentioned dominant interest in the study of impact and shock sensitivities within the mentioned period of nine years, and also with regard to the importance of these sensitivities in technical practice, the topic of splitting of EM molecules by the said stimuli have been included in Sect. 3 on Impact and Shock Sensitivities.

2.1
Low-Temperature Thermal Decomposition

The basic problem of defining the kinetics and mechanism of thermal decomposition of energetic materials lies in the strong dependence of the corresponding kinetic parameters on temperature, pressure, and construction materials in contact with the sample decomposed [9]. Hence the mutual compatibility of results obtained from thermal analyses of energetic materials using different methods and/or different types of apparatus of different origin is very rare [9, 10]. So far the most reliable results in this area have included both theoretical and practical findings obtained by Russian scientists on the basis of their manometric method (see [10, 11] and citations herein). The data obtained by this method are known to correspond to the non-autocatalyzed stage of thermal decomposition of the given material (i.e. to molecular structure [10]), and also to the absolute values of the corresponding Arrhenius parameters. The mechanisms of primary unimolecular fragmentation in low-temperature thermal decomposition (up to 600 K) of technologically attractive organic polynitro and polynitroso compounds in condensed state, specified mostly by the Russian authors, can be divided into the following basic classes:

- *Homolysis of* $C-NO_2$, $N-NO_2$, $O-NO_2$, $N-NO$ and $C-NF_2$ *bonds*. The homolytic fission of the first $C-NO_2$ bond, particularly that connecting a sterically hindered nitro group, is characteristic of the decomposition of polynitro paraffins [11] and unsubstituted polynitro arenes [12, 13]. Homolytic splitting of the $N-NO_2$ bond is typical of secondary nitramines [11–18, 20] (Scheme 1), that of the $O-NO_2$ bond is typical of nitric esters [11], that of the $N-NO$ bond is typical of nitrosamines [19], and that of the $C-NF_2$ bond is typical of *gem*-bis-difluoroamino paraffins [11, 20, 21]. The relationship between molecular structure and thermal stability of aliphatic, azaaliphatic and some heterocyclic high energy compounds containing the above-given types of bonds is presented in [20].

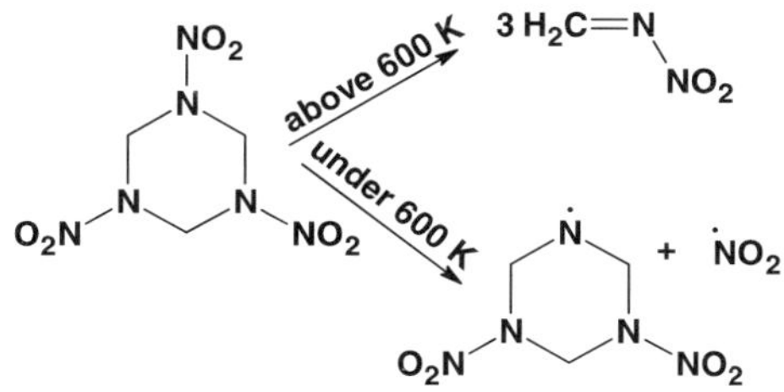

Scheme 1 Mechanism of high- and low-temperature primary thermolysis of RDX

- *Homolysis via a five-, six- or seven-membered transition state or aci-form.* A six-membered transition state (or aci form) is connected with the presence of a hydrogen atom at the γ-position with respect to the nitro group (Scheme 2) [7, 11] in derivatives of polynitro arenes exhibiting the so-called trinitrotoluene mechanism of thermal decomposition [7, 11]. The five-membered transition state might have significance in a group of polychlorinated derivatives of TNB—the primary step of the thermal decomposition of these derivatives could perhaps be connected with the chemical interaction between the chlorine atom and oxygen of the *ortho*-standing nitro group ([6] and references therein).

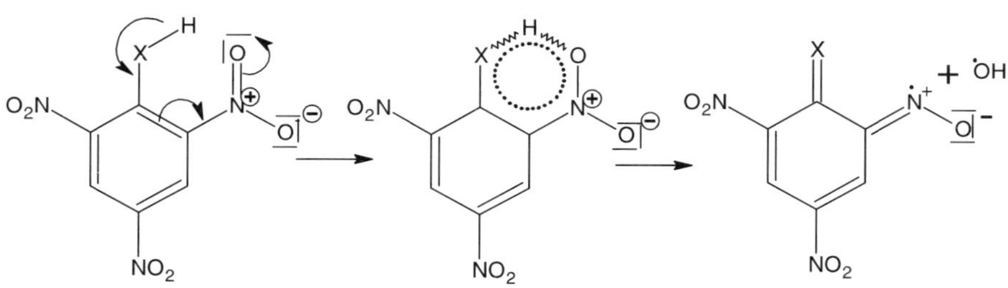

Scheme 2 Trinitrotoluene mechanism of thermal decomposition of polynitro arenes with a hydrogen atom in the γ-position towards the nitro group – here X can be CH, CH_2, O, N, or S; in the case of TNT, the last fragment forms 4,6-dinitro-2,1-benzoisoxazole and other decomposition products [11]; in the case of amino derivatives (X = NH_2) derivatives of 4,6-dinitrobenzofurazans and furoxans result

- *Homolytic fragmentation of another bond in a molecule* (see [6] and the references therein). In the case of HNAB it was assumed that its thermal decomposition is started by primary fission of the bond between the carbon atom and azo-bridge [13]. Similarly, DIPSO and DIPS (Scheme 3b in Sect. 4.3.1) should be liable to C – S bond homolysis [13].

The findings concerning mechanism and kinetics of thermal decomposition of nitric and perchloric acids and their salts (ammonium, hydrazinium, hydroxylammonium and metal salts) are presented in the monograph [11].

Sorescu et al. have published theoretical and computational studies of energetic salts [22]. Opinions concerning the mechanism of thermal decomposition of nitrates are approximately the same in both monographs. The decomposition mechanism of the best-known oxidizer, ammonium nitrate (AN), is given below. In the solid state, AN is dissociated according to the following equilibrium [11, 22]:

$$NH_4NO_{3(S)} \rightleftharpoons NH_{3(g)} + HNO_{3(g,l)} \,. \tag{1a}$$

Another decomposition of AN takes place through nitration of ammonia [11, 22]:

$$NH_{3(g)} + HNO_{3(g,l)}H^+ \rightarrow H_2O + NH_3NO_2{}^+ \rightarrow H_2O + H_3O^+ + N_2O \,, \tag{1b}$$

and the main decomposition can be accompanied by side reactions [11]:

$$NH_4NO_3 \rightarrow N_2 + 0.5O_2 + 2H_2O \tag{1c}$$
$$NH_4NO_3 \rightarrow 0.8N_2 + 1.8O_2 + 0.4HNO_3 \,. \tag{1d}$$

A relatively new problem is the study of the decomposition mechanism of extremely nitrogen-rich EMs. The mechanism of thermal decomposition of s-tetrazine derivatives has been outlined by Oxley et al. [23]. The authors of paper [24], dealing with thermal decomposition of 3,6-bis(2H-tetrazol-5-yl)-1,2,4,5-tetrazine (BTT) and 3,3′-azobis(6-amino-1,2,4,5-tetrazine) (DAAT), are somewhat more specific. In their opinion, the tetrazole derivatives depolymerize to azoimide and nitrile, and tetrazine derivatives decompose via a cyclic mechanism to give nitriles and nitrogen [24]. The data obtained from study of chemical kinetics and mechanism of thermal decomposition of furazanes, furoxanes, 5-monosubstituted tetrazoles, 1,5- and 2,5-disubstituted tetrazoles are presented in an accessible monograph by Manelis et al. [11]. The mechanistic ideas of thermal decomposition of tetrazoles in [24] agree with the data given in this monograph [11].

2.1.1 Quantum Chemical Simulations of the Thermal Decomposition

On the basis of UHF-SCF-AM1 MO calculation, studies on pyrolysis and sensitivity of nitro derivatives of aminobenzenes [28] and nitro derivatives of phenols [29] were carried out. Accounts of the thermal decomposition in the first study [28] do not agree with practical experimental findings (Scheme 2), while in the second case [29] they are in full accordance with experiment. The authors mention the sensitizing effect of the nitro group in nitro derivatives of aminobenzenes and the sensitizing effect of the hydroxyl group in polynitrophenols. The same MO calculation method was used for investigation of the decomposition of the azido derivatives of 1,2,3-, 1,2,4- and 1,3,5-trinitrobenzenes via breaking of $C-NO_2$, $C-N_3$ and $CN-N_2$

bonds [25]; it was shown that these compounds may be initiated by the cleavage of both $C-NO_2$ and $N-N_2$ bonds. The given conclusion does not agree with reality, because derivatives of *ortho*-nitroazidobenzenes primarily give derivatives of benzofuroxanes by thermal decomposition [11], therefore, only cleavage of the $N-N_2$ bond in the azido group might be important here. The same methodology was used to obtain the decomposition activation energies, E_a, of 1,2,3-, 1,2,4-, and 1,3,5-trinitrobenzenes (TNB) and their chloro derivatives [26]. The authors did not compare these results with experimental data. The same is true of analogous theoretical study of polymethyl derivatives of 1,3,5-trinitrobenzene and dinitrotoluene [27]. In this case, the authors calculated the E_a values for homolysis of the weakest bond $C-NO_2$ and for hydrogen transfer from the methyl group to the oxygen atom of the *ortho*-nitro group [26]. The splitting discussed does not correspond with the real course of the process, while the hydrogen transfer partially corresponds with Scheme 2. Energy of this transfer correlates to the impact sensitivity of studied polymethyl derivatives of TNB [27].

By using the B3LYP level of density functional theory, possible reaction pathways of 2,4,6,8,10,12-hexanitro-2,4,6,8,10,12-hexaazawurtzitane (HNIW) in the gas phase have been investigated [32]. In accordance with the findings presented in [11–18], the homolysis primarily concerns the $N-NO_2$ bond in the 2-position of the molecule. The same result was obtained by UHF-SCF-PM3 MO calculation [33].

Research based on quantum-chemical simulation of thermal decomposition of solid-state RDX using the Hartree–Fock method (combined with the cluster and periodic models) [15] showed that the dissociation energy of the $N-NO_2$ bond strongly depends on the environment of the molecule. The equatorial bond of the molecules inside a crystal is sensitive to the crystal field and is characterized by a notable increase in energy compared to the gas-phase dissociation energy. A significantly lowered energy barrier (by 8–15 $kcal\,mol^{-1}$) is obtained for a molecule placed near the surface [15].

Methods of quantum mechanics (quantum chemistry) were utilized to predict the products of thermal decomposition of triacetone triperoxide (TATP) [30], i.e. oxygen, acetone and unstable acetone monoperoxide. By means of a ReaxFF monomolecular molecular-dynamic cookoff simulation it was found [31] that the initiation of TATP decomposition in the condensed phase is dominated by a monomolecular process [31]. This simulation demonstrates that this initiation is entropy-driven, since the initial reaction is almost energy-neutral [31].

2.2
High-Temperature Thermal Decomposition

The authors of a large majority of papers published in the field of detonation initiation micro-mechanism of EMs presume that the primary fragmentation of molecules of these materials takes place only after adiabatic compression of EM by the front of the detonation wave, i.e. under conditions close to the Chapman–Jouget state (e.g., see [1, 3, 4, 6, 33]). This defines the conditions for molecular-dynamics simulations of decomposition of EM molecules. Such a simulation performed for decomposition of HMX at 3500 K showed that the primary splitting involves depolymerization of the molecule giving Digen [33, 34] (in principle, as in Scheme 1) or its dimers [34]. For RDX at 1505 and 1540 K it was found [36] that the depolymerization and homolysis of the $N-NO_2$ bond are processes of equal probability, while at 2685 K the depolymerization is dominant (Scheme 1). Analogously, it was found for 2,4,6-trinitroaniline [38] that at 2500 K the primary splitting process concerns the $C-NH_2$ bond, while in the temperature range of 2800–5000 K the $N-H$ bond is primarily split. A favoured nitrocompound in such simulations is nitromethane (NM): allegedly, at 30 GPa and 2000 K, the primary process consists of the formation of the *aci*-form of NM via a bimolecular mechanism [5, 34, 35]:

$$\begin{aligned} CH_3NO_2 + CH_3NO_2 &\rightarrow CH_3NO_2H + CH_2NO_2 \\ &\rightarrow CH_3NO_2 + CH_2NO_2H\,. \end{aligned} \tag{2}$$

It turns out that the primary fragmentation mechanism under these conditions is entirely different from that of the low-temperature variant. However, the results of molecular-dynamic simulations depend considerably on the starting conditions set for a given simulation [39].

2.3
Laser Initiation

In the past nine years, two review articles have been published about laser ignition and initiation [40, 41]. In this section, facts are presented that are connected with studies of initiation reactivity of explosives.

The energy density threshold of a beam in laser ignition is $1\ \mathrm{kW\,cm^{-2}}$ with a laser pulse over $1\ \mu s$. The majority of research work in the field of EMs studies using lasers utilizes the quantum generators based on neodymium glass or ruby, which operate in the mono-pulse regime [42]. The applied pulse times vary from fractions of a second to several tens of nanoseconds [42]. The intensity of laser irradiation can be high (up to $10^9\ \mathrm{W\,cm^{-2}}$ or even higher) [42]. The distribution of density of absorbed energy in irradiated EM is determined by its optical characteristics. The chemical changes in a laser-irradiated EM start in hot spots [42, 43]. The hot spots can be represented by structural non-homogeneities in the given EM (point defects, clusters of vacancies, interfacial boundaries), chemical admixtures etc. [42, 43]. Crystal defects can also be formed by action of a massive light wave. With sufficiently high energy of the initiation impulse, the number of hot spots can increase to such an extent that the reaction appears as quasi-homogeneous [43]. If the supplied energy fails to exceed the critical value needed to produce detonation, then pre-explosion phenomena (such as luminescence, formation of caverns) can be observed.

Luminescence and conductivity of crystals of heavy metal azides after their irradiation were successfully used in studies of their pre-explosion state [43–50, 52]. The authors of [49] summarized their results in the following way: The explosive decomposition of these azides is composed of two stages: a pre-explosive stage and the stage of the explosion products dispersion. The role of the initiating pulse is to create the necessary density of free electrons. At the pre-explosive stage the actual processes are of chain nature and provide a smaller portion of energy-release (up to 20%), which is connected with the non-radiative trapping and the recombination of electrons and holes, multiplying by chain mechanism. The objective of this stage is to provide the primary heating (up to 600 K) and the dispergation of the model, that means to set conditions for the second stage of the product dispersion. At this stage the majority of the energy connected with the exothermic reaction $2N_3 \rightarrow 3N_2$ is released. The reaction is more preferable on the surface or in the gas phase than inside a crystal. This model of the process of chain multiplication of holes due to their intraband transitions into various energy states [49] was also supported by quantum-chemical calculations using the Hartree–Fock method [45]. The said model, which is consistent with currently available experimental evidence, can be presented by the following quasi-chemical reaction [49]:

$$h + \mathrm{A} \rightarrow \mathrm{N_3}^0 + \mathrm{A} \quad (3a)$$

$$\mathrm{N_3}^0 + \mathrm{N_3}^- \rightarrow \mathrm{N_6}^- \quad (3b)$$

$$\mathrm{N_6}^- + \mathrm{Ag}^+ \rightarrow \mathrm{N_6}^0 + \mathrm{Ag}^0 \quad (3c)$$

$$\mathrm{N_6}^0 \rightarrow \mathrm{N_6}^- + h^* \quad (3d)$$

$$h^* \rightarrow h + h + e \quad (3e)$$

$$h + \mathrm{A} + \mathrm{N_3}^- + \mathrm{Ag}^+ \rightarrow 2h + e + \mathrm{N_6}^- + \mathrm{Ag}^0 , \quad (3f)$$

where h and h^* are thermal and hot holes, respectively. Here Eq. 3a represents hole trapping by a cation vacancy, Eq. 3b reconstruction of the centre as a result of the chemical reaction with the formation of a quasi-local hole state in the valence band, Eq. 3c hole detrapping from the reconstructed centre, Eq. 3d bearer multiplication as a result of impact ionization by hot holes and Eq. 3e reconstruction of a local state in the band gap. According to final Eq. 3f, an extra electron-hole pair is generated at the expense of the decomposition of lattice fragments next to A (the formation of N_6^- and Ag^0) [49].

According to the authors of paper [50], however, within the framework of the above-mentioned model it is difficult to explain the dependence of the ignition energy threshold on the excitation pulse parameters and excitation type as well as the occurrence of the induction period. They suppose [50] that the kinetics of initial processes of heavy metal azide decomposition upon exposure to laser pulses, pulsed electron beams, and individual microparticles have demonstrated that they possess some features in common. Severe local deformation of the lattice initiate reactions between lattice anions, and energy of ~ 10 eV is liberated as a result of each individual reaction act. When a high density of such reactions is observed simultaneously in a small volume, the resultant heat liberation in this region can be sufficient for the avalanche process development culminating in the formation of a shock wave with a pressure $> 10^{10}$ Pa and explosive decomposition of the given sample [50].

One of the ways of initiation by a laser beam consists of its action on EM placed in a metal foil sheath. The authors of [52] point out that wrapping of silver azide crystals in aluminium foil of 10 μm thickness results in disappearance of cathode luminescence, but has virtually no effect on the kinetics of explosion fluorescence (that corresponds with the kinetics of explosive decomposition in that part of the sample in which this decomposition is initiated). The authors of paper [53] have found that surface roughness and a thin metal layer are very effective at enhancing absorption of laser energy and produce a very strong shock wave in ambient media. They have succeeded in the detonation of a very slender PETN powder column almost in the steady propagation mode. A calculation of the laser initiation of thin PETN (density $1.0\ g\,cm^{-3}$), using one-dimensional hydrodynamic code to study of the effect of the initiation methods on the initiation process, is a topic of paper [54]. Three types of the initiation are modelled: utilizing direct absorption of laser energy, impact of the laser driven-flyer plate, and ablation of the aluminium film. In the last case the initial pressure in aluminium is extremely high, after 0.05 mm propagation of shock wave, the pressure at the shock front falls to 4 GPa. During propagation to intermediate particles, the decomposition of PETN can not affect the shock front, so the pressure decrease continues [54].

A presentation of direct experimental evidence for the initial fast decompositions of the more powerful military explosives, HMX, RDX and TNAZ, as well as of a non-explosive molecule, dimethyl nitramine (DMNA), is a topic of paper [55]. Experimental data indicate that electronic excitation, in the gas phase, of all the nitramines studied, with 248 nm, 15 nanosecond pulses from the excimer laser leads to the formation of electronically excited NO_2 ($^2B_2$), resulting from the primary N – N bond scission in the electronically excited nitramine molecule. Furthermore, electronically excited OH ($A^2\Sigma$) is also formed biphotonically, possibly through the intermediate formation of vibrationally excited nitrous acid (HONO). The acid is believed to be produced through a five-member ring formation in the electronically excited nitramine, and eventually HONO elimination. A subsequent absorption of a 248 nm photon by the vibrationally exited HONO leads to OH ($A^2\Sigma$) and NO [55].

As for the laser impulse sensitivity proper, it is generally accepted [42] that the initiation energy threshold of a given EM depends on wavelength and impulse action time. This threshold also depends on the size of EM crystals: decreasing of size of single crystals of metal azides is connected with increase in critical initiation energy density. There exists a limit size of crystals—samples with fine crystals decompose without explosion after laser mono-impulse of any force. The initiation process is the easier, the smaller escape of gaseous products is from the initiation zone. Very sensitive to initiation by laser mono-impulse are heavy metal azides (critical impulse density 15–100 μJcm^{-2}; for α-modification of lead(II) azide this value is 4–6 μJcm^{-2} [37]). The technologically most attractive light-sensitive EMs are complexes of d-metals, such as $[Co(NH_3)_4(Tz)_n](ClO_4)_m$, where Tz is 5-substituted tetrazolyl [42, 51].

A relatively new problem in science and technology is the study of nanostructured EMs. Dlott et al. have developed several methods for fabricating and rapidly initiating these materials ([56] and the references herein). Pulses of near-IR laser are used to suitably flash-heat the sample at a high repetition rate about 100 Hz. Each laser pulse ignites a region about 100 μm in diameter, so a single sample provides more than 10^5 spots that can be independently ignited and studied. The laser pulses flash-heat only the metal fuel particles. The authors demonstrate the application of this laser with two types of samples [56]: one of them consists of Al nanoparticles in the 30–100 nm diameter range, suspended in continuing oxidizing medium (NC or Teflon oxidizers). The Al particles have a native oxide coating and were used as provided. The second sample consists of a vacuum-evaporated manometric film of Al approximately 30 nm thick with a native oxide, covered with a much thicker slab of NC or Teflon oxidizer. In the first type of the EM, energetic reactions occur at the interface between the nanoparticles and the oxidizer, and each sample has a high surface area. In the second case, reactions occur at a well-defined planar interface between Al and oxidizer [56].

2.4
Electrostatic Discharge Initiation

This initiation has not attracted much attention in the past. Nevertheless, the electric spark sensitivity is an important characteristic of explosives, especially primers and propellants. Important results were published by Skiner et al. [61]. The authors adopted a needle-to-plate electrode for a capacitive discharge, electrostatic discharge sensitivity test procedure. With the EMs investigated, they determined the critical temperatures T_c according to the Frank–Kamenetskii relationship, and, furthermore, the temperatures T_3, for which the rate constant is equal to $10^3\ s^{-1}$. The electric spark energy values for 50 percent probability of ignition of the EMs investigated are in excellent correlation with theT_c and T_3 values [61]. From this fact, the authors deduce a conclusion that the electric spark initiation is primarily of thermal nature [61], i.e. it represents a thermal decomposition. This means that the mechanism of primary splitting of EMs molecules by action of electric spark should be identical with their primary splitting induced by heat. The centre of the problem of this initiation continues to be a focus of interest today and it will be commented on further in Sect. 4.

2.5
Effect of the Hydrostatic Compression

Sorescu et al. have investigated [58] the hydrostatic compression effect on RDX, HMX, HNIW and PETN through crystal packing and isothermal-isobaric molecular dynamic simulations (assuming the rigid-molecule approximation). In the case of RDX, HMX and HNIW crystals, the results indicate that the proposed potential model is able to reproduce accurately the changes in the structural crystallographic parameters as a function of pressure for the entire range of pressures that has been investigated experimentally. Kuklja and Kunz have studied [59] the effect of a hydrostatic compression (up to 71% of equilibrium volume) on the electronic structure of RDX by means of the ab initio Hartree–Fock method for a periodic system combined with the many-body perturbation theory. An external pressure causes a significant decrease in the optical gap for both the perfect material and the crystal with dislocations. Edge dislocations produce local electronic states in the optical gap, whereas the external pressure moves these states deep within the band gap. This creates favourable conditions for the $N-NO_2$ bond rupture due to exciton formation [59]. However, according to Manaa's study [60], the applied uniform and uniaxial compression to the nitromethane crystal, with and without molecular vacancies, demonstrated only a slight reduction of the HOMO-LUMO gap for hydrostatic pressure up to 150 GPa or higher. This small reduction essentially prohibits any considerations of electron excitations, and hence of non-radiative transitions due to the crossing of adjacent energy surfaces of molecules. The band-gap drop is most likely attributed to the relatively weak intermolecular interactions, as the molecules tend to maintain positions that yield the lowest possible pressure in the crystal. Accordingly, no appreciable bending of the nitro group is observed for this range of volume strains [60]. Manaa calls attention to distinct differences between his results [59] and those of Kuklja and Kunz [59].

Manaa's conclusions, however, have a connection with the dynamic pressure effect.

A study of RDX behaviour under static ultrahigh pressure (up to 65 GPa), generated by using diamond anvil cells (DACs) using FT-IR spectroscopy and UV-VIS absorption spectroscopy, is a topic of paper [61]. RDX changed its colour into dark red when compressed up to 20 GPa with caesium iodide (CsI), filled as a pressure medium. In this case the intensities of characteristics of IR absorption peaks of RDX decreased as the pressure increased, and did not return to the intensities measured at ambient pressure, after the pressure was unloaded. However, when RDX was compressed alone, its colour changed into yellow at a pressure above 60 GPa. In its UV-VIS absorption spectra the peak shifted from 243.5 nm at ambient pressure to 410 nm at 65.5 GPa. The authors assume that the HOMO-LUMO band gap of this nitramine decreases with increasing pressure [61] which is in line with the opinion of Kuklja and Kunz [59].

Rice et al. [62] used quantum-chemical calculations to treat the experimental data about unit cell dimensions for nitromethane (NM) in a pressure range from 0.3 to 15.3 GPa. These calculations generated the electronic band structures, from which the energy gap between the valence and conduction bands of the crystalline NM is found. Metallization (loss of covalency) results when this energy gap is closed. By analysis of data, the authors arrived at a conclusion that the metallization of explosive due to compression cannot be dismissed as the initiating step in the sequence of reactions leading to detonation [62].

2.6 Effects of Crystal Defects

Crystal lattice defects (vacancies, impurities, voids, pores, dislocations) play an important part in all types of initiation of crystalline EMs (generally, they increase the sensitivity of these materials to stimuli). The most significant types of defects in thermal decomposition [11, 63] and initiation of thermal explosion [63] are dislocations. Manelis et al. [11] documented the effect of dislocations in thermal decomposition using ammonium perchlorate (AP) as an example: the decomposition of AP is initiated and proceeds predominantly in the dislocations of its crystal. The liberated decomposition products produce strain and deformations around the increasing reaction centre. This leads to movement of dislocations by sliding along the surface (plastic flow), their mutual interactions as well as interactions with other crystal lattice defects. The result is a generation of further dislocations and an increase in decomposition rate. The rate of this type of reaction is to a considerable extent determined by the factors that are connected with the rate of proliferation of dislocations [11]. The value of the real rate constant then depends slightly on the rate of chemical change. The same is true of the temperature dependence of the rate of process, which follows the Arrhenius law in the limit. However, among the parameters of the Arrhenius equation also values, that determine the temperature dependence of generating of dislocations, are predominant. The concentration of dislocations and degree of transformation depend exponentially on the distance from the front of reaction propagation. This means that the width of the zone of chemical reaction increases with temperature more weakly as the velocity of propagation of the reaction front [11]. In this light a paper by Kunz et al. [59, 65] can be mentioned: on the basis of their study of laser pulse initiation of azides [54] they suggest that chemical decomposition in crystals with a low density of dislocations develops noticeably more slowly than that in the highly defective samples, showing a strong correlation between density of dislocations in the crystal and chemical reactivity [59, 65]. Kuklja and Kunz [66] have experimentally determined configurations for edge-type dislocations with the Burgers vector (001) in crystalline cyclotrimethylene trinitramine (RDX) and pentaetythritol tetranitrate (PETN), and theoretically simulated them. It is shown that a shear stress, induced by the dislocations, produces local electronic states in the fundamental band gap of the crystal. These states are mainly formed by molecular orbitals of critical bonds (which are the $N-NO_2$ group in RDX and the $O-NO_2$ group in PETN) responsible for the stability of the materials. Kuklja et al. [67] also studied structural and electronic features of shear strain in the molecular crystal of 1,1-diamino-2,2-dinitroethylene (FOX-7) and investigated its effect on decomposition of the material. The authors demonstrate that shear strain lowers the decomposition barrier and narrows the band gap of the solid and thus facilitates thermal chemistry in molecular materials. The dynamic behaviour of the band gap contains a rich variety of information about the details of the decomposition processes at the initiation stage.

It is generally known that if the thermal effect of exothermic decomposition of a given EM exceeds the thermal losses of this process, then the given material will explode. In this context, Armstrong's review article [64] presents linear relationships between the size of the above-mentioned centres of initiation of reaction (generally, defects of crystal lattice that are also called "hot spots") in RDX and PETN crystals and reciprocal critical temperature of these explosives. Armstrong also states [63] that local concentration of strain energy within the relatively more stable energetic crystals is required to induce intramolecular decomposition, relating to (intermolecular-type) hydrogen and (intermolecular) nitrogen/oxygen bond breakages. Dislocation pile-ups provide an explanation of crystal size effects in energetic crystal decomposition. Such molecular crystals give enhancement of local heating from plastic work because of their insulating character [63].

According to Coffey [68, 69] an initiation of chemical reaction in explosive crystals occurs due to the local energy dissipated in the crystals during plastic flow arising from shock or impact. The dislocations created during the flow are held responsible for producing lattice and molecular distortions within the crystals. A distorted lattice potential is developed through which the dislocations move by quantum tunnelling [69]. These processes together with energy localization are likely to be responsible for detonation. If somehow the plastic flow in the explosive crystals is stopped, even during full detonation, the energy dissipation due to plastic flow and the chemical energy release also stop regardless of the applied pressure (this is Dead Pressing) [68]. Coffey developed the theory of plastic flow in detail and summarized his results in a chapter [68] of monograph [2].

A significant role during initiation is ascribed to dislocations also in the theory of electronic initiation developed by Kuklja et al [15, 45, 59, 65, 70–72]. The authors used ab initio simulation (Hartree–Fock level) [15, 45, 59, 65, 70–72]. They found [70] that low formation energy characterizes defects in the solid RDX. This indicates that a significant number of vacancies should be present in the crystal even at low temperatures. Narrowing of the fundamental gap (about 1 eV) is caused by the presence of vacancies in the material. This fact demonstrates anisotropy of the RDX crystal with respect to the vacancy distribution [70]. In another paper [72], Kuklja et al. deal with structure and properties of molecular vacancy dimers in the RDX crystal. The three nearest-neighbour vacancy dimers have positive binding energies, i.e. vacancy interactions are attractive. Higher vacancy association may also be energetically favoured. High RDX anisotropy manifests in the different binding energies and optical gap narrowing for the different mutual orientations of vacancy dimers [72]. These authors also stated [71] that a key role belongs to the edges of dislocations, which promote dramatic changes in the electronic structure, primarily a reduction of the optical gap due to the splitting off local electronic states from both valence and conduction bands.

The crystal lattice defects also include cracks, either present in the EM crystal or produced in it during initiation of ongoing decomposition. Holmes et al. [73] presented a model for time and spatial dependences of the "heating of molecular vibrations" and the possible initiation of chemical reaction from heat dissipated in the vicinity of a propagating crack in a molecular crystal. Energy from a moving crack tip is released as phonons in proximity to the crack. Initially, the phonons and the molecular vibrations are not in thermal equilibrium. Subsequently, there is a competition between excitation of molecular vibrations by multiphonon up-pumping and diffusion of phonons from the crack region. If the coupling between the locally hot phonon bath and the molecular vibrations is sufficiently large, a transitory high vibrational temperature will be achieved prior to eventual thermal equilibration with the bulk of the crystal. The authors found that the peak vibrational temperature can be sufficiently high for a significant time period for chemical reaction to occur. This temperature change is connected with a production of chemical reaction in EM [73].

3
Impact and Shock Sensitivities

3.1
Model of the Multiphonon Up-pumping in Energetic Materials

Some authors approach the study of impact and shock sensitivities as if it were a single type of initiation. However, from the standpoint of the way the impulse energy is transferred to the reaction centre of the molecule there are two different initiation processes (Sect. 4). On the basis of his multi-phonon up-pumping model of initiation of EMs, Dlott considers the impact initiation to be a kind of low velocity initiation (LVI) [1, 74].

The above-mentioned model by Dlott [1, 74] for both the types of initiation represents the best at developed theory so far trying to unify the approaches to the solution of the problem of initiation of detonation, namely (as already mentioned in the Introduction) the approaches of continuum mechanics, chemistry and quantum chemistry. Like most authors of papers in this scientific field, he relates his model to the processes taking place in the layer of EM molecules compressed by impact or shock wave, i.e. to decomposition under extreme conditions. This is understandable, and he consults a number of specialists (supporters of this idea—see [1, 74]) about his clearly and intelligibly formulated considerations. The introductory reasoning of the model starts from the idea [1, 74] that mechanical perturbation of intramolecular vibrations and intramolecular stretching (the latter gives rise to collective delocalized excitations—phonons) here leads either to energy transfer vibrations to phonons (vibrational relaxation), or energy transfer from the phonons to vibrations (multiphonon up-pumping) [1, 74]. This consideration is complemented by an extensive set of data and discussions mainly obtained by analysis of outputs of molecular dynamics simulations of the decomposition of EMs, mostly under extreme conditions, and spectroscopic methods. Dlott submits two scenarios where up-pumping could have a significant effect on chemical reactivity other than a brief delay between phonon generation and chemical reaction [74]. The first (connected with a crack surface or edge moving dislocations) involves the non-linear thermal conductivity catastrophe. The catastrophic nature results from the combination of more phonons and less thermal conductivity, so that at a presumably rather sharp threshold rate for phonon generation, the ignition temperature will abruptly be exceeded at the surface. The second involves non-equilibrium vibrational populations caused by ultra-fast up-pumping that is faster than the intramolecular vibrational redistribution (IVR). In this case, if the point is either coherent or incoherent phonon pumping, the mechano-initiation consists either of primary splitting of the weakest bond or of splitting of stronger bonds in the molecule of a given EM, respectively [74]. However, missing experimental verification is a shortcoming of this model.

3.2 Impact Sensitivity

3.2.1 Low Velocity Initiation

In the case of low velocity initiation Dlott [74] arrived at a preliminary conclusion that the sensitivity to this impulse might increase in materials characterized by fast up-pumping (a connection between fast up-pumping and poor thermal conductivity was suggested in this case). In Dlott's opinion [1] the phonon generation and up-pumping under conditions of LVI could be investigated, where one obstacle is accurately describing the localized generation of phonons under LVI conditions. McNesby and Coffey developed a simple theory [75] for predicting impact sensitivities in crystalline explosives from Raman vibrational spectra measured at room temperature. They suggest that the initial rate of energy transfer, from states within a low-energy phonon and near-phonon to a state within higher vibrational levels, plays a role in determining impact sensitivities of the explosives examined in their paper [75]. Shuji Ye et al. [76] estimated frequencies of the normal mode of 12 different individual explosives by means of density functional theory calculations at the B3LYP/6-31G(d) level of theory and evaluated the number of doorway modes (i.e. the lower frequency vibrations with significant bandwidths) in the three doorway regions. They compared the number of doorway modes with the impact sensitivities (from the hammer test). A qualitative correlation found indicates that the number of vibrational states in the doorway region can be an important parameter for the intrinsic impact sensitivity of solid molecular explosives.

3.2.2 Aspects of the Characteristics of Molecular Structure

Important papers are those studies of relationships between molecular structure and impact sensitivity. Politzer and co-workers [77–82] established correlations between the features of the quantum chemically determined electrostatic potential (ESP) surrounding an isolated molecule and many of its condensed-phase properties. The molecular EPS reveals a striking distinction between energetic organic polynitro compounds and the great majority of other organic molecules [81]. Especially the surface ESPs of the two classes are distinctly different. In contrast to the second class of compounds, in energetic molecules the positive surface potentials still cover the larger areas but now tend to be much stronger then the negative one. Politzer and Murray have proposed that the metastabilities of energetic compounds are associated with an anomalous imbalance between positive and negative surface potentials and that it can serve as a basis for correlating and predicting sensitivity [80, 81]. It is fully understandable that the above-mentioned imbalance depends on mutual effects of electron-withdrawing and electron-donating functional groups present in the energetic molecule. From the standpoint of organic chemistry, mesomeric effects in aromatic systems will stabilize the molecule (i.e. will act in opposition to this imbalance), while in aza-aromatic systems this stabilization could be weaker, and in aliphatic and/or aza-aliphatic compounds such stabilization is virtually missing. Therefore, the correlations found by Politzer and Murray [81] are restricted to specific classes, i.e. nitroaromatics, nitroheterocycles and nitramines.

Rice et al. [83], using density functional theory, calculated the dissociation energies for rupture of the weakest $C-NO_2$ bond in some nitroaromatic molecules. They tried to find a relationship between these energies and impact sensitivity of the nitro arenes studied. They state that the results indicate the said relationship. The polynitro arenes studied represent either "genuine" non-substituted polynitro derivatives of benzene (the primary splitting is the homolysis of the $C-NO_2$ bond [12, 13]) or substituted polynitro benzenes with a hydrogen atom at the γ-position to the nitro group (Scheme 2). In this way, however, the authors mix two different primary splitting mechanisms into one, their approach being justified only in the case of the first class of benzene derivatives. Fried et al. [84] made an attempt at inter-correlating the dissociation energies of $C-NO_2$, $N-NO_2$, $O-NO_2$ bonds and logarithm of drop heights $\log h_{50\%}$ of the respective classes of polynitro compounds with the same result. The said authors, using Cheetah 2.0 code, calculated "energy for decomposition" (perhaps heat of explosion) for these compounds, and correlated the values with $\log h_{50\%}$ with the same result as in the previous case. However, a linear dependence resulted from the relationship of ratios of "energy of decomposition" and dissociation energy with $\log h_{50\%}$.

Rice and Hare [85] continued the above-mentioned findings of Politzer et al. and developed a hybrid model of prediction (four models) of impact sensitivity of CHNO explosives. They adopted the approximation to the electrostatic potentials and bond midpoints, statistical parameters of these surface potentials, and generalized interaction properties function or calculated heats of detonation. The authors showed that patterns of charge on the electrostatic potentials for iso-surfaces of electron densities surrounding energetic molecules are useful guides in assessing the degree of sensitivity of explosive to impact. The most sensitive molecules have regions of very positive ESPs localized over covalent bonds. This localized region of electron deficiency is not apparent in the insensitive explosives. The approach given should not depend upon chemical classification of the 54 individual energetic materials of various chemical structures. However, the diagram of dependence between heat of explosion and drop height $h_{50\%}$, which is presented by Rice and Hare as an exponential function [85], can be specified by a partial linear dependence associating the EMs studied according to the primary chemical mechanism of splitting (see also Sect. 4). A linear dependence between log $h_{50\%}$ and the heat of detonation, obtained by B3LYP/6-31G* level of theory (no carbon monoxide path), was published by Edwards et al. [86]. These authors also published relationships between drop height $h_{50\%}$ and the lowest unoccupied orbital molecular energy (B3LYP/6-31G*), between the said height and the highest occupied molecular orbital energy (B3LYP/6-31G*), and between the said height and other quantities, all that without any substantiation or other interpretation.

The PM3-SCF-MO calculation was used to study the impact sensitivity of 5-substituted tetrazoles and their salts [87]. The authors found a parallel relationship between the activation energy and the sensitivity and a good linear relationship between the experimental impact sensitivity of metal salts and the reaction activation energy. However, no parallel relationship was found between activation energies and the bond orders between nitrogen atoms in positions 1 and 2 [87]. This bond should be primarily split during thermal decomposition of the said heterocycles [11]. The relative magnitude of sensitivity for most tetrazole compounds is determined mainly by the activation energy of the opening ring in the thermal decomposition. The sensitivity found order for 5-substituted tetrazoles was as follows (order of substituents): $CH_3 \approx Ph < NH_2 < NHNO_2 < Tz < Cl < NO_2 < N_3 < N_2^-$ [87]. From the same workplace also comes UHF MO PM3 and PM1 calculation of pyrolysis mechanism of nitromethane, methyl nitrate, nitroguanidine, nitramine, methylnitramine and dimethylnitramine [89]. The mechanisms of primary homolysis of the given compounds stand in accordance with experimental findings [11]. The found activation energies of this process are consistent with the experimental values of impact sensitivity [89]. The authors found a parallel linear relationship between the bond orders of $N-NO_2$ bonds in the molecules of studied nitramines and the activation energies of their initiation reactions breaking these bonds [89]. Türker investigated structure-impact sensitivity of some substituted 1,3,5-trinitrobenzenes by using a topological approach [88]. He found that the sensitivity increases with decreasing of a defined upper bound of the deepest lying molecular orbital energy, or the fourth coefficient of the secular polynomial of the molecular graph of the iso-conjugate hydrocarbon system within the Hückel MO framework [88].

3.2.3
Specification of Centres of Initiation Reactivity in a Molecule

Recent trends in research focus on the finding of centres of initiation reactivity in molecules of energetic materials (see also Sect. 4). Chaoang et al. [90, 91] calculated the Mulliken net charges, Q_{NO_2}, of the nitro group by means of a new DFT BLYP/DNP method. The authors found that the charges can be regarded as structural parameters to estimate the impact sensitivity. A dependence is indicated between this sensitivity and the Q_{NO_2} values with the most negative Q_{NO_2} being connected with the lowest sensitivity [90, 91]. According to the authors' opinion the compound may be sensitive ($h_{50\%} \leq 0.4$ m for a hammer weight of 2.5 kg) when its nitro group has a lower negative charge than about 0.23 e. The method is applicable to almost all nitro compounds in contrast to the methods of length of the $C-NO_2$, $N-NO_2$, or $O-NO_2$, molecular electrostatic potential, and oxygen balance [90, 91]. However, in some cases the selection of the most reactive groups in a molecule differs from experimental findings: especially in the cases of TNT and 2,4,6-trinitroaniline, the authors choose for the correlation the most negative Q_{NO_2} values, i.e. the nitro groups at the 4-position of the given polynitro arenes, which contradicts experimental findings (Scheme 2). Nevertheless, these are seminal papers inspiring further activities in this area of specific chemical (initiation) reactivity of EMs.

3.2.4 Effect of Desensitizing Admixtures

Xiao et al. studied the intermolecular interaction of 2,4,6-triamino-1,3,5-trinitrobenzene (TATB) with HMX and RDX using the PM3 method [92]. This interaction can strengthen the $N-NO_2$ bond. The binding energies of TATB with HMX or RDX are much larger than those between graphite and HMX or RDX. Therefore, TATB can wet and desensitize both nitramines more effectively than graphite can [92]. Technologically, this is an important piece of knowledge. An experimental investigation of effect of desensitizers (phlegmatization agents) upon sensitivity of phlegmatized EMs to mechanical stimuli was realized by Loginov and Surkova [93]. The model EM used was HMX desensitized by addition of *N*-phenylaniline (DFA). Samples of this HMX after exposure to impact, friction or vibration were investigated by means of thin-layer chromatography, IR and UV spectroscopy and X-ray-phase analysis. In this way, they proved the presence of chemical interaction between DFA and active particles from decomposition of HMX by action of short-term mechanical strains. The authors point out [93] that the efficiency of action of a desensitizing agent upon the sensitivity of EMs to mechanical stimuli is connected with not only the change in physical and mechanical properties of desensitized EMs (as compared with the original ones) but also the ability of a desensitizing agent to trap active products of decomposition of the given EM and inhibit this decomposition.

3.2.5 Methods of Prediction

From among the prediction methods of impact sensitivity, three methods can be mentioned here. Vaullerin and Espagnacq discuss a criterion that aids in such screening for all families of EMs [94]. This criterion is based on the maximum heat of reaction (perhaps heat of explosion) calculated by means of thermochemical calculation code CHETAH. The authors have introduced a sensitivity criterion (the product of square of heat of explosion and molecular weight divided by the number of atoms in 1 gram) which correlates with experimental drop energies. On the basis of this relationship, the set of EMs studied by them falls into three subsets, namely sensitive, fairly sensitive and insensitive explosives [94]. Cho et al. [95] have performed neural networks (NN) studies to predict the impact sensitivity of energetic molecules. A set of 234 individual EMs has been taken from a single database, and 39 molecule descriptors were computed for each energetic molecule. Optimization of NN architecture was carried out by examining seven different sets of molecular descriptors and varying the number of hidden neurons. The best result was obtained from 17(S5)-2-1 architecture ($r^2 = 0.8148$, and SEP is 0.190) [95]. Keshavarsz and Jaafari published a method of prediction of impact sensitivity using an artificial network [96]. As compared with the previous method, the outputs of their calculations are loaded with very large error (RMS deviation of 56 cm of the $h_{50\%}$ quantity [96]).

3.3 Initiation by Shock

3.3.1 Physical Kinetics Model

In 1998 one of Wolker's most recent papers [97] argues for his theory of physical kinetics of detonation. The theory is described as a non-equilibrium, non-thermal process elicited by the exceedingly high kinetic energy of shock and/or in the detonation front. This process leads to massive mechanical fracture of covalent bonds at and near to the detonation front. According to this theory, the reaction rate is determined by the medium vibrational energy of free electron pairs (radicals) formed by the fracture of a molecule within a time period of 10^{-14} to 10^{-12} s [40]. This theory can be considered as an example of views of some physicists on chemical reactions, but in its time it was a considerable contribution to development of detonation theory.

3.3.2 "Accumulation Mechanism" Application

In 1999 Zeman [98] summarized the published significant views on initiation of detonation (except for the model of physical kinetics) as follows: the influence of shock on energetic materials results in adiabatic compression of the molecular layer struck. According to Klimenko's and Dremin's "accumulation mechanism" [99–101], the kinetic energy of the shock in this compression is accumulated through translational-vibrational relaxation processes by translational and vibrational modes of molecular crystals of the material within 10^{-13} to 10^{-12} second. This causes a considerable quasi-overheating (20 000 to 40 000 K [99, 100]) especially of vibration modes. A non-equilibrium state is established with concomitant primary splitting of the energetic material into ions and radicals [99–101]. These active particles, by chemically interacting with each other, evoke a process of spontaneous spreading of the shock front in the starting substance, i.e. evoke a second equilibrium stage of detonation behind the front. Zeman wrote [98] that this or similar ideas of transformation of low-frequency vibrations of crystal lattice (*acoustic phonons*) into high-frequency vibrations (*vibrons*), with subsequent spontaneous localization of vibrational energy in the explosophore groupings [74, 102], had been applied by a number of authors in their studies of shock reactivity of energetic materials.

3.3.3 The Multidimensional Reactive Flow Models (NEZND)

The multidimensional reactive flow models of shock initiation and detonation of solid explosives was developed by Tarver et al. [5, 103, 104] for the large, fast, tetraflop computers currently available. This Non-Equilibrium Zeldovich–von Neuman–Döring (NEZND) theory was developed to identify the non-equilibrium chemical processes that precede and follow exothermic chemical energy release within the reaction zones of self-sustaining detonation waves in explosives. The model has been developed in the ALE3D and LS-DYNA2D/3D hydrodynamic codes, in which heat transfer and detailed chemical kinetics are directly coupled into hydrodynamic equations. This model creates hot spots by various mechanisms (void, collapse, friction, shear), determines whether these hot spots are large and hot enough to react and grow, and then calculates the spreading rates of the growing reaction sites. The reactions in the model are propagated as they are in nature: by heat transfer using the Arrhenius kinetics. The chemical decomposition mechanisms are based on thermal explosion experimental data [5, 103, 104]. Tarver starts from the multiphonon up-pumping, induced by action of the shock front, and predicts the subsequent reaction course as follows [103]: Early unimolecular reactions are endothermic or weakly exothermic, but they channel most of the available energy into excited vibrational states of intermediate product species. The intermediate products transfer some of their vibrational energy back into the transition states, accelerating the overall reaction rates. As the decomposition progresses, the highly vibrationally excited diatomic and triatomic molecules formed in very exothermic chain reactions are rapidly vibrationally equilibrated by "supercollisions", which transfer large amounts of vibrational energy between these molecules. Along with vibrational-rotational and vibrational-translational energy transfer, these excited vibrational modes relax to thermal equilibrium by amplifying pressure wavelets of certain frequencies. These wavelets then propagate to the leading shock front and reinforce it. This is the physical mechanism by which the leading shock front is sustained by the chemical energy release [103]. The induction period for the initial endothermic bond breaking reaction can be calculated using the high-pressure, high-temperature transition state theory [5]. Experimental unimolecular gas-phase reaction rate at the low temperature (< 1000 K) shock conditions obey the usual Arrhenius law [5]. The authors in their papers present a scheme of NENZD theory of detonation that demonstrates their view of the processes following the adiabatic compression of a tiny layer of EM by a shock wave [5, 103, 104]. However, it must be remembered that appearance of the induction period of the EM decomposition in the front of detonation wave makes the front kinetically unstable and pulsating [101]. In this case the CJ pressure influences the kinetic of the EM decomposition rather than the chemical spike [101].

3.3.4
Application of Raman Spectroscopy to Study

Like McNesby and Coffey in the case of impact sensitivity [75], Koshi et al. [105] have also used Raman spectra (but only for very low temperatures) in the characterization of relationships between the shock/impact sensitivities and molecular structure. The authors start from Dlott's ideas (see above) of the transition of energy of excited lattice vibrations (phonons) to molecular vibrations before bond breaking can occur. This energy transfer process is expected to be the rate-determining step for the explosion. The rate of this transfer has been evaluated by the authors on the basis of a simple theory coupled with experimental measurements of line width of Raman spectra at 0 K (the line was obtained by extrapolation of the spectra recorded within the temperature interval of 3.8–300.0 K). A good correlation was found between the sensitivities derived from drop-hammer tests and the rate of energy transfer. However, no correlation was obtained for the data of TNT (for its primary splitting see Scheme 2) [105]. This is understandable since the authors did not include it among those EMs whose primary step of initiation consists of homolysis of $C-NO_2$ bonds (1,3-dinitrobenzene), $N-NO_2$ bonds (HMX, RDX, NQ) and $O-NO_2$ bonds (PETN and propyl nitrate). The authors have compared also the rates of energy transfer with wedge test results (as a measure of the shock sensitivity) [105]. In this case, loading pressures were very high, and the effect of pressure on this transfer has to be considered. Taking the increase of anharmonic coupling into account, the rates of energy transfer behind the shock can be compared with the induction distance in wedge tests [105]. Apart from the above-mentioned effect of chemical mechanism of primary fragmentation, there is a clear difference between the initiation by impact and that by shock in this case; the authors [105] do not point out these two important facts.

3.3.5
The Excitonic Model

Also significant are the already-mentioned results of the work by Kuklja et al. [15, 45, 59, 65–67, 70–72, 106]. Ab initio studies were performed on the initiation chemistry in high explosive crystals from a solid-state physics viewpoint. Specifically, they were looking for the relationship between the defect-induced deformation of the electronic structure of solids, electronic excitations and chemical reactions under shock conditions. Band structure calculations by means of the Hartree–Fock method with correlation corrections were done to model an effect of a strong compression induced by a shock/impact wave on the crystals with and without edge dislocations (using the example of RDX, the authors point out the strong dependence of dissociation energy of the $N-NO_2$ bond on the environment of molecule [15]). On the basis of the results obtained, an excitonic mechanism of the earliest stages for initiation of high explosive solids was discussed as follows [106]: The necessary condition for the chemical reaction to occur in liquids and gases is known to be the real migration of reactants towards each other. In solids, a completely different situation is possible. Here electronic excitations migrate across the crystal. Localization of these excitations at definite sites of the crystalline lattice (structure of impurity defect) leads to the appearance of actual radicals at the necessary site. Thus, a sufficiently slow process of migration of real heavy particles (usually a diffusion process) is replaced by much faster migration of electronic excitations (see Sect. 3.1—pre-explosion processes in heavy metal azides [49, 71]). Experimental verification of the validity of the proposed model is reported for RDX and heavy metal azides. Thus, the key role of electronic excitations facilitated by edge dislocations in explosive solids was established [15, 45, 59, 65–67, 70–72, 106]. A paper by Mathieu et al. [107] focuses on another mechanism involving shock-induced excitations, with the idea that the nuclear dynamics on excited potential energy surfaces might favour molecular decomposition. The relevance of non-adiabatic electronic effects to shock initiation of EMs is discussed by the authors on the basis of simulation results. The latter is obtained using a classic path approach to the electron dynamics in shock waves. The time-dependent electron wave-function is propagated coherently, and attempts are made to introduce environment effect. The results are analyzed in terms of a parameter η defined as the fraction of electrons promoted into virtual orbitals as a result of non-adiabatic transition. The authors have mentioned [107] that current investigations of pressure-induced chemical changes suggest further mechanisms that could make the early decomposition processes easier [108]. It is also alleged [109] that when chemical decomposition occurs, excited states may eventually play a more significant role as the system goes through transition states associated with avoided crossing.

3.3.6 Influence of Crystal Characteristics

White et al. [109] report that crystalline structure has little effect on the initiation of EMs by a shock wave. van der Heijden et al. [110], using the example of RDX, HMX and HNIW, state that the following crystal parameters play a role in determining the sensitivity towards a shock stimulus: (a) internal product quality, (b) mean particle size, (c) surface smoothness/shape of the explosive particle. Like the impact sensitivity, also the shock sensitivity is affected by the density defect content (dislocations, grain boundaries, voids, impurities, inclusions). The shock initiation tests with HMX, for example, have clearly demonstrated a relationship between the average crystal density and shock initiation pressure. The findings from this area are very important for practice, because a modification of quality and crystal shape of, for example, RDX can give a product with increased resistance to impact and shock (I-RDX [110]).

The initiation depends on crystal orientation in energetic materials such as PETN [109, 111, 112]. Using the anharmonic potentials [74], Jindal and Dlott have calculated the directional compressibility, the mode Grüneisen parameters, and the bulk Grüneisen parameters along different crystallographic directions for naphthalene [111]. They show that the temperature increase is quite a bit greater for a shock of given pressure along the *a* or *b* axes than along the *c* axis, mainly because the compressibility along the *c* axis is smaller. With the use of thermochemical data for EMs it is shown that the temperature increase anisotropy is large enough to lead to reaction rates which differ by several orders of magnitude, when typical initiating shocks are incident along different crystallographic directions [111]. PETN is sensitive along the shock propagation perpendicular to the (113) plane, but highly insensitive along the shock propagation perpendicular to the (102) plane [112].

3.3.7 Effect of Desensitizing Admixtures

From the point-of-view of technological practice it is important to study the effects of inert admixtures added to EMs upon the sensitivity of resulting mixtures (see also Sect. 3.1, papers [92, 93]). Desensitization of detonable materials by diluent was studied by Rice et al. [113] on the basis of molecular-dynamics investigation. An inert diluent, a heavy material, was inserted into a crystalline explosive in two ways. The first series of simulations investigates the attenuation of the energy of a detonation wave in a pure explosive after it encounters a small layer of crystalline diluent that has been inserted into the lattice of the pure explosive. After the shock wave has passed through the diluent layer, it re-enters the pure explosive. Unsupported detonation is not re-established unless the energy of the detonation wave exceeds a threshold value. The second series of simulations investigates detonation of solid solution of different concentrations of the explosive and diluent. For both types of simulations, the key to re-establishing or reaching unsupported detonation is the attainment of a critical number density behind the shock front. Once this critical density is reached, the explosive molecules make a transition to an atomic phase. This is the first step in the reaction mechanism that leads to the heat release that sustains the detonation. The reaction fragments formed from the atomization of the heteronuclear reactants subsequently combine with new partners, with homonuclear products formation exothermally favoured [113].

3.3.8 Some Types of Chemical Behaviour of Energetic Materials Under Shock Wave

The effect of a shock wave upon initiation of self-sustaining chemical decomposition (detonation) of EMs has already been mentioned. However, several specific cases call for separate comment. Engelke et al. [114], using time-of-flight (TOF) mass spectroscopy, observed formation of TNT dimer resulting from the action of high-pressure shock waves on this polynitro arene. The dimer is thought to arise from a Diels–Adler cross-linking of two TNT molecules. It is noteworthy that the cross-lining processes are strongly pressure enhanced [114].

Andreev studied decomposition of benzotrifuroxane brought about by mechanical and shock-wave stimuli [115]. This compound is a high-energy heterocycle with relative low impact sensitivity but high shock sensitivity. Andreev tries to explain this strange behaviour within the framework of existing theories of hot spots and the difference in the course of kinetic processes depending both on physical state of the compound and on the intensity of intermolecular interactions in its crystal. The author presumes that the specific chemical capability of benzotrifuroxane results from the possibility of decomposition of various bonds in furoxane cycles. According to monograph [11], the decomposition of furoxanes proceeds via a diradical transition state formed by homolysis of one of the N – O – N bonds of the furazane cycle.

Wu et al. [116] have carried out semi-empirical quantum mechanical molecular dynamic simulation involving collisions of two PETN molecules at different molecular orientations and at several intermolecular separations. The common features of reactive scattering among all molecular orientations are: the dissociation mechanism of PETN remains unimolecular, and the dominant reaction channel is the breaking of an $O-NO_2$ bond. These findings fully agree with those about primary splitting of these bonds in thermal decomposition of nitrate esters [11]. The probability of collision-induced decomposition of this nitric ester depends strongly on initial conditions, in agreement with the experimentally observed sensitivity of shock-initiated detonation in bulk PETN along different crystalline orientations [116].

Dreger et al. [117] investigated the decomposition mechanism in shocked PETN using time-resolved emission spectroscopy. PETN single crystals were subject to stepwise loading along (100) and (110) planes to peak stresses between 2 and 13 GPa. Because of concurrent changes in the optical transmission of this nitric ester, emission spectra were analyzed using the absorption data acquired separately under the same loading conditions. The emission was identified as chemiluminescence from the nitronium ion, NO_2^+. Although the authors state that this ion is an intermediate observed experimentally, their suggested mechanism of PETN decomposition consists of the primary splitting off of the nitrate anion, NO_3^-, which by subsequent reactions with the organic moiety should produce the nitronium cation. It need not be stressed that this idea diametrically differs from the knowledge of splitting of nitric esters [11, 116].

Raikova and Likholatov [118] studied the chemical kinetics of detonation reactions in some liquid nitric ester mixtures (nitroglycerine, nitroglycol, diethylene glycol dinitrate). The main sources of information on the chemical kinetics were the experimental dependence of failure diameter on composition of mixtures. Calculations were carried out in terms of classic theory of Dremin [101]. The authors have found [118] that the obtained rate constants for nitric ester mixtures differed slightly from each other. The corresponding values of activation energies, E_a, and pre-exponent A (i.e. Arrhenius parameters) were about 140–150 kJ mol^{-1}, and 10^{14} s^{-1}, respectively. It may be assumed that the decomposition of this ester mixture in the detonation wave includes only one monomolecular reaction. Values of these Arrhenius parameters correspond to those from low-temperature thermal decomposition of the said liquid esters in general (see in [11]). This is an important finding.

3.3.9 Problems of Nitromethane Initiation

It has already been mentioned that nitromethane (NM) is very frequently used as a model EM for MO calculations and simulations of various processes connected with initiation reactivity of EMs. Therefore, from the point-of-view of initiation by shock wave, it is given separate attention here.

The problems of unimolecular isomerization and dissociation of NM were probed theoretically by means of different MO calculations in two papers [119, 120]. In both cases, the isomerization to methyl nitrite seems to be improbable, the homolysis of the $C-NO_2$ bond being dominant. The isomerization of NM during gas-phase decomposition is also discussed in monograph [11] and labelled as little probable. The dominant primary splitting is the above-mentioned homolysis. Photolytic decomposition of NM also starts by homolysis of the $C-NO_2$ bond [121]. Studies of electronic excitations in shocked crystals of NM showed [122] that dynamic actions have a more significant effect on the band gap than static actions, but relative molecular velocities in excess of $6\,km\,s^{-1}$ are required to produce a significant thermal population of excited states. Some authors [5, 34–36] consider the hydrogen transfer in the sense of Eq. 2 to be the primary process in shocked NM. Applying the DFT-binding method to crystalline NM at 0 K [123], the authors arrived at the conclusion (surprising—in their opinion) that uniaxial compression of about 25–40 GPa along the *b* lattice vector causes the $C-H$ bond to be highly stretched and leads to proton dissociation.

Zhang and Bauer studied decomposition of nitromethane in the gas phase over a temperature range of 1000–1100 K in reflected shock waves (shock heating) and they gravitate to homolysis of the $C-N$ bond as the first fission process [124]. The authors state that this pyrolysis is found to be a chain process—about 40% of the nitromethane is lost through secondary reactions. This paper was continued by Glarborg et al. [125], who found out that at high temperatures and pressures the methyl radical reacts with the nitro radical to give a methoxy radical and NO (secondary reactions), this reaction being more than one order faster than the recombination of these two radicals giving the starting NM.

Monitoring of the shock exposure of nitromethane by means of Raman spectroscopy [126] or detonation of nitromethane by means of the MS spectroscopy [126] led to the conclusion that condensation reactions of two or more nitromethane molecules are a part of its initiation (see schemes, for example, in [126]). However, arrangement of both the experiments had to detect extinguishing detonation (in the first case explosive decomposition in a very thin layer—0.20 to 0.35 mm [126], in the second case an expansion of the explosive decomposition of the very small sample into high vacuum [127]). The critical diameter of NM is in the range of 10–20 mm [101, 129], which also supports the idea of extinguishing detonation at the conditions used in the two experiments. That means the secondary termination reactions should be detected in both the cases. By means of time-resolved emission spectroscopy measurements a shock-to-detonation transition of NM was studied [128]. Corresponding radiant spectra show that the detonation front and detonation products are semitransparent. Shocked NM reaches at least 2500 K, showing the existence of local chemical reaction after shock entrance. This last finding corresponds to Klimenko's model of multi-process detonation [100] (a combination of focal and frontal mechanisms of detonation).

In primary and secondary nitroparaffins, hydrogen transfer reactions take place involving a 1,3-H shift [11, 130]. This produces *aci*-forms of these compounds, which are also called nitronic acids. According to Manelis et al. [11], the existence of the *aci*-form of NM represents a more probable cause of lowering of activation energy of thermal decomposition of liquid NM as compared with the gas-phase thermolysis (i.e. in relation to the theories of bimolecular course of this decomposition [11]). Paper [130] discusses both possible ways and intermolecular mechanisms of formation of the said acids. The mentioned transfer of hydrogen atom can also be evoked by amines [130], whose presence sensitizes both primary and secondary nitroparaffins to stimuli [129, 131]. In the case of nitromethane sensitized with amines, Gruzdkov and Gupta suggested the following mechanism of initiation [132]:

$$\begin{gathered} H_3C-NO_2 + RNH_2 \rightarrow [H_2C=NO_2]^- + RNH_3{}^+ \\ [H_2C=NO_2]^- + H_3C-NO_2 \rightarrow H_3C-NO_2{}^{\bullet -} + H_2C=NO_2{}^{\bullet} \\ H_3C-NO_2{}^{\bullet -} \rightarrow CH_3{}^{\bullet} + NO_2{}^- . \end{gathered} \tag{4}$$

As shown, the first stage should be, according to the authors' opinion, an acid-base equilibrium. This conclusion was made on the basis of application of time-resolved optical spectroscopic techniques and action of shock (12–17 GP) upon an NM layer of 0.2 mm thickness. Similar conclusion was arrived at by Woods et al. [133] on the basis of laser-initiated decomposition of single aerosol particles of NM and NM-diethyl amine mixture in the source region of a time-of-flight spectrometer, followed by vacuum UV laser ionization (products identification by mass spectroscopy). In comparison to the above-mentioned enhancement of detonation parameters, an influence of amines on the failure diameter of nitromethane [129] is better known and more studied [127, 136]. Also this effect is connected with the increase in reaction rate in the reaction zone of detonation (i.e. with reduction of the zone length) due to influence of amines on nitromethane [136]. This fact might indicate that the primary fragments entering the reaction zone of detonation wave in the case of neat NM are different from those in the case of NM sensitized by addition of amines. It has been known for quite a long time that the concentration of the nitromethyl *aci*-anion ($CH_2NO_2{}^-$) in the NM is increased not only with the addition of bases [130] but also by application of static high pressure [134] (see also the above-mentioned paper [123]). However, evidence for an increase in concentration of the anion in neat nitromethane under shock conditions (dynamic high pressure) has been found lacking [135].

Sensitivity of NM is also increased by additives that are inert to it and which can represent active EMs (RDX, ammonium nitrate) [137], or by non-active hard powders (aluminium, copper, fumed silica, boron carbide etc.) [131, 137]. The amounts of the additives are 3 and 40% w/w in [131] and [136], respectively. The sensitivity of the mixtures is higher than that of neat NM [131, 137]; however, for example, the sensitivity increase achieved by addition of 3% w/w of boron carbide is less efficient than the effect of the same amount of added diethylenetriamine [131]. The increase can be due [131] to difference in the number density of each in the mixture (difference in the number of available hot spots) or to an impedance mismatch difference, or to a function of both variables.

3.4
Relation Between Performance and Sensitivity of EMs

Performance of explosives can be defined on the basis of heat of explosion, Q, by equation [138]:

$$A = Q \cdot \left(1 - \left(\frac{V_1}{V_2}\right)^{\gamma-1}\right), \tag{5}$$

where V_1 is the initial volume of gaseous products of detonation and V_2 is their final volume, or also by equation

$$A = Q \cdot \left(1 - \left(\frac{p_2}{p_1}\right)^{\frac{\gamma-1}{\gamma}}\right), \tag{6}$$

where p_1 is the initial pressure of detonation products and p_2 is their pressure after fulfilment of work [138]. This means that if a relationship exists between the Q values and the sensitivity characteristics, then these two equations indirectly relate the sensitivity with performance. The relationships mentioned have so far been presented in this text as the results of work done by Rice and Hare [85] and Edwards et al. [86] (Sect. 3.2.2). These results are in accordance with a result of Licht's analysis [139] which showed that high performance is usually accompanied by an enhanced sensitivity and that an insensitive explosive will not exhibit a top performance. The author has stated that this result cannot be proved by theory. Therefore, also exceptions are possible, i.e. explosives which approach the ideal high explosive (high performance combined with high security), which is exemplified in [139].

4
Sensitivity of EMs from the Standpoint of Physical Organic Chemistry (POC Model)

Explosive decomposition of conventional EMs is based on chemical processes. In spite of this fact, very few authors in the past dealt with, and today are dealing with, the initiation mechanisms from the point-of-view of chemistry. As follows from Sect. 3, most authors of papers published in this field try to solve the problem of micro-initiation by means of quantum chemistry [1, 3–5, 39, 56, 60, 74, 81, 85, 105, 106, 109]. The simulations are predominantly restricted to a few selected molecules of energetic substances (NM, RDX, HMX and TATB are the dominating ones) and their crystals [1, 3, 5, 56, 60, 74], or in some cases only to hypothetical two-atom molecules [109].

Hence, from the standpoint of sensitivity to stimuli (i.e. the micromechanism of initiation) attention is predominantly focused on organic polynitro compounds [1–4, 6, 37, 39, 60], and more recently of heavy metal azides [43–50]. Also the development in the area of new individual EMs mostly advances with orientation to organic polynitro compounds [51, 84, 140–143], and more recently it is intensively directed also to organic high nitrogen EMs (see, e.g., [142–144]). Thus, this special area of organic chemistry plays a highly significant part in the study and development of EMs. In this sense, studies of sensitivity of EMs, mainly organic polynitro and polynitroso compounds, are also conducted at the workplace of the author of this review article.

4.1 Approach to Solution of the Initiation Reactivity

It is a generally accepted idea that nitro groups represent the primary cause of sensitivity (initiation reactivity) of polynitro compounds (see, e.g., [6–20, 59, 90, 91]). Sensitivity of a given polynitro compound to stimuli results from the type of bonds to nitro groups in the molecule, their number, the type and intensity of their interaction with the rest of the molecule, and intermolecular interactions. Recently relationships have been found between outputs of non-isothermal differential thermal analysis (DTA, i.e. low-temperature thermolysis) and characteristics of detonation of polynitro arenes [145, 146] and also nitramines, nitrosamines and nitrate esters [146]. A more detailed analysis of these results for polynitro arenes shows that their classification in the sense of the relationships found is given primarily by (a) steric conditions and (b) electron configuration in the ground state of the reaction centre of the molecule [145, 146]. In this context, the term reaction centre means the grouping of atoms and/or functional groups in the molecule whose primary chemical changes initiate decomposition of this molecule (e.g., see Schemes 2, 3a, 3b, 7 and 8). At the same time, the said facts represent one of the basic principles of approach of organic chemistry to dealing with reactivity problems in general.

4.1.1 Methods of Specification of the Reaction Centre in Molecule

Both the electron configuration and steric situation in the reaction centre of the molecule can be characterized from the standpoint of organic chemistry on the basis of:

- NMR chemical shifts of key atoms of the reaction centre of the molecule (semi-empirical determination of the nitro groups that are first to react in the initiation) [6, 7, 10, 16, 98, 148],
- direct correlation of reaction characteristics with electron charges at the nitrogen atoms of these first-reacting nitro groups [7, 12, 152, 160, 161] or with net charges of these nitro groups [90, 91, 160].

For the first time, the approach given was demonstrated on a group of secondary nitramines [7, 10, 98], because these compounds have relatively simple molecular structures, and the mechanism of their primary homolysis is well known [11, 14] (see also Scheme 1). On the basis of these findings, the approach was extended to the application of ^{13}C NMR [6, 16, 148] and ^{15}N NMR [150] chemical shifts in polynitro arenes. It can be argued that in this approach, an NMR study in solution, neglects important crystal-lattice effects that are vital to the determination of explosive properties [149]. From our several recent papers [6, 7, 10, 16, 98, 148] and the following text it will be seen that this objection has no fundamental significance for studies of chemical micro-mechanism of initiation of energetic materials.

The calculation of electronic charges, q^{N}, at nitrogen atoms and net charges, Q_{NO_2}, of nitro groups of the individual EMs investigated was carried out by means of the Mulliken population analysis of electron densities obtained by the ab initio DFT B3LYP/6-31G** method, under total geometry optimization and without subsequent frequency analysis [158].

4.1.2
Selection of Sensitivity (Reaction) Characteristics of EMs

The thermal decomposition characteristics used included the Arrhenius parameters obtained by means of the Russian manometric method ([9, 11, 151] and citations herein) or by means of methods that give comparable results (especially differential scanning calorimetry [9, 151]). Another source of these characteristics was the simple non-isothermic differential thermal analysis (DTA) whose outputs were evaluated by means of the Kissinger method [152–154] in the sense of relationship [154]:

$$\ln\left(\frac{\phi}{T^2}\right) = -\left(\frac{E_{\mathrm{a}}}{R}\right)\frac{1}{T} + \ln\left(\frac{AR}{E_{\mathrm{a}}}\right), \tag{7}$$

where ϕ is the rate of temperature increase and T is the peak temperature of the exothermic decomposition. The reactivity was expressed as the E_{a}/R slopes of the relationship Eq. 7.

In the case of electric spark sensitivity, the available data were obtained by application of two instruments [8] whose design and construction were realized with contribution from the Institute of Energetic Materials. These data are spark energies, E_{ES}, for 50% probability of initiation.

As for the impact sensitivity, the data considered were the drop energies, E_{d}, obtained by conversion of drop heights values, $h_{50\%}$, both published and determined by us. Attention was given both to the impact sensitivity as "the first reaction" (detection with respect to the first decomposition) [6, 16, 148, 155], and impact sensitivity detected by sound ([6, 148, 155] and references herein).

The studies of shock sensitivity were based on the detonation velocities, D, calculated by the Kamlet and Jacobs method [156], and in some cases they also were determined experimentally. The heat of explosion, Q_{real}, was calculated by means of the semi-empirical relationships by Pepekin et al. [157]. Both these quantities characterize the shock wave due to chemical transformation of EM, i.e. the detonation wave.

Also polarographic half-wave potentials, $E_{1/2}$, of 24 polynitro arenes were applied; these potentials were determined in aqueous medium buffered at pH 7, the final concentration of their solutions being 0.5×10^{-4} M [159].

By means of the above-mentioned methods and data of described nitramines, basic data were predicted [16, 98, 148, 176] for three substances not yet prepared, i.e. for 1-nitro-1-azaethylene (Digen), 1,3-dinitro-1,3-diazetidine (Tetrogen) and 1,3,5,7,9-pentanitro-1,3,5,7,9-pentazecane (Decagen), which are of considerable theoretical interest in the chemistry of nitramines.

4.2 Initiation by Heat

The application of ^{15}N NMR chemical shifts, δ_N, of nitrogen atoms in nitro groups to the analysis and prediction of kinetic data of thermal decomposition of nitramines [10] is presented in Fig. 1; a non-isochronous molecule HNIW correlates through the δ_N value in the position 2.

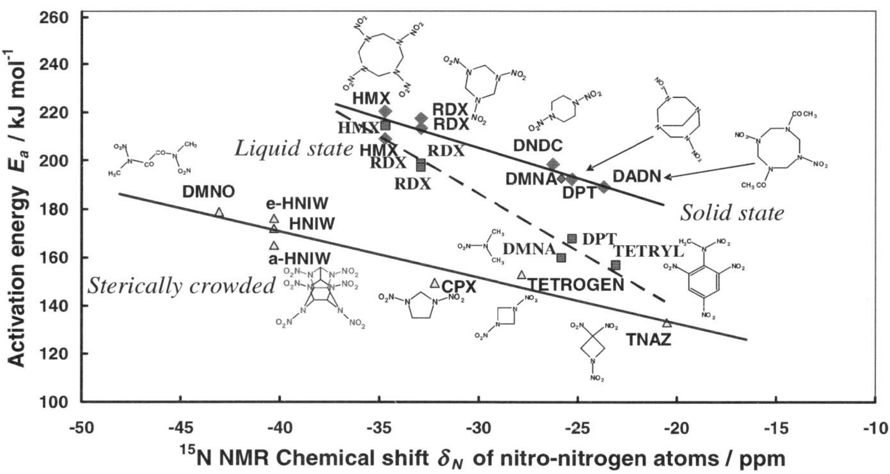

Fig. 1 Relationship between the activation energies, E_a, resulting from the Russian manometric method, and the ^{15}N NMR chemical shifts of nitrogen atoms of the nitro groups (non-isochronous molecule HNIW correlates through the shift of the nitro–nitrogen atom in the position 2)—taken from [7]

Figure 1 shows that in the case of HMX the effect of the liquid phase on its thermolysis should be absent even if the thermolysis is realized within a temperature range above 553 K (see discussion in [10]). An analogous dependence for polynitro arenes is presented in Fig. 2 [151].

Figure 3 presents the relationship between E_a/R (from Eq. 7) and electronic charges, q^N, at nitrogen atoms of the first-reacting nitro groups for nitramines taken from [152]. The discussion of this graph confirmed the differences between linear (line *D*) and cyclic nitramines (lines *A*, *B* and *C*) [152] from the standpoint of the initiation energy transfer to the reaction centre. The quantum chemical calculation itself revealed non-equality of nitrogen atoms in nitro groups from the point-of-view of their potential participation in primary splitting processes of initiation [152], which was only further clearly confirmed by the dependences in Fig. 4. A similar approach applied to highly thermally stable polynitro arenes resulted in the relationships between E_a/R and q^N values depicted in Fig. 4 [160]. Also in this case one can clearly see very close molecular-structure dependence and, in the cases of TNB and HNB, also effect of state of aggregation (both these polynitro arenes decompose in the liquid state).

On the basis of relationship between electron charges, q^N, at the nitrogen atoms of the first-reacting nitro groups and DTA onsets, T_D, of exothermic decomposition of polynitro arenes in Fig. 5, a set of these substances are di-

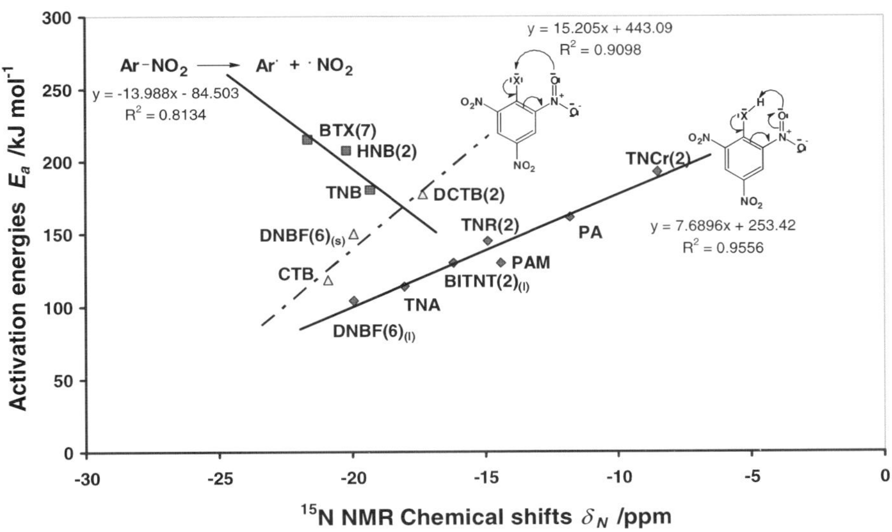

Fig. 2 Graphical representation of the relationship between activation energies, E_a, of low-temperature thermal decomposition and ^{15}N NMR chemical shifts of polynitro arenes (the numbers in parenthesis denote the position in the molecule): a general scheme of the primary step of chemical decomposition is presented with each line [150]

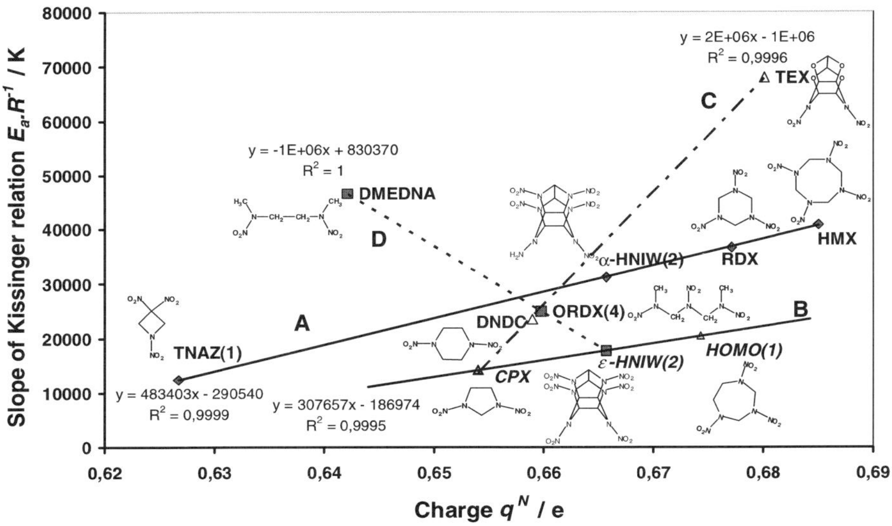

Fig. 3 Relationship between the slopes of the Kissinger relation Eq. 7, E_aR^{-1}, and Mulliken B3LYP/6-31G** charges, q^N, at the nitrogen atoms of the primarily reacting nitro groups in nitramino groupings in the molecule (given in *parenthesis* are the positions in the molecule which correlate with the non-isochronous molecules). Taken from [152]

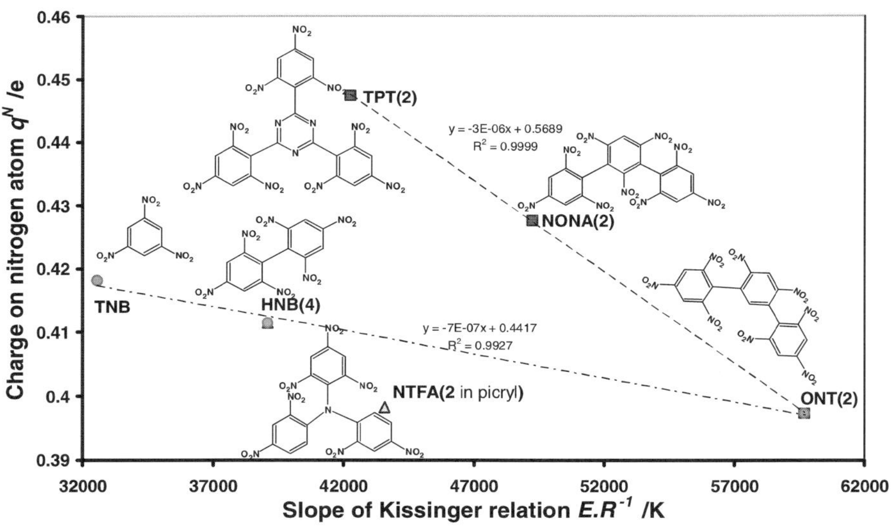

Fig. 4 Relationship between the slopes of the Kissinger relation Eq. 7, E_aR^{-1}, and Mulliken B3LYP/6-31G** charges, q^N, at the nitrogen atoms of the primarily reacting nitro groups in molecules of high-thermostable polynitro arenes (decomposition of TNB and HNB in the liquid state; given in *parenthesis* are the f positions in the molecule). Taken from [160]

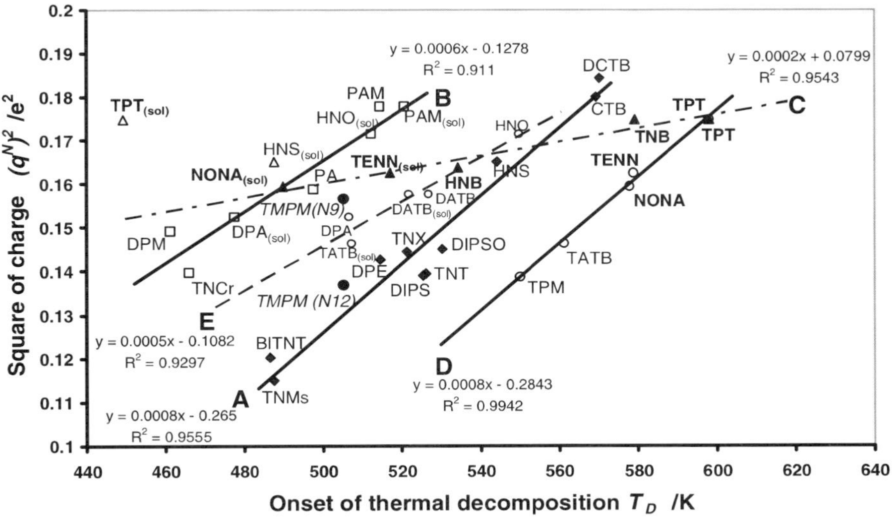

Fig. 5 Relationship between the q^N values of the nitrogen atoms of the primarily reacting groups and onsets, T_D, of exothermic decomposition. Codes of the genuine polynitro arenes are written in *boldface type*. Taken from [161]

vided into several logical subsets [161]: one with the dominant stabilizing effect of the crystal lattice (line *D*), one in which the effects of the mechanism of primary homolysis (lines *A*, *C*, *E*) dominates, and one characterized by combination of both above-mentioned effects (line *B*). An addition of solvent (TNB) to substances of class *D* will suppress this stabilizing effect: TENN and NONA move to class C (with primary homolysis of the $C-NO_2$ bond), TATB moves to class *E* (splitting in the sense of Scheme 2) and the stability of TPT decreases so markedly that it does not correlate with any of the classes in Fig. 5.

From the results in paper [6] it follows that a molecule of energetic material can contain more than one potential reaction centre of initiation. As an example, we can give *N*,*N*′,*N*″-tris(3-methyl-2,4,6-trinitrophenyl)-1,3,5-triazine-2,4,6-triamine (TMPM) with potential reaction centres in the vicinity of nitrogen atoms N9 (position 6 in trinitrophenyl group) and N12 (position 2 in the same group) [6]; the data of the second reaction centre mentioned approach those of substances of class *A*, which means that the primary splitting by heat should go by the classic trinitrotoluene mechanism (see also Scheme 8). The existence of this type of relationship (i.e. Figure 5) might be connected with the electrostatic interaction of "instantaneous point dipoles" at the reaction centre of their molecules (discussion of this problem in [161] is unfolded from the Einstein–Nernst relationship [162]).

4.3
Impact Sensitivity

4.3.1
Impact Sensitivity as the First Reaction

The problem of this type of impact sensitivity was experimentally studied at the author's workplace [6, 16, 148]. Figure 6 presents a relationship between ^{15}N NMR chemical shifts, δ_A, of aza (amino) nitrogen atoms carrying the most reactive nitramino groups and drop energies, E_{dr}, of the first reaction [6]. This figure was taken from papers [6, 16] and shows that the impact reactivity of linear polynitramines (more than two nitramino groups in a molecule) is connected with primary homolysis of "inner" $N-NO_2$ bonds in their molecules (Scheme 6). This finding stands in accordance with the mechanism formulated for this process on the basis of molecular dynamics by Kohno et al. [163].

No analogous relationships between E_{dr} values and ^{15}N NMR chemical shifts, δ_N, of nitro-group nitrogen atoms were found [6] due to lower correlation of the shifts with structural details. It must be stated that the aza (amino) nitrogen chemical shifts, δ_A, in the nitramino groups are expected to be influenced by the nitrogen hybridization, the size and conformation of

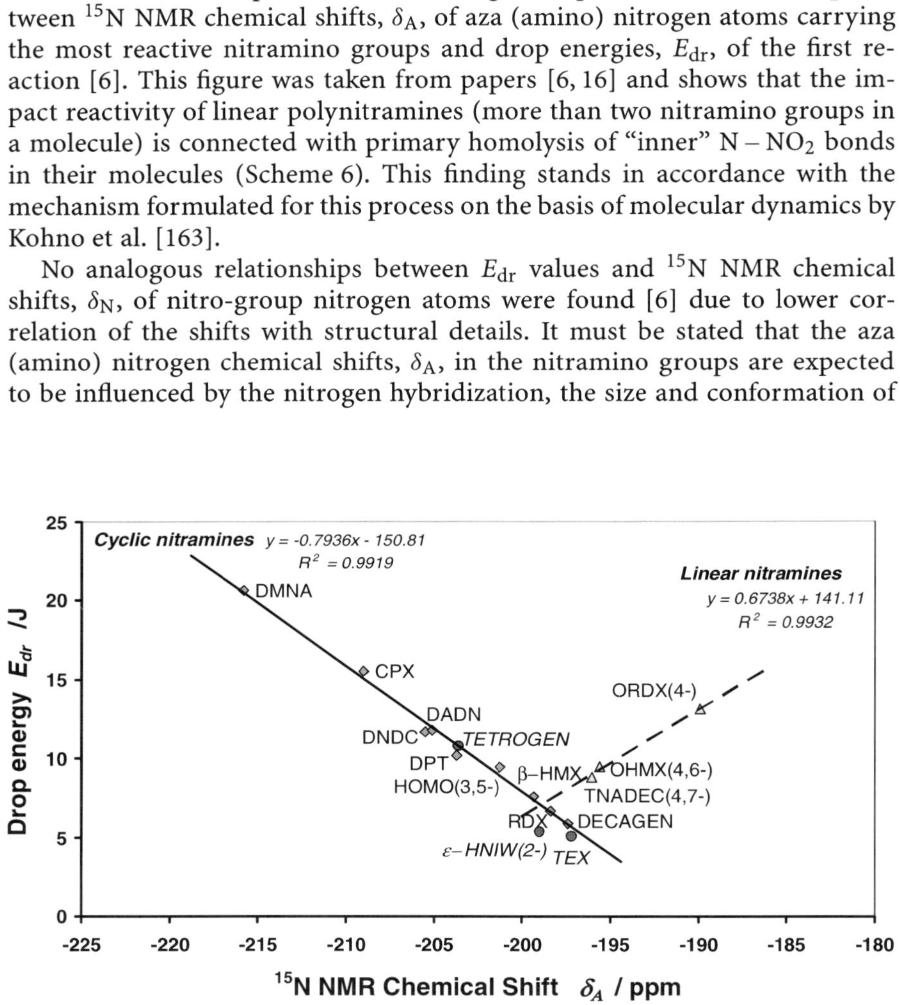

Fig. 6 Relationship between the drop energy, E_{dr}, of "the first reaction" and ^{15}N NMR chemical shifts, δ_A, of the aza nitrogen atoms carrying the nitro groups that are first to react in the initiation reaction (given *in brackets* are the positions in the molecule). Taken from [16] and complemented by data for Tetrogen, HNIW and TEX [6]

the molecule, and the extent to which the nitrogen lone pair is involved in π-bonding with the NO_2 group. As the conformation and size of the molecule plays a dominant role in the intermolecular interactions in the corresponding crystal, the interactions should have a significant influence on the impact sensitivity.

This hypothesis is supported by Fig. 7 (taken from papers [6, 16]), which represents a relationship between the E_{dr} values and heats of fusion, $\Delta H_{m,tr}$, of nitramines (the heats of fusion $\Delta H_{m,tr}$, defined in the sense of monograph [164] as the sum of heat of melting and heats of all polymorph transitions). As the heat represents the work needed for formation of defects in the crystal lattice, the relationships found seem to stand in accordance with the ideas about the decisive role of dislocations and plastic deformations of crystal played in the initiation of energetic materials (Sect. 2.6).

Likewise, for nitramines a relationship exists between E_{dr} values and ^{13}C NMR chemical shifts, δ_C, of the most reactive nitro group "bearers" (i.e. carbon atoms in the reaction centres of the molecule) in polynitro arenes [148].

As shown in Figs. 8a and 8b the relationship has once, again, the same form as that for nitramines. The more complex molecular structure of polynitro arenes (particularly the mesomeric and steric effects) and the two presumed basic mechanisms of primary splitting, as compared with nitramines, result in a more varied assortment of shapes of this relation-

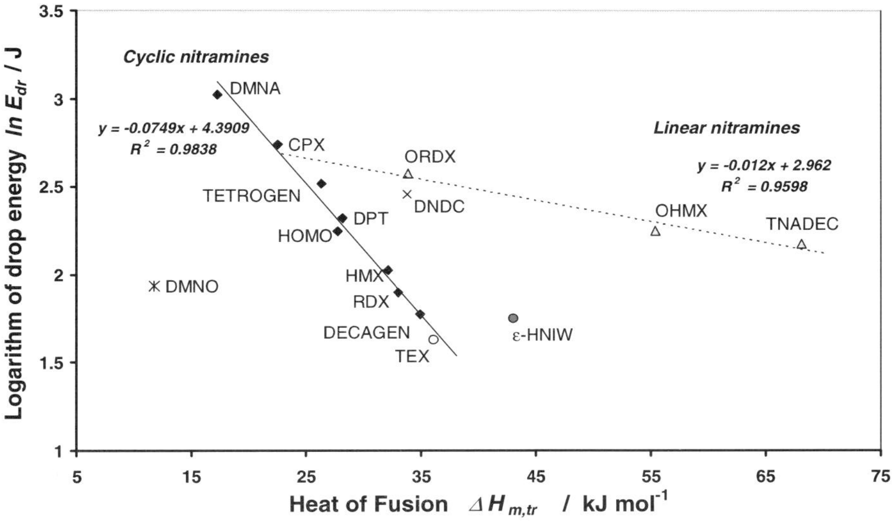

Fig. 7 Relationship between the drop energy, E_{dr}, of the first reaction and the heat of fusion, $\Delta H_{m,tr}$, of nitramines; taken from [16] and complemented by data for TEX [6]

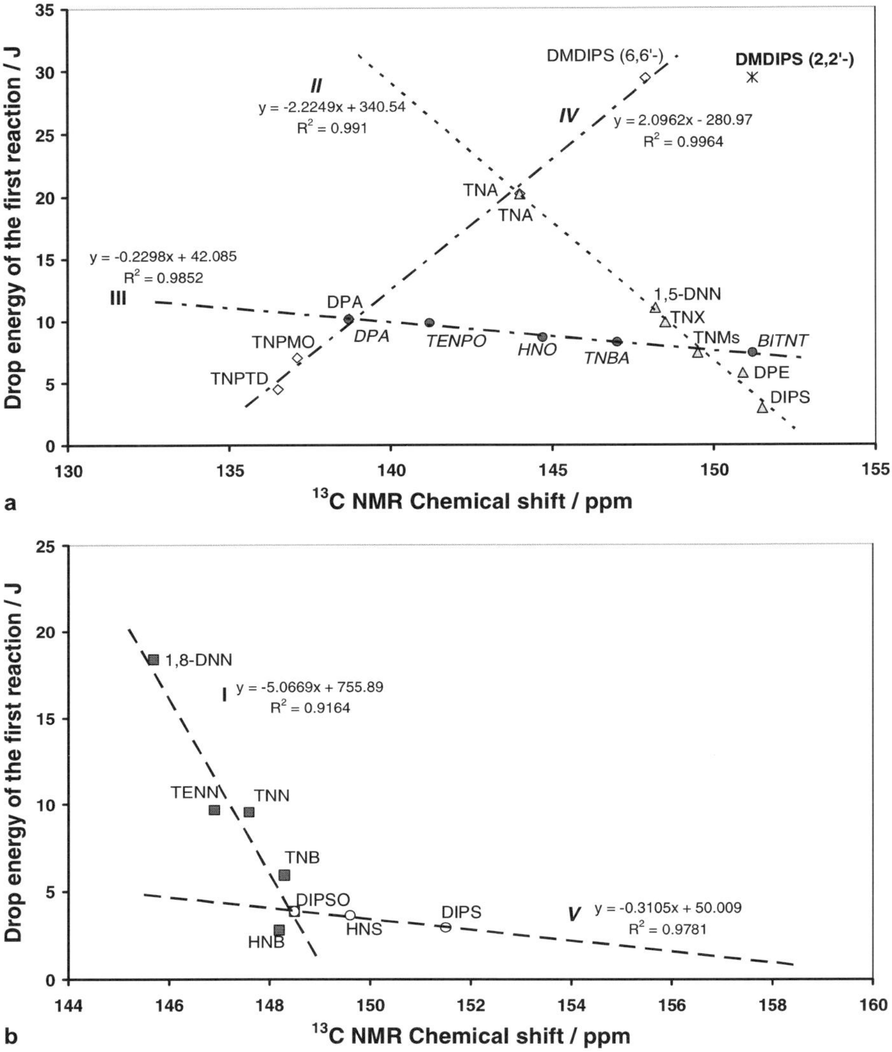

Fig. 8 **a** Relationships between the impact sensitivity (drop energy of the first reaction, E_{dr}) and ^{13}C NMR chemical shifts of the carbon atoms which bears the primarily reacting nitro groups in the polynitro arenes studied (compounds with primary thermal decomposition via a five-, six- or seven-membered transition state or aci-form) [148]. **b** Relationships between the impact sensitivity (drop energy of the first reaction, E_{dr}) and ^{13}C NMR chemical shifts of the carbon atoms which bears the primarily reacting nitro groups in the "genuine" polynitro arenes (group I) and compounds (group V) with primary thermal decomposition via a five- or six-membered transition state [148]

ship. The straight line IV in Fig. 8a predicts a likely primary mechanism of splitting of 2,4,6-trinitro-3-[(3-methyl-2,4,6-trinitrophenyl)thio]toluene (DMDIPS), which possesses two potential reaction centres in its molecule (see Schemes 3a and Scheme 3b); as this straight line correlates with the δ_C values of the "bearer" of the nitro group at position 2(2′), the DMDIPS primary splitting in the sense of Scheme 3b [6, 148] is most probably. In the case of low-temperature thermal decomposition of 1,3,5-trinitro-2-[(2,4,6-trinitrophenyl)thio]benzene (DIPS), as a result of C – S bond homolysis, 2,2′,4,4′,6,6′-hexanitro-1,1′-biphenyl (HNB) is formed as a stable intermediate [167].

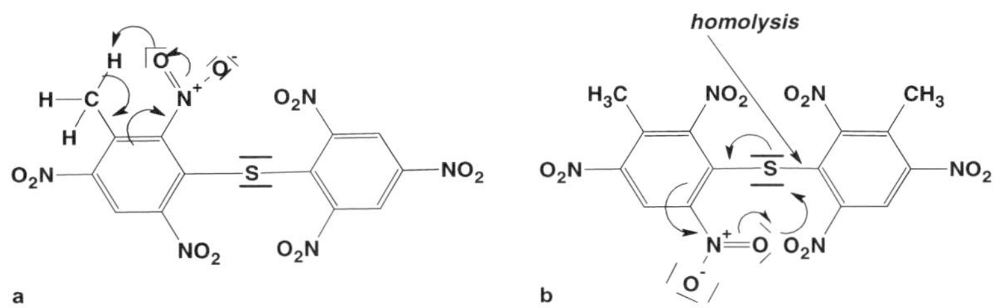

Scheme 3 **a** Theoretical, less probable mechanism of electron shifts at the beginning of the low-temperature thermal decomposition of 2,4,6-trinitro-3-[(3-methyl-2,4,6-trinitrophenyl)thio]toluene (DMDIPS) with participation of the nitro group at the 2-position in the reaction centre (typical "trinitrotoluene mechanism") [6, 148]. **b** Presumed mechanism of more probable electron shifts at the beginning of the low-temperature thermal decomposition of 2,4,6-trinitro-3-[(3-methyl-2,4,6-trinitrophenyl)thio]toluene (DMDIPS) with participation of the nitro group at the 6-position in the reaction centre [6, 148]. In the low-temperature thermal decomposition of 1,3,5-trinitro-2-[(2,4,6-trinitrophenyl)thio]benzene (DIPS), as a result of C – S bond homolysis, 2,2′,4,4′,6,6′-hexanitro-1,1′-biphenyl is formed as a stable intermediate [167]

The relationship valid for class II in Fig. 8a is excellently followed by the δ_C data of 1,5-dinitronaphthalene (1,5-DNN), which means that the thermal decomposition of 1,5-DNN should begin by the interaction of the oxygen atom of the nitro group with the hydrogen atom at the *peri*-position [148], as contrasted with the presumed homolysis of the C – NO_2 bond.

The relationships between impact sensitivity in the sense of the first reaction of polynitro arenes and their heats of fusion on the one hand (Fig. 9), and the ^{13}C NMR chemical shifts at the reaction centres of their molecules (Figs. 8a,b) on the other hand, are of the same types as those found for nitramines.

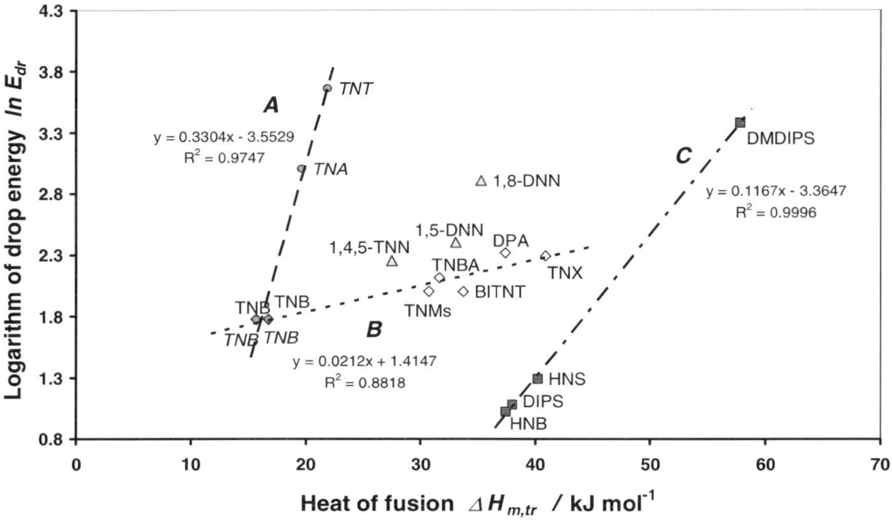

Fig. 9 Semilogarithmic relationships between the impact sensitivity (drop energy of the first reaction, E_{dr}) and heat of fusion ($\Delta H_{m,tr}$) of polynitro arenes [148]

4.3.2 Impact Sensitivity Detected by Sound

In this case, the impact sensitivity data were mostly taken from the literature. Figure 10 represents the relationship between the impact sensitivity detected by sound and ^{15}N NMR chemical shifts of nitramines [6]. This relationship, in comparison with Fig. 6, is mathematically different. The difference could be interpreted by the different mechanisms of transfer of drop energy to the reaction centre of the molecule in the case of the first reaction as compared with the impact sensitivity detected by sound. This difference is evident also in the relationship between the sensitivity and the $\Delta H_{m,tr}$ values (Fig. 11 taken from [147]).

The same shape of relationship as shown in Fig. 9 is valid for polynitro arenes in Fig. 12: here the impact sensitivity detected by sound correlates with ^{13}C NMR chemical shifts of the bearers of the most reactive nitro groups in the molecule. According to this diagram, they are the nitro groups in 2-positions of the DATB, TNX, TNR and BITNT. These groups are also most sterically crowded [6]. However, in the N,N'-bis(2,4,6-trinitrophenyl)-3,5-dinitropyridine-2,6-diamine (PYX) molecule there exist two potential reaction centres, namely that with participation of the nitro group at position 2 (Eq. 2) in the picryl group and that in position 3 (Eq. 1) of the pyridine ring (see also Scheme 7).

The dependences given for polynitro arenes in Fig. 13 are analogous to those given for nitramines in Fig. 10. They are discussed in [147]; we should

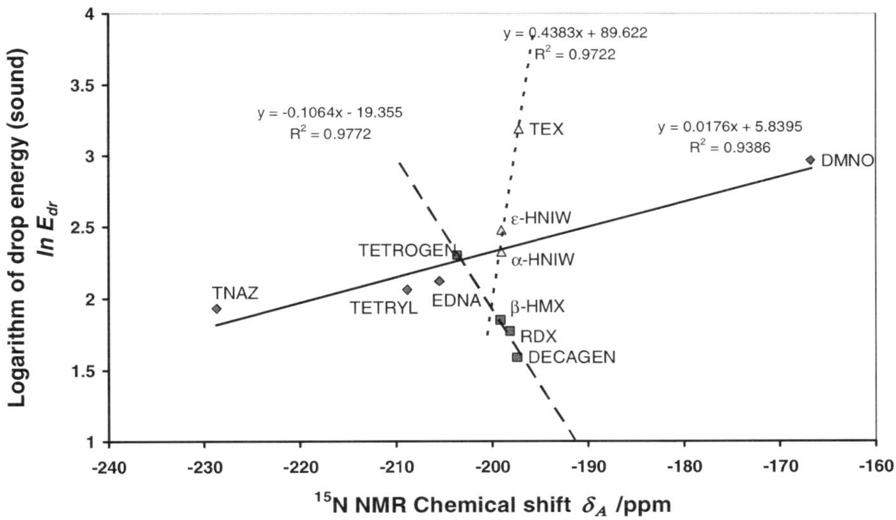

Fig. 10 Relationship between the drop energy, E_{dr}, (impact sensitivity detected by sound) and ^{15}N NMR chemical shifts, δ_{A}, of the aza nitrogen atoms carrying the first reacting nitro groups in the molecules of nitramines studied [6]

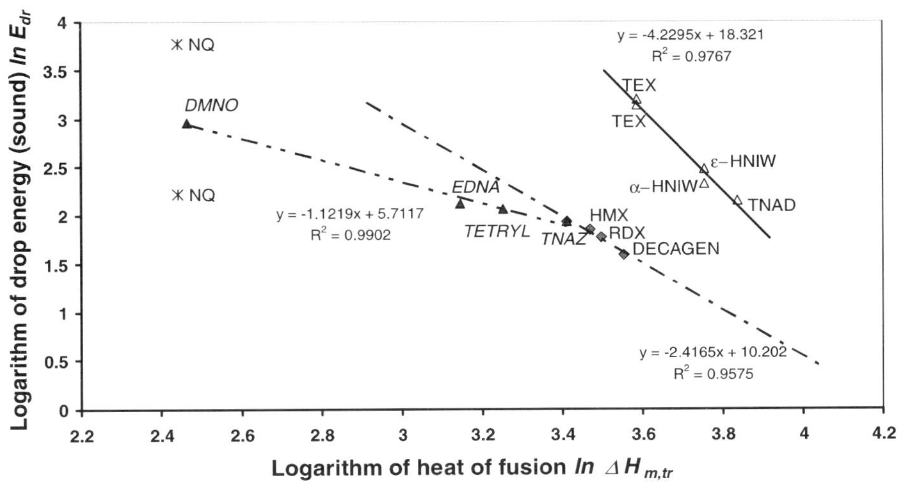

Fig. 11 Relationship between the drop energy, E_{dr}, (impact sensitivity detected by sound) and heat of fusion, $\Delta H_{\mathrm{m,tr}}$, of nitramines; taken from [147]

draw attention here to the effect of the introduction of the hydroxyl group into the TNB molecule (line *D*), amino group (line *A*), alkyl groups (line *B*), or "doubling" of trinitrophenyls (line *C*).

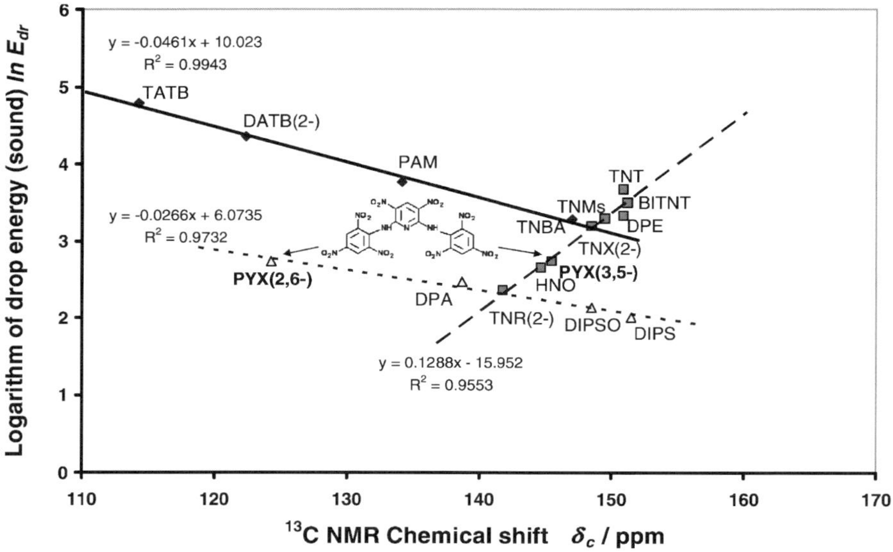

Fig. 12 Relationship between the energy, E_{dr}, (impact sensitivity detected by sound) and ^{13}C NMR chemical shifts, δ_C, of the carbon atoms of the aromatic system carrying the most reactive nitro groups (the *numbers in brackets* denote the positions in the molecule) [165]

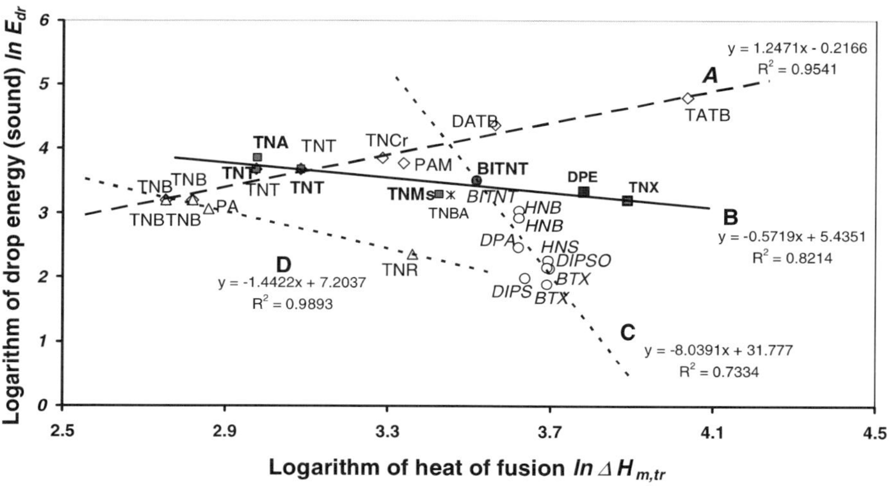

Fig. 13 Logarithmic relationships between the impact sensitivity, E_{dr}, (impact sensitivity detected by sound) and heat of fusion, $\Delta H_{m,tr}$, of polynitro arenes [147]

Figure 14 represents very logical relationships between impact sensitivity and the net charges of the primarily reacting nitro group: here linear condensed polynitro (poly)arenes are sharply separated from spatially con-

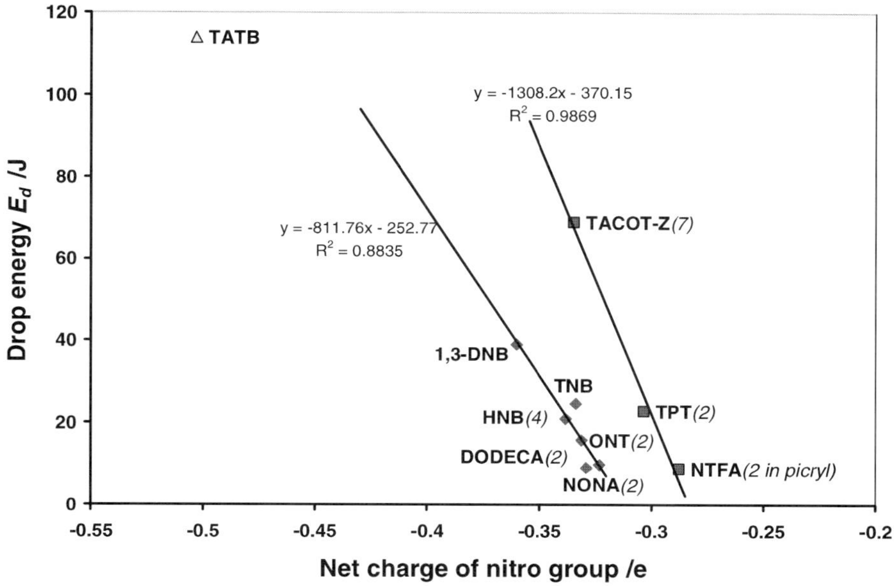

Fig. 14 Relationship between impact sensitivity and net charges of primarily reacting nitro groups in molecules of highly thermally stable polynitro arenes (the *numbers in brackets* denote the positions in the molecule) [160]

densed analogues (i.e. NTFA, TPT and TACOT-Z). The data of TATB do not correlate with either group due to the above-mentioned singularity of primary fission of its molecule.

4.3.3 Allocation of Polynitro Compounds on the Basis of their Impact Sensitivities

The mechanism of transfer of drop energy to the reaction centre of the molecule in the case of the impact sensitivity as the first reaction should differ from that in the case of impact sensitivity detected by sound. Comparison of the drop energy values, E_{dr}, of the two types of impact sensitivity leads to relationships shown in Fig. 15 [166].

The polynitro compounds in Fig. 15 fall into three classes. Each of them includes compounds with a characteristic range of values for thermolysis rate constants at their onset temperature of exothermic decomposition [166]. The given temperatures result from differential thermal analyses of the said compounds. This finding corresponds with the known relationships between impact sensitivities and thermal decomposition characteristics of polynitro compounds [16, 89, 148]. This means that some relationship exists between

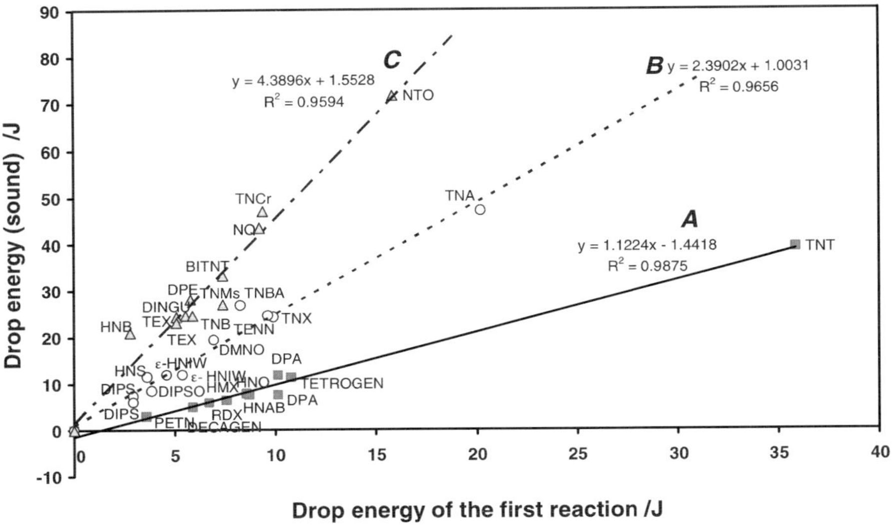

Fig. 15 Relationships between the drop energies of the impact sensitivity detected by sound and the drop energies of the first reaction [155]

vibrational excitation by impact, on the one hand, and thermal activation of the molecules of energetic materials, on the other [166].

4.4
Initiation by Shock

By means of an analogous approach to that used in the case of impact sensitivity mutual relationships between the characteristics of detonation and ^{15}N NMR chemical shifts of nitramines [26] were studied. In contrast to the already-mentioned Figs 6, 8a, 8b, 10 and 12 the characteristics of detonation (i.e. heat of explosion, Q, or square of detonation velocity, D^2) correlate with the ^{15}N NMR chemical shifts, δ_N, of nitrogen atoms of the first reacting nitro groups in the initiation process [6, 7, 98]. Corresponding relationships for the square of detonation velocity, D^2, for nitramines are presented in Fig. 16 [6, 7, 98]. It therefore follows that the primary fission of the nitramine molecule is the same as in the case of initiation by impact (Scheme 1) [6, 7]. Again this fact agrees perfectly with the conclusions drawn by Kohno et al. [163] from the molecular dynamics study of impact and shock reactivity of nitramines. Replacement of squares of detonation velocities, D^2, of nitramines by their heats of explosion, Q, gave as good correlations as in the case of Fig. 16 [98].

A closer relationship between Q_{real} and δ_N values (Fig. 17) exists for polynitro arenes and aza-arenes [150]. The more complex intramolecular interactions in these molecules results in more complex relationships: in

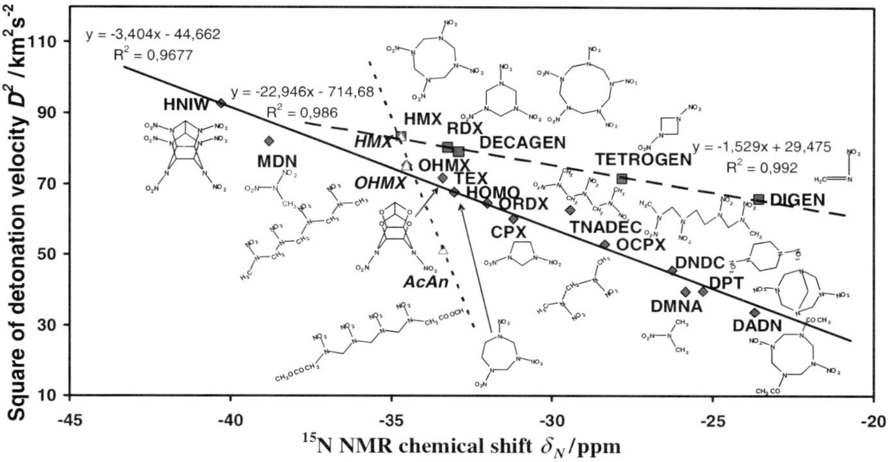

Fig. 16 Relationship between the square of the detonation velocity, D^2, and ^{15}N NMR chemical shifts of nitrogen atoms of the primarily reacting nitro groups [6, 7, 98]

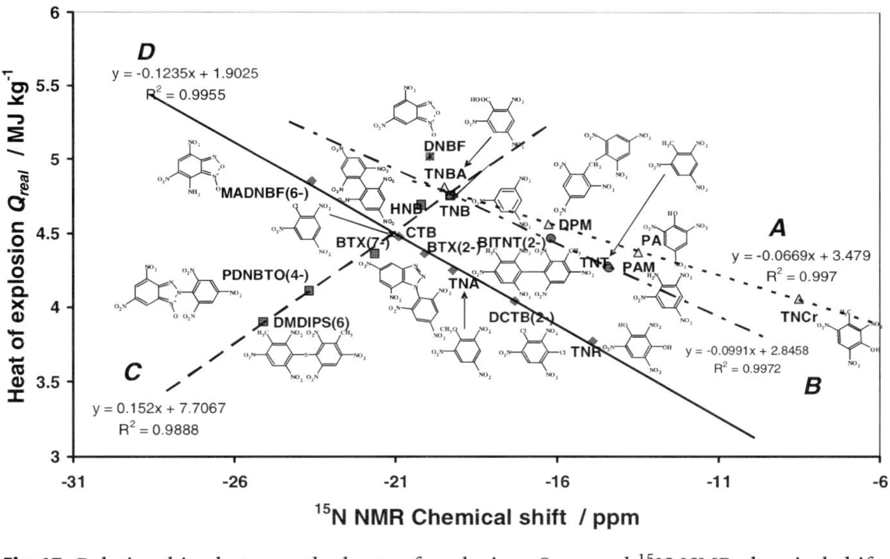

Fig. 17 Relationships between the heats of explosion, Q_{real}, and ^{15}N NMR chemical shifts of the nitrogen atoms of the primarily reacting nitro groups in polynitro arenes [150]

Fig. 17 the derivative groups of *A* and *B* mutually differ by the presence of intramolecular hydrogen bonds in the first group (the splitting according to Scheme 2). Class *D* includes polynitro derivatives characterized besides a more distinct steric effect probably also by the interaction between the *ortho*-nitro group and substituent (mediated by the electron pair at the oxygen atom) as the primary chemical process of thermolysis. Class *C* in Fig. 17

represents substances with primary homolysis of the $C-NO_2$ bond, and the $C-S$ bond in the case of DMDIPS.

4.5 Relationship Between Sensitivity and Heat of Explosion

As already stated above, Rice and Hare [85] found an exponential dependence, and Edwards et al. [86] found a semi-logarithmic dependence between impact sensitivity and heat of explosion, Q, without trying to interpret these facts. Also reported is a relationship between the ratio $Q.DH^{-1}$ and logarithm of drop height $h_{50\%}$ [84] (in this case DH stands for the dissociation energy of the weakest bond of the nitro group in the molecule).

From what has been said so far, and from the published papers [6, 7, 9, 14, 16, 98, 145, 151–153] it follows that there exist logical relationships between the characteristics of low-temperature thermal decomposition and those of initiation and detonation, respectively. The homolytic character of primary fission in both the detonation and low-temperature thermal decompositions of energetic materials (for relevant quotations, see [9]) was a motive for Zeman to use the Evans–Polanyi–Semenov equation (E–P–S) [168] to study the chemical micro-mechanism governing initiation of energetic materials [9]. A relationship formally similar to the E–P–S equation can also be obtained by mutual comparison of the dependences shown in Figs. 2 and 17; it has the following form:

$$E_a = \alpha \cdot Q + \beta \,, \tag{8}$$

and application of the relationship between detonation velocities and explosion heats Q [138] in the form

$$Q = \frac{D^2}{2\left(\gamma^2 - 1\right)} \,, \tag{9}$$

where γ is the polytropy coefficient, transforms Eq. 8 into the following form

$$E_a = a \cdot D^2 + b \,, \tag{10}$$

which can also be derived by mutual comparison of the dependences in Figs. 1 and 16. Equations 8 and 10 were called a modified E–P–S equation by Zeman [9]. The original E–P–S describes a relationship between activation energies E of most substitution reactions of free radicals and corresponding heats of reaction ΔH of narrow sets of substance structures. The equation shows that the strength of the bond being split is a decisive factor in the given reaction. In both Eqs. 8 and 10 the energy, E, can be the activation energy of thermal decomposition, E_a [7, 9, 98], the slope $E_a \cdot R^{-1}$ of the Kissinger relationship Eq. 7 [153, 169, 170], the energy of electric spark, E_{ES} [175, 176, 179], drop energy, E_p [160], it may be substituted by the charge, q^N, at the nitrogen atom of the most reactive nitro group [12] in the molecule, by the net charge of this nitro group [160], or by half-wave polarographic potential [159]. The following text presents several typical examples.

Figure 18 presents the relationship (Eq. 8) for nitramines, in which the E_a values for cyclic structures correspond to their decomposition in the solid state.

The validity of Eq. 8 was also successfully verified for the thermal decomposition of inorganic azides and fulminates [9]—see Fig. 19. For azides the literature gives activation energy values, E_a, for thermal decomposition within various temperature ranges. In the sense of Eq. 8, however, only those E_a values correlate that correspond to the lowest experimental temperature ranges [9]. This is a very important finding.

Figure 20 has been taken from [9] and presents Eq. 8 in which the E_a values resulted from the Russian manometric method for polynitro arenes and Q_{real} values were calculated according to the Pepekin et al. semi-empirical method [157]. It must be stated that the E_a value of TATB, obtained on the basis of the Russian manometric method (i.e. in vacuum, see [151]) in a temperature region above its hypothetical melting point [171], correlates with the data for thermal decomposition in the liquid state (this E_a value, which is very close to the heat of sublimation of TATB, is discussed in [171]). The data of TATB obtained from DSC (see [151], i.e. from measurements at atmospheric pressure) logically correlate with DATB and PAM data in Fig. 20.

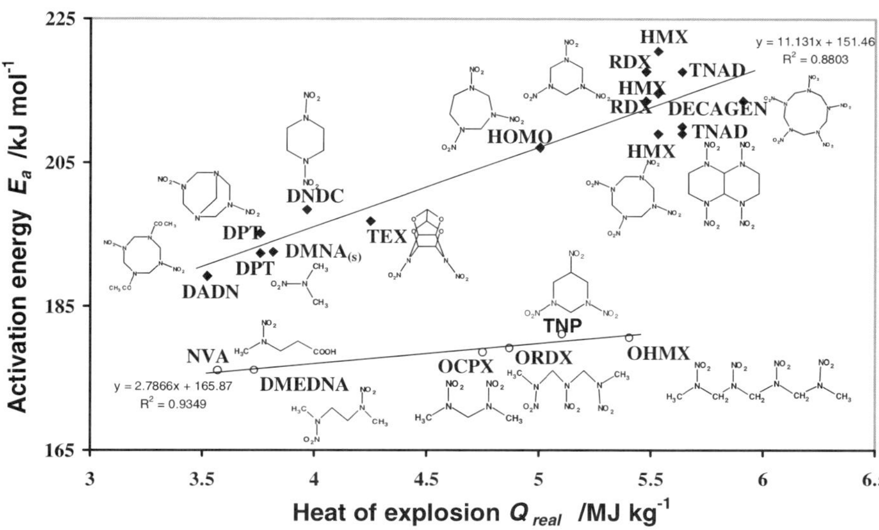

Fig. 18 Modified E–P–S equation (Eq. 8) for the relationship between the activation energies, E_a, corresponding to the outputs of the Russian manometric method for low-temperature thermolysis of nitramines, and the real heats of explosion, Q_{real} [9]

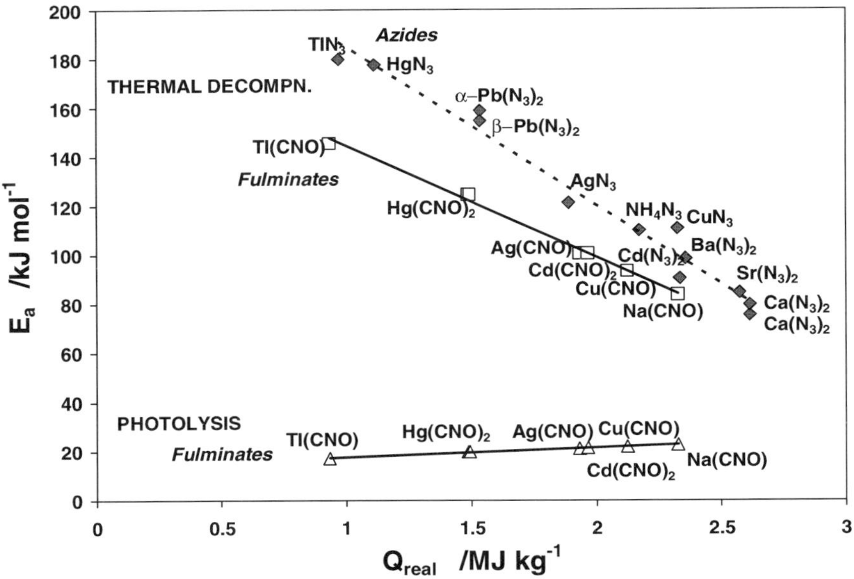

Fig. 19 Modified E–P–S equation (Eq. 8) for azides and fulminates [9]. For azides the literature gives activation energy values, E_a, for thermal decomposition within various temperature ranges; however, the E_a values that correlate correspond only to the lowest experimental temperature ranges [9]

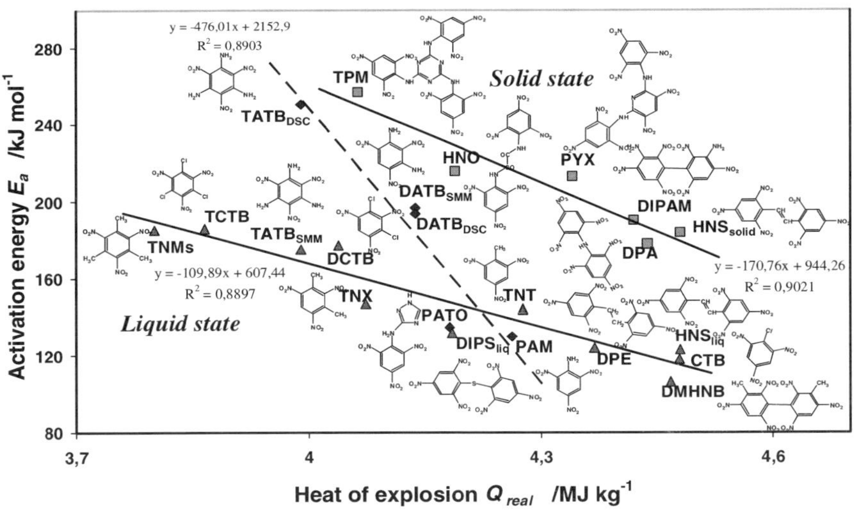

Fig. 20 Graphical representation of the modified E–P–S equation (Eq. 8) [9] for polynitro arenes with a hydrogen atom at the γ-position with respect to the nitro group (Scheme 2); exceptional are the polychlorinated derivatives of 1,3,5-trinitrobenzene—TNB (i. e. CTB, DCTB, and TCTB) whose thermal decomposition, however, represents a certain analogy with the decomposition of polymethyl derivatives of TNB [9]

The relationships in Fig. 21 exhibit very close molecular-structural correlations [172], but is very interesting in regard to HNIW molecules. According to this figure, in the series CPX-HOMO-α-HNIW the nitro groups at positions 2,6,8,12 of the last nitramine are seemingly "counted" twice; at these positions the $N-NO_2$ bonds are the longest and α-HNIW is more sensitive to stimuli ($E_d = 10.2$ J [173]; ignition temp. 220 °C; Zeman, 2006, personal communication) than ε-HNIW ($E_d = 11.9$ J [173]; ignition temp. 225 °C; Zeman, 2006, personal communication), in which similarly the less reactive positions 4 and 10 are seemingly counted twice.

Substitution of the activation energies E_a in Eq. 8 for highly thermostable polynitro arenes by drop energies results in the dependence presented in Fig. 22: here the increasing EM energy content corresponds to increasing sensitivity (Sect. 3.4) [160]. Equation 10 for net charges of primarily leaving nitro groups for the same group of polynitro arenes is presented in Fig. 23 [160].

Equation 10 with the E_a replaced by the slope, E_aR^{-1}, of the Kissinger relation (Eq. 7) was applied also to a study of explosive mixtures (Fig. 24); a thermal reactivity of several oxidizing systems of emulsion explosives and some commercial explosives with different nitrate ester content were specified in paper [153]. The main component of the oxidizing systems in all the studied materials was ammonium nitrate (AN). In Fig. 24 the line A corresponds to classical emulsion explosives W/O, the line A to W/O emulsions fortified by addition of high explosives in the amount of at least 30% w/w, and the line C to gelatinous explosives with the nitrate esters content be-

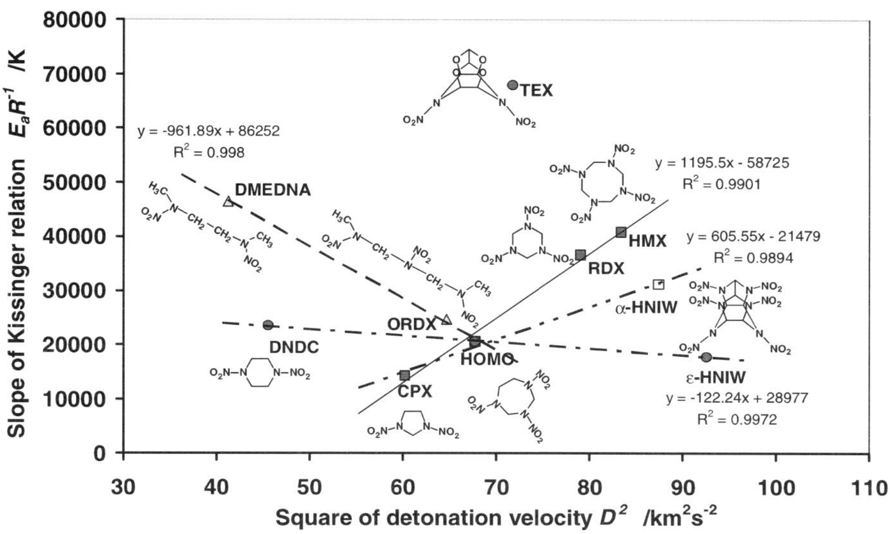

Fig. 21 Graphical representation of the modified E–P–S relation (Eq. 10) between the slope, E_aR^{-1}, of the Kissinger relationship (Eq. 7) and the square of the detonation velocity of nitramines [172]

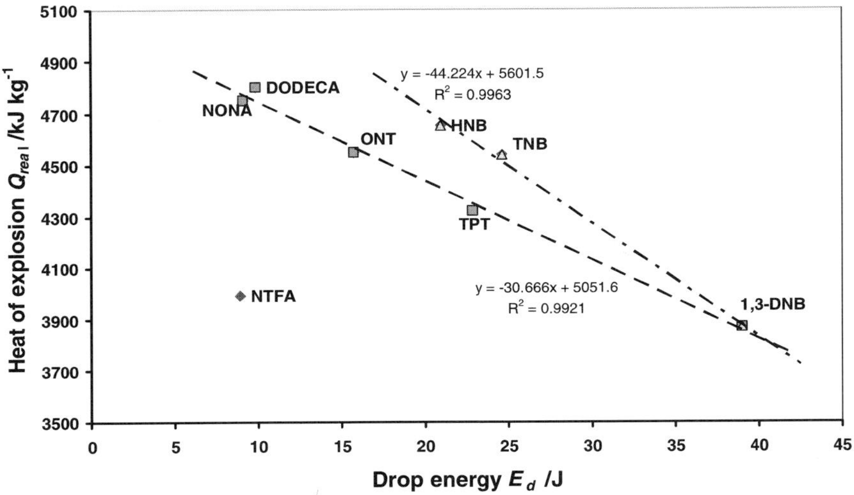

Fig. 22 Graphical representation of the E–P–S relation (Eq. 8) between the heat of explosion and the impact sensitivity of highly thermally stable polynitro arenes [160]

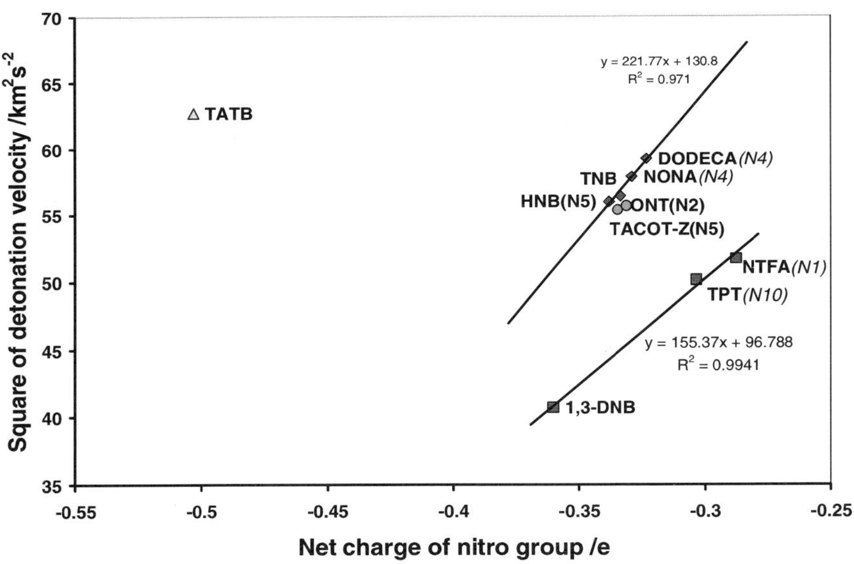

Fig. 23 Relationship between the square of the detonation velocity and the impact sensitivity of highly thermally stable polynitro arenes [160]. Data for TATB do not correlate due to discrepant primary fission (Scheme 2)

low 30% w/w. In particular, the group around line *B* indicates that a critical amount of admixtures of nitrate esters or nitramines exists for the mixtures (i.e. about 30% by wt.): the decomposition of explosive mixture is not influ-

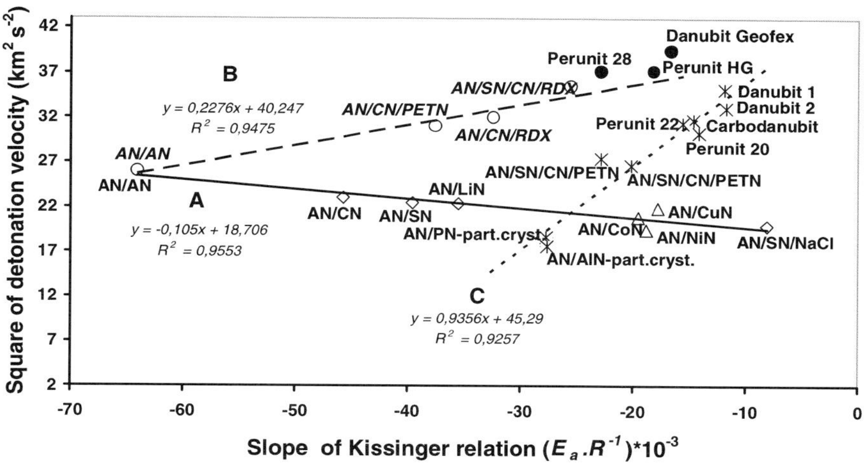

Fig. 24 Graphical representation of the E–P–S relation (Eq. 10) between the squares of the experimental detonation rates of explosive mixtures and the slopes, E_aR^{-1}, of Kissingers relationship (Eq. 7)—taken from [153]. *Black points* correspond to dynamites with a nitric ester content of over 30 wt %. Oxidizing systems of all of these explosives are based on ammonium nitrate (AN) and its mixtures with sodium nitrate (SN), potassium nitrate (PN), calcium nitrate (CN), aluminium nitrate (AlN), lithium nitrate (LiN), and sodium chloride–sodium nitrate (NaCl/SN), nickel(II) nitrate (NiN), copper(II) nitrate (CuN) and cobalt(II) nitrate (CoN)

enced by the physical state (i.e. solution or crystalline) of the oxidizing system of thus fortified mixtures only in the case when the content of the admixtures is above the critical value. With the line *B* also data of commercial gelatinous explosives correlate whose nitrate esters content exceeds of 30% (black points in Fig. 24).

This effect is analogous to the well-known influence of the shock pressure on initiation and propagation of the detonation in which the critical pressure plays a decisive role [101]. If the initiation pressure is lower than the critical value, then the initiation and growth of detonation themselves will depend on the micro-structure of the given charge (hot spots). If the pressure is higher than the critical value, then the initiation will depend only on the chemical composition of the charge [101]. The pressure during the detonation process is also given by the amount and reactivity of the molecular fragments which come into the reaction zone of the detonation wave. It was stated that in studied explosive mixtures the thermal reactivity of the oxidizing system and/or its mixture with a high explosive (for example: AN increases the thermal reactivity of RDX) replaces the primary thermal reactivity of explosophore groups in the individual energetic materials [153].

4.6
Electric Spark Sensitivity

Two different instruments are available for studies of electric spark sensitivity [6, 8, 61, 174–180] at the University of Pardubice: they differ by configuration of electrodes and structure of circuit [6, 8, 174–180]. Scheme 4 presents the spark gap of the older instrument with code RDAD; Scheme 5 depicts the newer one with code ESZ KTTV.

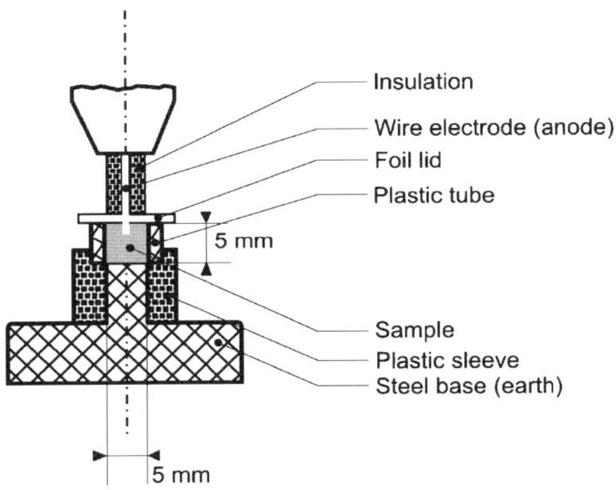

Scheme 4 A spark gap in the RDAD instrument [8, 174]

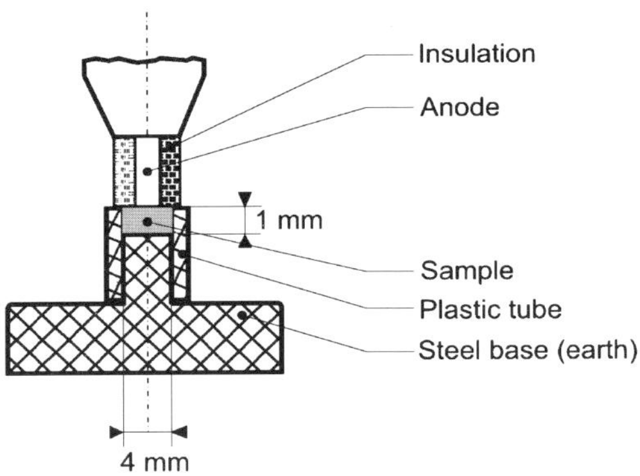

Scheme 5 A spark gap in the ESZ KTTV instrument [8]

The values of measurements carried out with the instrument RDAD were comparable with those of LANL in Los Alamos [174]. In the papers [175, 176] the electric spark sensitivities of 23 polynitro arenes and 13 nitramines were determined as the spark energy, E_{ES}, required for 50 per cent initiation probability. The relationship between the E_{ES} values and the squares of detonation velocities of the polynitro arenes resembles a certain form of Evans–Polanyi–Semenov equation (Eq. 10) (it is limited by molecular structure as documented by the example given in Fig. 25 [175]).

For nitramines relationships (Eq. 8) or (Eq. 10) are much more unambiguous [176] (their molecular structure is simpler). The relationship itself signalizes an identity of chemical mechanisms of the primary fragmentation processes in molecules of polynitro compounds during their detonation and in initiation of their decomposition and/or explosive transformation by action of electric spark. The relationship was used to predict the E_{ES} values of nitramines not yet synthesized, particularly 1-nitro-1-azaethylene (Digen), 1,3-dinitro-1,3-diazetidine (Tetrogen) and 1,3,5,7,9-pentanitro-1,3,5,7,9-pentazecane (Decagen). Application of the approach by Skiner et al. [61] (Sect. 2.4) to solving the relationship between electric spark sensitivity and thermal reactivity of individual EMs gave a relationship between $\ln E_{ES}$ and reciprocal values of temperatures for thermolysis rate constants $k = 10^3\ s^{-1}$ [177]. In contrast to paper [61], these relationships were strongly limited by molecular structure [177] (as in the case of the E–P–S equation). The conclusion was made [177] that both the chemical mechanism of primary thermolysis of energetic materials and intermolecular interaction factors in their crystals play decisive roles in their initiation by electric spark.

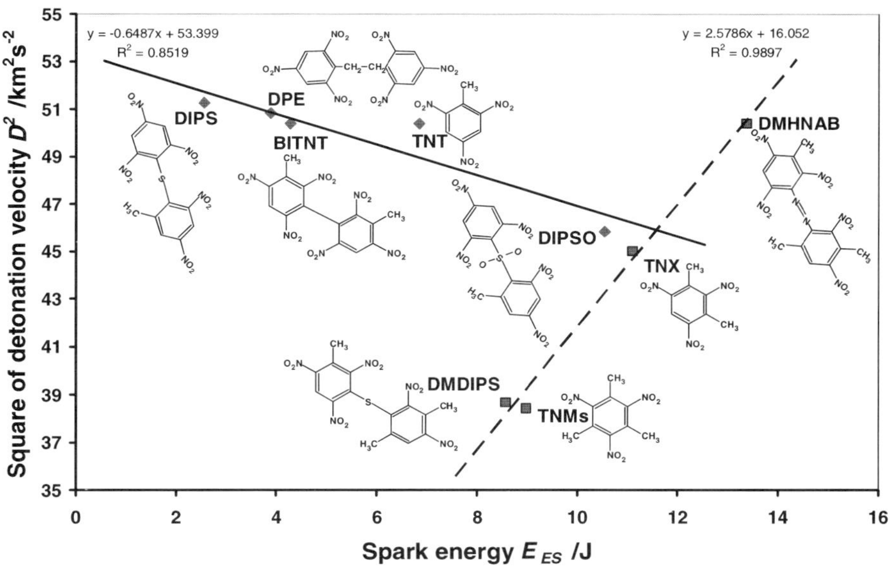

Fig. 25 Graphical representation of the E–P–S relation (Eq. 10) between the electric spark sensitivity and the square of the detonation velocity for compounds which thermolyze via a five- or six-membered transition state involving a direct participation of an *ortho*–nitro group (Schemes 2 and 3b) [175]

Attention was also focused on the relationship between the E_{ES} values and impact sensitivity, expressed as drop energies, E_d, of the first reaction. It was concluded that, depending on intermolecular interaction factors in crystals of energetic materials, the mechanism of impact energy transition to the reaction centre of their molecule can differ from that of transition of energy of electric spark [178]. In an attempt specify the reaction centre in molecules of polynitro arenes by means of the ^{13}C NMR chemical shifts of bearers of primarily leaving nitro groups logical dependences (Fig. 26) were observed especially in the case of derivatives whose primary reacting nitro group is conjugated with the rest of the molecule (Scheme 2) [6].

According to Fig. 26 it was confirmed that the most reactive nitro groups in the TMPM molecule are bound at the 2-position, those in the DMDIPS molecule at the 6-position, those in the BITNT molecule at the 2-position, and those in PYX at the 3(5)-position of the pyridine ring [6].

Xiao et al. [179], using the DFT-B3LYP6-31G* approach, re-evaluated the results of [175–177] for nitramines. They confirmed the linear relationship between D^2 and E_{ES} values and, in addition, specified another linear relationship between the logarithm of detonation pressure and E_{ES} values. Both relationships show, that increasing performance is connected with increasing electric spark sensitivity in technologically attractive nitramines [179].

In comparison with the instrument RDAD, the measurements carried out with the instrument ESZ KTTV give E_{ES} values lower by as much as two orders of magnitude. The outputs obtained from the two instruments for nitramines are compared in Fig. 27 [8].

Figure 27 indicates that the initiation mechanisms in the spark gaps of the two instruments are different. With regard to the considerably higher E_{ES}

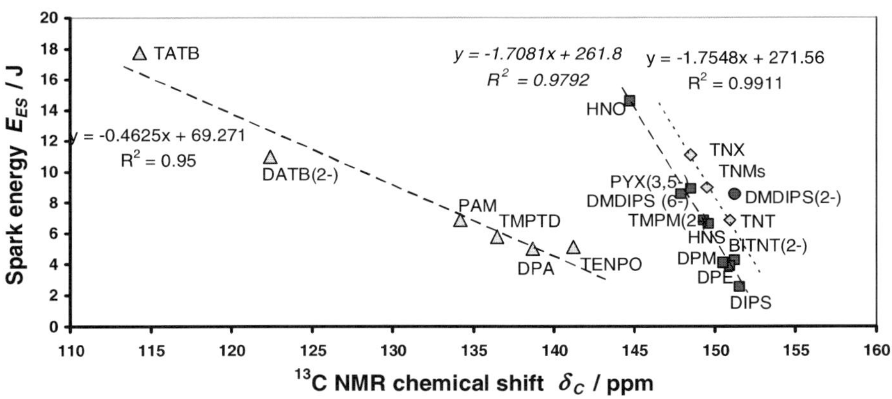

Fig. 26 Relationship between the spark energy, E_{ES}, and ^{13}C NMR chemical shifts, δ_C, of the carbon atoms of the aromatic system carrying the most reactive nitro groups (the *numbers in brackets* denote the positions in the molecule) in polynitro arenes containing amino, hydroxy and alkyl substituents [6]

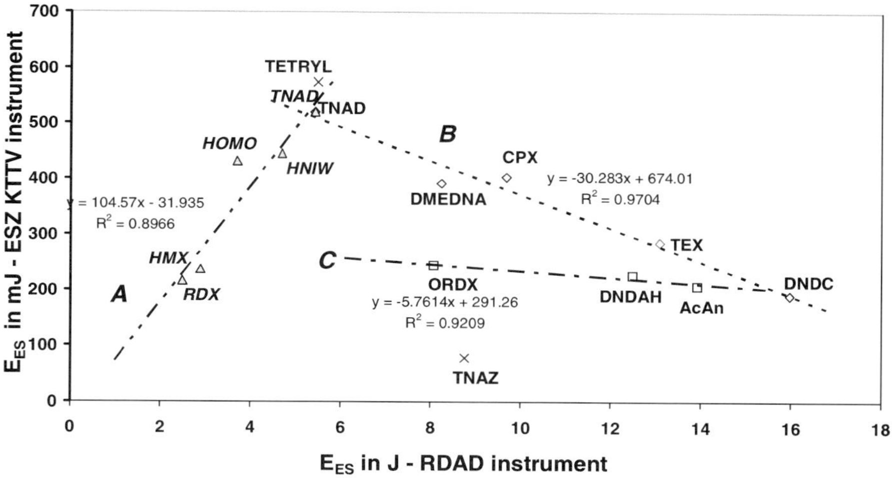

Fig. 27 Relationship between the outputs from ESZ KTTV and RDAD instruments (see also Schemes 4 and 5)

values observed with the instrument RDAD, a substantial part of discharge energy is realized as thermal energy, which can become engaged in thermal decomposition of the sample investigated. Different ways of energy transfer through the spark gap to the reaction centre of molecule are also indicated by the opposite trends of the dependences of $\ln E_{ES}$ on the dissociation energies of the weakest $N-NO_2$ bonds obtained from the two instruments [8]. The E_{ES} values obtained from the instrument ESZ KTTV logically correlate with ^{15}N NMR chemical shifts of nitrogen atoms of the primarily reacting nitro group and with the charges on these atoms [8]. Their relationship to the heat of fusion (Fig. 28) [8] is also interesting.

Figure 28 reminds us of analogous relationships between impact sensitivity (taken as the first reaction) and the $\Delta H_{m,tr}$ values of nitramines (Fig. 7). However, these relationships in Fig. 7 strongly differentiate cyclic and linear nitramines while according to Fig. 28 this differentiation is only slightly visible. This might mean that plastic deformations of the crystal should not have significance in the initiation by electric spark. But dislocations should be a reason of existence of the relationships in Fig. 28. This means that the results obtained from the measurements by means of the SEZ KTTV instrument should be more dependent on the shape and size of crystals than those in the case of RDAD. Satisfactory correlation of the data from the first-mentioned instrument with characteristics of nitramine molecular structure might be caused by the wide spectrum of their granulometry (given by the method of samples crystallization) which can lead to the "averaged" E_{ES} values [8]. In a continuing paper [180], samples of RDX and HMX were screened to obtain narrow fractions, and it was found that an increase in

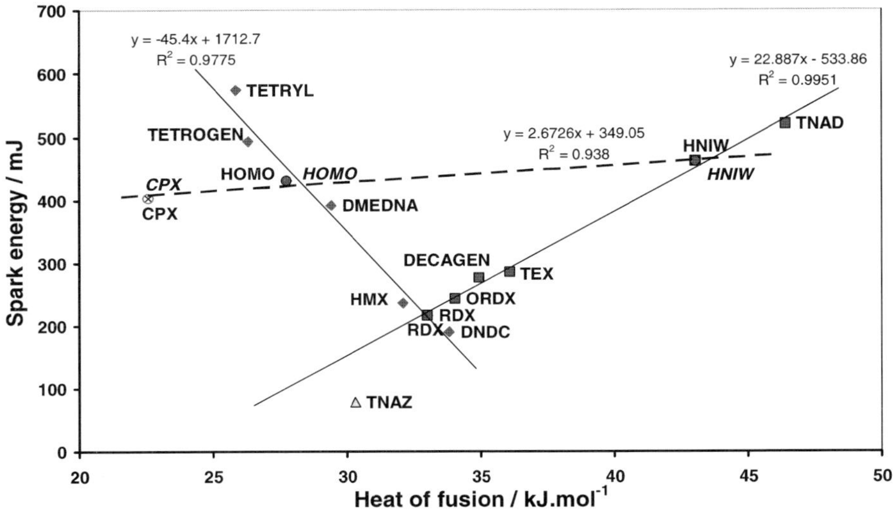

Fig. 28 Relationship between the E_{ES} values from the ESZ KTTV instrument and the heats of fusion of nitramines [8]

grain size of these nitramines is connected with an increase in E_{ES} values. This corresponds with the opinion of Auzanneau et al. [181] about the mechanism of spark energy transfer into the powdered reactive solid, i.e. in this

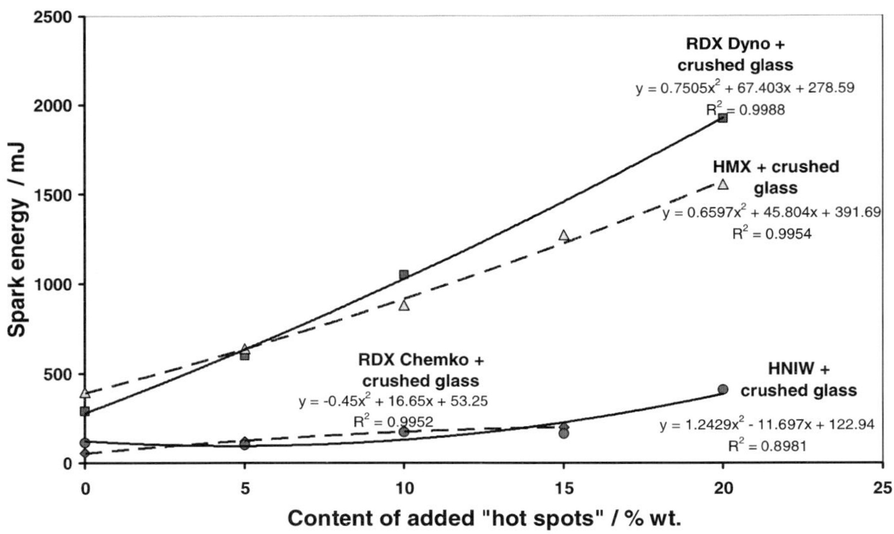

Fig. 29 Electric spark sensitivity of mixtures of the RDX, HMX and HNIW with crushed glass

case the decrease in the number of inter-grain points per volume unit. However, also dislocations in the crystals should have some effect on this type of EMs. This theory was also confirmed by the electric spark sensitivity of some nitramines with admixtures of fine particles of hard additives (crushed glass, diamond powder [180]). The results are documented, for example, by Fig. 29.

Artificial introduction of fine and hard particles into crystalline nitramines decreases the electric spark sensitivity of the resulting mixtures, i.e. the particles behave as a flegmatizing additive, here the number of inter-grain contact points of the nitramine grains is decreased by incorporation of the foreign particles [180].

4.7 Application of Polarography

The electrochemical reduction of the polynitro arenes in aqueous medium (concentration less than 10^{-4} M) at pH 7 gives the values of resulting half-wave potentials, $E_{1/2}$, which markedly depend inter alia on the solvation of these compounds [159]. However, it is possible to find logical linear dependences between the $E_{1/2}$ values and squares of detonation velocities of polynitro arenes. These relationships are another form of modified Evans–Polanyi–Semenov equation (Eq. 10) for energetic materials. The strong dependence of $E_{1/2}$ values on the conditions and way of carrying out the polarographic measurements, along with the very limited solubility of the polynitro compounds studied in media containing water, considerably restrict the application of polarography to studies of chemical micro-mechanism of initiation of organic energetic materials [159].

4.8 Inequality of Nitramine and Nitro Groupings in Molecules of EMs

From what has been written so far it follows that the initiation of nitramines by impact should be dominated by the key role of the aza-atoms carrying the primarily leaving nitro groups [6, 7]. The dominant factor in the initiation by shock, electric spark, and in low-temperature thermolysis should be the electronic structure and surroundings of the primarily leaving nitro group [6, 7]. For nitramine molecules, whose nitrogen atoms are not isochronous, the said fact is documented in Scheme 6.

In the case of the 2,4,6,8-tetranitro-2,4,6,8-tetraazanonane (OHMX) molecule, the primary reactivity of inner nitramino groups (positions 4, 6) was confirmed by Kohno et al. [163] using molecular-dynamic simulation. Their paper tends to be quoted without mentioning one of its important merits, i.e. the selectivity in the initiation reactivity of nitramino groups [163]: "in connection with the action of impact or shock on a nitramine crystal, the in-

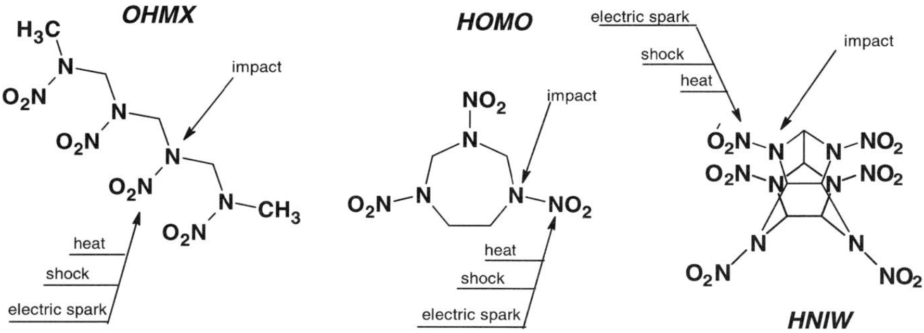

Scheme 6 Summary of the findings concerning initiation reactivities, taken from [7], can be presented as follows: the dominating reactivity of the "inner" nitramino groups of the OHMX molecule in impact and shock was confirmed by molecular dynamics simulation [163]; the most reactive nitramino group in the HOMO molecule is that at the 1-position and that in HNIW at position 2 (this is the longest N – N bond in the molecule of the ϵ-polymorph—see [7])

tramolecular vibration energy is transferred into nitramino groups. There is a difference between linear and cyclic nitramines in transfer of the energy". According to Kohno et al. [42], the linear OHMX molecule shows only low probability of transfer of surplus energy from the inner to outer nitramino groups. The longest N – N bond in ε-HNIW is that of the nitramino group at the 2-position of its molecule [163] (143.6 nm, the single bond N – N length being 142.5 nm [173]). Therefore, this bond should be the first to undergo homolysis in initiation processes. This statement is supported by MO calculations of these compounds and here also by the findings presented in Figs. 1, 3, 10 and 16.

A similar situation is encountered with polynitro arenes. Apart from the MO calculations, the inequality of nitro groups in the molecule is also indicated especially in Figs 2, 4, 8a, 8b, 12, 14, 17, 23 and 26. All these inequalities are potential sources of initiation, i.e. one of the causes of sensitivity of individual EMs to stimuli.

4.9 Several Reaction Centres in a Molecule

The complexity of molecular structure of polynitro arenes (aside from nitramines) makes the problem of their primary fission somewhat complicated: if a molecule of these compounds contains several types of substituents, it can contain several potential reaction centres [6]. This likelihood was already mentioned in Sect. 4.3.1 (Scheme 8a,b), Sect. 4.3.2 (here Fig. 12) and Sect. 4.6 (here Fig. 26). Two potential reaction centres exist in the molecules of DMDIPS (Schemes 3a and 3b), PYX and TMPM. These possibilities are shown in Scheme 7 for the PYX molecule.

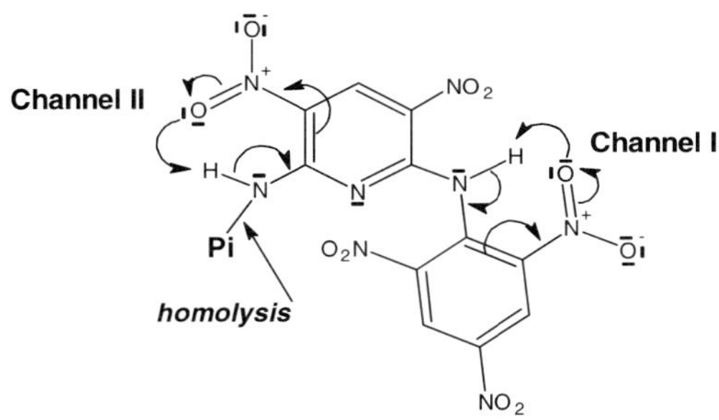

Scheme 7 Presumed reaction mechanisms (centres) of primary fission in the initiation process of the *N*,*N*-bis(2,4,6-trinitrophenyl)3,5-dinitropyridine-2,6-diamine (PYX) molecule (taken from [165, 183]); here Pi- is 2,4,6-trinitrophenyl whose break away (a secondary process) as a radical and subsequent reaction with other fragments of the PYX molecule gives DPA (see Scheme 9)

More probable pathways of initiation of the PYX molecule were estimated [165] from considerations involving the following relationships: (1) between impact sensitivity and the ^{13}C NMR chemical shifts of some polynitro arenes (Fig. 12); (2) between electric spark sensitivity and these shifts (Fig. 26), (3) between Mulliken charges on nitrogen atoms of the primarily reacting nitro group and onsets of thermal decompositions from differential thermal analyses of the said compounds; and, (4) with computations obtained using the DFT-B3LYP/3-21G methods in the GAUSSIAN 98/03 program. It was found that both reaction centres in the PYX molecule probably participate in the first fission caused by initiation stimuli although primary participation of the picryl nitro group in position 2 (together with the amino bridge) in this process seems to be more probable (i.e. Channel I) [165].

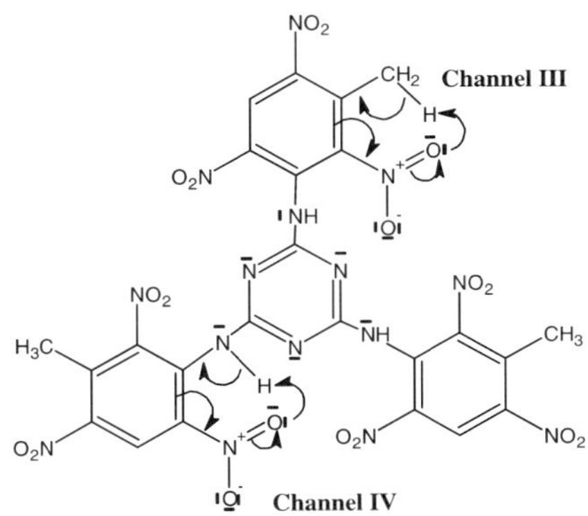

Scheme 8 Presumed reaction mechanisms (centres) of primary fission in initiation processes of the *N*,′,*N*″-*tris*(3-methyl-2,4,6-trinitrophenyl)-[1,3,5]triazine-2,4,6-triamine (TMPM) molecule (channel III is preferred) [165]

The same approach to the TMPM molecule (Scheme 8), i.e. taking into account the data in Figs. 5 and 26 with application of the same computation apparatus showed [165] that primary participation of the nitro group in position 2 together with the methyl group in position 3 during initiation processes in this molecule should be a unique reaction centre (i.e. only Channel III).

4.10
Comparison of Splitting of Polynitro Arenes by Heat and by Shock

In order to verify the hypothesis of the identity of the primary fragmentation reactions in low-temperature thermolysis and detonation reaction of EMs, a group of polynitro arenes [182, 183] (with a presumed mechanism of primary decomposition such as that given in Scheme 2) was submitted to a comparison study of splitting by heat and shock wave. The chromatographic records of TNT after thermal exposition (Fig. 30a) and after exposure to shock wave (Fig. 30b) and its post-detonation residues (Fig. 30c) show [182] that the main intermediates identified in the decomposition of TNT by shock and heat are identical. 4,6-Dinitro-2,1-benzoisoxazole and 2,4,6-trinitrobenzaldehyde are the most reactive. The first finding excellently agrees with the experimental results obtained recently by Bulusu and Autera [185] in the field of application of DKIE to the study of initiation of TNT. The results, as well as those found by Bulusu, signalize that the chemical micro-mechanism of the primary fragmentations of shock-exposed TNT molecules and/or its detonation transformation should be the same as in the case of their low-temperature thermal decomposition. The formation of products on the basis of reaction of the methyl group in TNT can be connected with the action of the hydroxyl radical (Scheme 2) and nitrous gases upon the TNT itself [11]; these reactions are in progress namely during thermal decomposition of TNT and in the phase of extinguishing of its detonation.

In the second group of substances, the same investigation method was applied to *N*,*N*′-bis(2,4,6-trinitrophenyl)-3,5-dinitropyridine-2,6-diamine (PYX), 2,4,6-trinitroaniline (PAM), 2,4,6-trinitro-*N*-(2,4,6-trinitrophenyl)aniline (DPA), and *N*,*N*′,*N*″-tris(2,4,6-trinitrophenyl)-1,3,5-triazine-2,4,6-triamine (TPM) [183]. Figures 31a and 31b present the results obtained from the investigation of PYX, which also seem to confirm the suggested splitting process in Scheme 6.

Fig. 30 **a** LC-UV chromatogram of thermally exposed TNT at 316 °C [182]; **b** LC-UV chromatogram of shock-exposed TNT at the limit of its initiation ability [182]; **c** LC-UV chromatogram of residues after the TNT detonation [182]; here DNAt is 4,6-dinitro-2,1-benzoisoxazole, TNB is 1,3,5-trinitrobenzene, TNBAld 2,4,6-trinitrobenzaldehyde, TNBAlc 2,4,6-trinitrobenzylalcohol and TNBA 2,4,6-trinitrobenzoic acid ▶

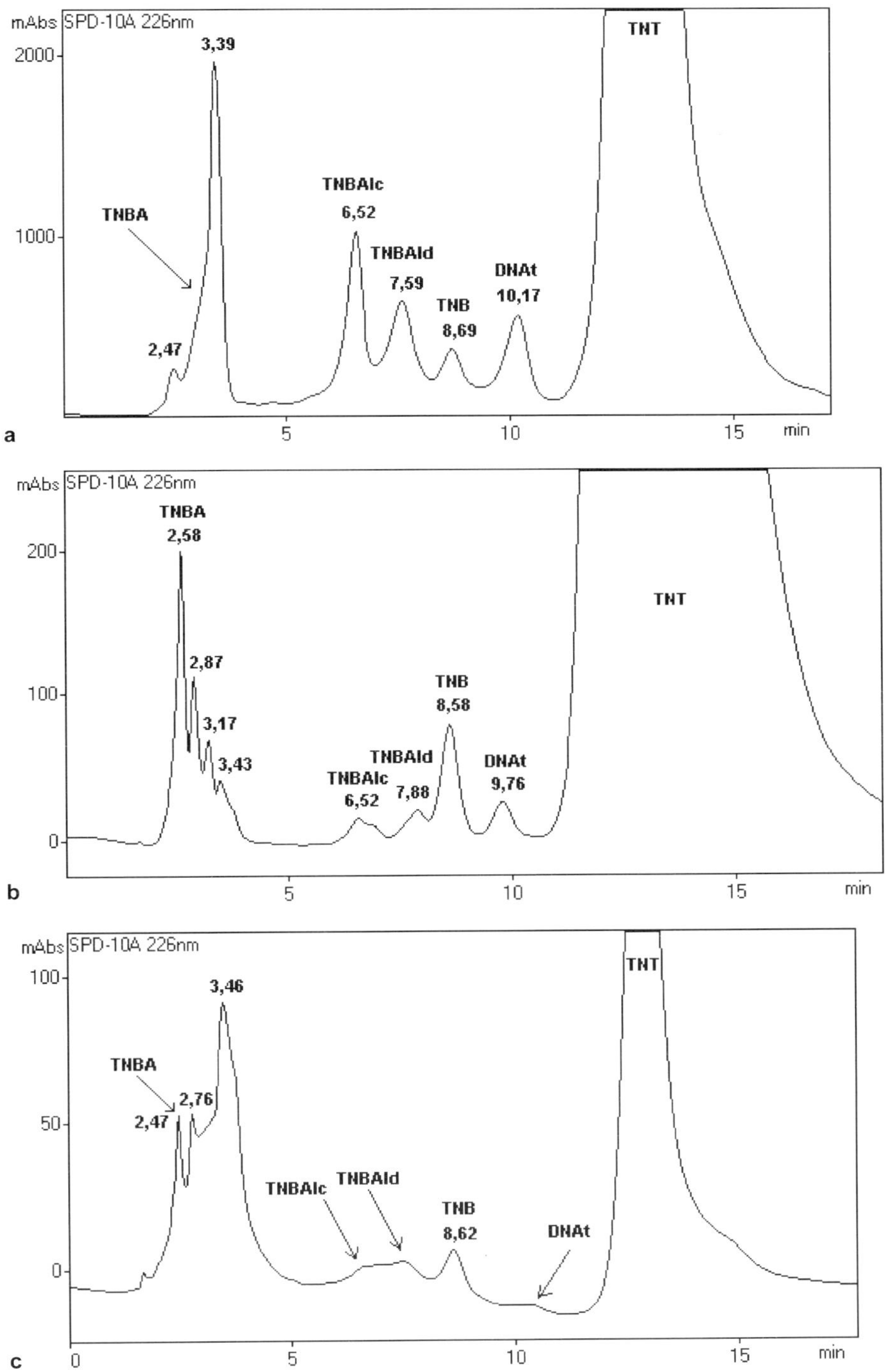
mAbs SPD-10A 226nm
2000
1000
3,39
TNBA
TNBAlc
6,52
TNBAld
7,59
TNB
8,69
DNAt
10,17
2,47
TNT
5
10
15
min
a
mAbs SPD-10A 226nm
200
100
0
TNBA
2,58
2,87
3,17
3,43
TNBAlc
6,52
TNBAld
7,88
TNB
8,58
DNAt
9,76
TNT
5
10
15
min
b
mAbs SPD-10A 226nm
100
50
0
3,46
TNBA
2,47
2,76
TNBAlc
TNBAld
TNB
8,62
DNAt
TNT
0
5
10
15
min
c

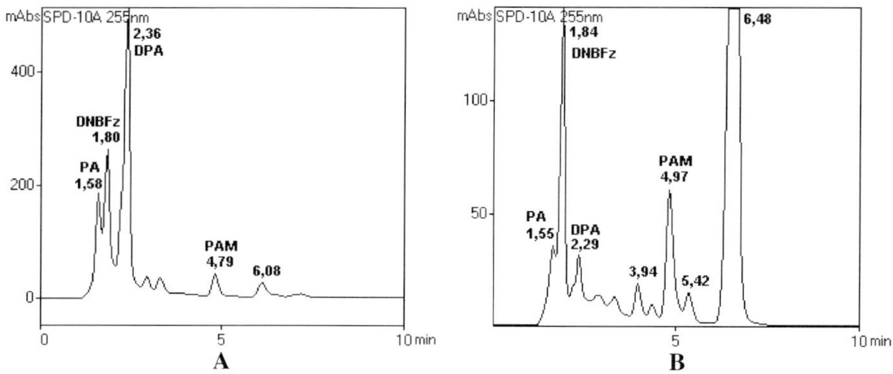

Fig. 31 LC-UV chromatograms of thermally exposed PYX at 367 °C (**A**) and of shock-exposed PYX (**B**) [183]; here the following intermediates are: PA 2,4,6-trinitrophenol, PAM 2,4,6-trinitroaniline, DPA N-(2,4,6-trinitrophenyl)-2,4,6-trinitroaniline and DNBz 4,6-dinitrobenzofurazane

On the basis of the results obtained in the paper [183], it is possible to present the mutual relationship between the main stable products from incomplete initiation of PAM, DPA, PYX and TPM by heat and shock in Scheme 9.

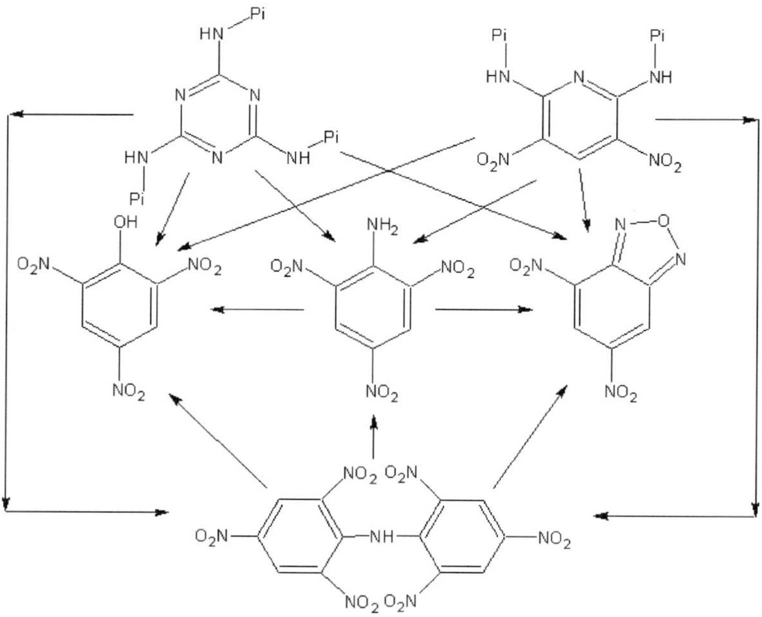

Scheme 9 Mutual relationship of the main products of the incomplete initiation of PAM, DAP, PYX, and TPM by heat and shock; here Pi- is 2,4,6-trinitrophenyl [183]

4.11 Survey of Results from Application of Physical Organic Chemistry (Outputs of POC Model)

A common feature of impact, shock, electric spark and thermal sensitivities of organic polynitro compounds is the reactivity proper of $C-NO_2$, $N-NO_2$ and $O-NO_2$ groupings in their molecules and/or molecular crystals. Besides the relationships between the characteristics of the said sensitivities [6–9, 16, 89, 118, 145, 146, 148, 153, 155, 159, 160, 169, 170, 172, 175–178] it is also possible to study the micro-mechanisms of their primary chemical processes with help of the ^{15}N and ^{13}C NMR chemical shifts of key atoms in reaction centres of their molecules [6, 7, 9, 98, 147, 150, 165] with the help of charges on nitrogen atoms of the primarily reacting nitro groups [7, 12, 152, 160, 161, 172] and/or of net charges of these nitro groups [8, 90, 91, 160]. Application of polarography to such studies is limited by solvation effects of the polynitro compounds studied in aqueous solutions [159]. So far the NMR chemical shifts have appeared to correspond best to the real electron configuration and steric conditions of the key atoms in the reaction centre of the given molecule. From the point-of-view of physics of explosion these atoms can be taken as hot spots [7, 172]. In the case of technologically attractive explosive mixtures with the oxidation system based on ammonium nitrate, these chemical hot spots represent a part of this system (or its mixtures with high explosives, as the case may be) [153].

The relationships between detonation characteristics of energetic materials and characteristics of their low-temperature decomposition and other characteristics of their mechanical and electric spark sensitivities [6–9, 16, 60, 89, 98, 118, 145, 146, 148, 153, 155, 159, 169, 170, 172, 175–178] means that:

- the primary fragmentation processes of molecules of energetic materials in the low-temperature thermal decomposition should be identical with those in the impact, electric spark and shock initiations, which, in the case of shock, can also be documented by experimental results from initiation of TNT [182] and some of *N*-(2,4,6-trinitrophenyl)-substituted polynitrated aminoarenes [183];
- the effect of temperature (i.e. *thermal decomposition*) in the classic sense is not applicable to the process of detonation initiation by shock, impact or electric spark;
- the primary fragmentation of EMs in their detonation transformation should proceed at milder conditions than those present at the front of the detonation wave or in its reaction zone, which means that the detonation transformation itself of the given substance should be preceded by an induction period (whose necessity is, however, also considered in Non-Equilibrium Zeldovich–von Neuman–Döring theory of self-sustaining detonation [103, 104]).

Electron configuration and steric effects at C or N-atoms, which are bearers of the primarily reacting nitro groups in a molecule, should play a key role in the initiation by impact. The dominant factor in the initiation by shock, electric spark, and in low-temperature thermolysis should be the electron structure and close neighbourhood of nitrogen atoms of the primarily leaving nitro group [6, 7].

The results mentioned are best demonstrated on secondary nitramines [7], which in their molecular structure are relatively simple polynitro compounds, and the mechanism of primary homolysis of their molecules is well understood [11, 14] (Scheme 1). Polynitro arenes, on the other hand, have a more complex structure and intramolecular effects in their molecules: here the mesomeric, inductive and steric effects on reactivity operate simultaneously. This fact makes the problem of their primary fragmentation somewhat complicated too [6, 9] (Schemes 2, 3a, 3b, 7 and 8). If a molecule of these compounds contains several types of substituents, it can contain several potential reaction centres (e.g., the PYX and TMPM molecules, see Schemes 7 and 8). The initiation proper can then be realized by the molecule simultaneously participating by several centres or always by a single centre in a given type of initiation (the initiation of PYX by impact or shock versus its initiation by electric spark, see Figs. 12 and 26) [6].

As for the electric spark sensitivity, which has also been assessed from the standpoint of development of the respective instruments [8], this characteristic is also dependent upon the configuration of electrodes and structure of the circuit, which complicates the specification of energy transfer from the electric spark to the reaction centre of the molecule [8]. In instruments exhibiting large energy loss during electric discharge the thermal principle of initiation predominates [6, 8, 61]. On the other hand, instruments with minimized energy losses still need to be investigated from the point-of-view of the energy transfer. So far, the operation of hot spots in this initiation has been confirmed (i.e. the views of Auzanneau et al. [181]), i.e. the role of inter-grain contact points of the EM grains [8]. On the basis of this view it is understandable that addition of hard fine particles to EMs has a desensitizing effect in this type of initiation [8] (Fig. 29), while the same additives are well known to have a strong sensitizing effect in the initiation of EMs by mechanic stimuli.

5
Comments

It has already been mentioned that the study of energetic systems by theoretical methods has accelerated dramatically over the past 16 years. This study has provided a considerable insight into the understanding of some factors affecting their behaviour, but, on the other hand, it also has brought a great deal of opacity and guesswork. The main problem is that quantum chemical calculations and simulations start from certain opinions held by their authors, and the results correspond to this fact.

The discrepancy between some starting ideas of pro quantum chemical simulation of the thermal decomposition and practical findings from this area of chemistry of polynitro compounds is pointed out in Sect. 2.1.1 for concrete cases. It is illogical, for example, to consider the dissociation energy of $C-NO_2$ to be a stability criterion, if experimental evidence proves the participation of an oxygen atom of the nitro group in the primary fragmentation process (Schemes 2, 7–9 but also [11] and [166]). What has been said is also relevant to the trends that try to describe molecular-structure dependences of sensitivity connected with various mechanisms of primary fragmentation by means of a universal relationship, as mentioned in Sects. 3.2.2 and also 3.3.4. The large variety of intramolecular and thus connected intermolecular interactions, especially in polynitro arenes and azaarenes, makes these dependences relatively varied (in particular, see Sect. 4.5). Therefore, it should be stated that, for example, the well-known correlations by Politzer and Murray (Sect. 3.2.2 and [8] and [81]) must be restricted to specific classes (i.e. nitroaromatics, nitroheterocycles and nitramines). This is also the reason why Price and Hare [85] encounter problems when trying to describe by a single relationship the sensitivities of a larger set of structurally heterogeneous individual EMs.

Whereas in the area of simulation of high-temperature thermal decomposition of RDX and HMX there are no distinct differences between individual authors' results (Sect. 2.2), in the area of simulation of the hydrostatic compression effect again some of the outputs disagree with practical findings. Thus, for example, the author of paper [60], dealing with uniform and uniaxial compression of the nitromethane crystal, calls attention to the distinct difference between his results and those by Kuklja and Kunz [59]. It is axiomatic because dynamic pressure is under consideration in the paper [60]. However, the latter authors using simulation of an effect of a hydrostatic compression on the electronic structure of RDX, obtained very good agreement with both experimental results from the action of static pressure on this nitramine [61] (Sect. 2.5) and the findings about primary fragmentation of this molecule (homolysis of the $N-NO_2$ bond [11, 14]). The heterogeneity of views of the fragmentation mechanism under extreme conditions is well documented by the example of initiation of nitromethane (Sect. 3.3.9), which is a favoured model nitro compound in this field.

Studies of the sensitivity of EMs are mostly directed to impact and shock sensitivities (Sect. 3). A primary fragmentation of the given EM until adiabatic compression of a thin layer of its molecules by impact or shock wave is the unambiguously dominant view in this case. Physical Kinetics Model [97] was a typical representative of this view of mechanical splitting of the molecule at any deliberate bonds between its atoms by shock wave. One of the best developed models is Dlott's Model of the Multiphonon

Up-Pumping [1, 74]. As already mentioned in Sect. 3.1, the author consulted a number of specialists in advance about his ideas. However, in this way he introduced into his model the ideas of primary fragmentation of EMs under extreme conditions. Nevertheless, he enriched the theory of initiation of explosive transformations of EMs by introducing significant ideas about transfer of initiation impulse to molecules in the crystal lattice of these materials. Experimental verification of this model is absent to date. In Dlott's opinion [1, 74] the initiation by impact is a particular case of initiation by mechanical impulses (Low Velocity Initiation). Also [105] (Sect. 3.3.4) clearly documents the difference between the initiations by impact and by shock, though the authors do not call attention to this fact. Both the cited facts agree with study results of initiation reactivity of polynitro compounds by means of physical organic chemistry (Sects. 4.7 and 4.11), where the initiation by impact is treated separately from the set of initiations by shock, electric spark and heat.

Important to the study of sensitivity of individual EMs are the findings obtained by Politzer and Murray [77–82]. Their surface electrostatic potential sharply differentiates between organic compounds of explosive nature and the non-explosive ones. Politzer and Murray's approach [81] delimitates the part of the EM molecule in which primary fragmentation can take place by action of impulse. A similar result was also obtained by Rice and Hare [85], who improved the method devised by Politzer and Murray. Both of these approaches give results that agree relatively well with the outputs from the POC model; however, the last mentioned method provides much more detailed information. It is a disadvantage that the Politzer and Murray and Rice and Hare approaches do not consider experimental findings from kinetic studies and the mechanism of low-temperature thermolysis of individual EMs [9, 11, 14, 151].

Interesting is the Multidimensional Reactive Flow model developed by Tarver et al. [5, 103, 104]; it is based on the Non-Equilibrium Zeldovich–von Neuman–Döring theory. This model starts from the primary chemical changes occurring in the adiabatically compressed thin layer of molecules of the given EM and multiphonon up-pumping due to shock, but in the mathematical description it works with experimental data of thermal explosion of EM [5, 103, 104]; it considers the induction period of initiation of detonation. However, the induction period of the EM decomposition in front of the detonation wave makes the front kinetically unstable and pulsating [101].

The results of the Excitonic model by Kuklja and Kunz [15, 59, 65–67] are in concordance with experimental findings obtained by low-temperature decomposition of polynitro compounds. The outputs from this model also correspond with results from the application of physical organic chemistry to the study of initiation reactivity of EMs (POC model). Both these models are also supported by the results obtained by Raikova and Likholatov [118], who found that the Arrhenius parameters in both the detonation transformation and low-temperature thermolysis of nitrate esters have virtually the same values (Sect. 3.3.8). Kuklja and Kunz also significantly contributed to studies of pre-explosion states of heavy metal azides (Sect. 2.3). This problem is intensively dealt with by authors from Kemerovo State University [43–49] on the basis of an idea about electronic excitations facilitated by edge dislocations in explosive solids which was obtained from the application of the approach of solid-state physics to the study of this pre-explosion state (Sect. 2.3). However, there are also critics of this approach; see [50, 52]. Kuklja and Kunz are also successful when simulating the effect of dislocations upon the fragmentation (sensitivity) of RDX and PETN (Sect. 2.6); also in this case, their results correspond with those in [11, 14]). It should be noted that also Kuklja's simulation of thermal decomposition of RDX in the solid state [15] agrees well with the published findings of Russian authors [11, 14]. According to the opinion expressed by Fried et al. [3], the exited electronic states are not accessible in perfect crystals under realistic pressures and temperatures, but may still be accessed through defect or other energy localization mechanisms.

Good accordance with Excitonic and POC models from the point-of-view of the primary expirable bond and knowledge about the mechanism of low-temperature thermolysis of polynitro compounds [9, 11, 14] is achieved in the new methods of specifying reaction centres of initiation in molecules of polynitro compounds by means of the net charges of primarily reacting nitro groups [90, 91, 160], charges on nitrogen atoms of these nitro groups [7, 152, 160, 161], but mainly by means of ^{13}C and ^{15}N NMR chemical shifts of key atoms of these reaction centres [6–8, 98, 147, 148, 150, 165]; however, applications of the ^{15}N NMR spectroscopy are limited, because its present standard does not allow differentiation between individual nitrogen atoms in high nitrogen EMs, such as, for example, in azapentalene cycles of TACOT-Z. Also the results obtained by using the above-mentioned charges can slightly differ from reality (see the note in Sect. 3.2.3), hence also from the results of application of NMR chemical shifts [160], because these charge calculations are provided by total optimization of the equilibrium geometry of an isolated molecule in the gas phase at 0 K. Hence, ^{15}N NMR spectroscopy will hardly be applicable to the study of initiation reactivity of high nitrogen EMs. So far in this field, successful studies have been developed on the basis of investigation of their thermal decomposition [11, 24] and quantum chemical calculations [87]. It was found for tetrazole derivatives that the relative magnitude of their sensitivity is determined mainly by the activation energy of the opening ring in the thermal decomposition [87, 184].

Evaluation of the initiation mechanism of EMs on the basis of monitoring of products of their incomplete detonation transformation has its pitfalls. Action of impact or shock on EM brings about primary fragmentation of its molecule; due to the pressure drop of primary impulse (i.e. during extinguishing of this transformation) the products of these primary reactions are not further decomposed, and they end up as stable compounds. Thus, for example, in the case of TNT the primary stable intermediate is 4,6-dinitro-2,1-benzoisoxazole [11], which is formed in the sense of Scheme 2 in the extinguishing of the process or in low-temperature decomposition; the other products in Fig. 30a–c result from reaction of the OH-radical and nitrous gases with the methyl group in the starting TNT molecule during detonation extinguishing and/or low-temperature decomposition. For these reasons it is impossible to make conclusions about the multi-molecular course of initiation of nitromethane in the sense of experiments described in [126, 127] (Sect. 3.3.9)—in this case the process is finalization of primary fragments of NM.

According to the POC model, primary fragmentation of EM should take place before arrival of the detonation wave front in its detonation, which might be caused by electron excitation in the sense of the model by Kuklja and Kunz. In this connection also the finding by Price et al. [62] should be mentioned that the metallization of explosive due to compression cannot be dismissed as the initiating step in the sequence of reactions leading to detonation (physicists of explosion do not take account of metallization [101, 129, 138] because generation of this state during detonation of the convectional explosives is not possible).

6 Conclusions

Study of the sensitivity of energetic materials is inseparably connected with the understanding of the primary chemical processes of their initiation. Over the last nine years the study of problems of this sensitivity (i.e. initiation reactivity) has accelerated thanks to theoretical methods based on quantum chemistry. Impact and shock sensitivities have been the centre of interest. A beginning of chemical reaction in EM after its exposure to impact or shock (i.e. at high temperatures and pressures) has been a fundamental idea of this interest. Therewith connected are activities in the field of simulation of behaviour of individual EMs, most frequently nitromethane, RDX, HMX and sporadically TATB, at high temperatures and pressures. However, some outputs of such simulations for high pressures are contradictory. Rather confusing are the attempts to find relationships between dissociation energies especially of $C-NO_2$ bonds and the sensitivity parameters regardless of whether the primary process here is the homolysis of this bond or only participation of the oxygen atom of the nitro group in this reaction (this set of compounds tends to include also substances with $N-NO_2$ and $O-NO_2$ bonds). Particularly varied are the views of the chemical mechanism of fragmentation of nitromethane, which represents a favoured model EM in the field of physics and chemistry of explosion.

One of the best-developed models of sensitivity (initiation) of EMs is Dlott's Model of the Multiphonon Up-Pumping, which starts from the above-mentioned primary fragmentation of EM at high pressures and temperatures and presents a complex idea of impulse energy transfer into the EM molecule proper. However, exemplification of this model by experimental results does not exist yet. Significant is the Multidimensional Reactive Flow model based on the Non-Equilibrium Zeldovich–von Neuman–Döring theory, which within the mathematical description works with experimental data of thermal explosion and, therefore, considers the induction period of the process.

Sharp distinction between organic energetic materials and other compounds, using the surface electrostatic potential defined by Politzer and Murray, provided new possibilities of study of sensitivity of EMs from the standpoint of molecular structure and specification of the molecular moiety in which primary fragmentation could take place during initiation. The same possibility is offered by an innovation of the approach of the mentioned authors developed by Rice and Hare. More concrete and more realistic, from the point-of-view of both the primary fragmentation mechanism and the way of energy transfer to the reaction centre of the molecule, is the Excitonic model by Kuklja and Kunz. With regard to the primary fragmentation mechanism, the outputs of this model agree very well with experimental findings about primary fragmentation of polynitro compounds in low-temperature thermolysis which were mostly obtained by Russian authors.

The analysis of initiation reactivity of EMs from the standpoint of physical organic chemistry (POC model) showed that the primary fragmentation processes of molecules of energetic materials in the low-temperature thermal decomposition should be identical with those in the impact, electric spark and shock initiations. This means that the primary fragmentation of EMs in their detonation transformation proceeds at milder conditions than those present at the front of the detonation wave or in its reaction zone, which means that the detonation transformation itself of the given substance should be preceded by an induction period (already during the EM compression — before the front of detonation). As for the primary fragmentation mechanism of EMs by action of initiation stimuli, there exists accordance between the Excitonic and POC models. It would be desirable to evaluate the POC model also from the point-of-view of present knowledge of the physics of explosion (i.e. in the sense of [101]).

Acknowledgements The author is indebted to Prof. Zdenìk Friedl from the Chemical Faculty of the Brno University of Technology for his participation in solving some partial problems of the initiation reactivity of EMs, to Assoc. Prof. Pavel Vávra from the Institute of Energetic Materials and to Assoc. Prof. Josef Panchártek from the Department of Organic Chemistry, both from the University of Pardubice, for their valuable remarks on this paper. This material is based upon work supported partly by the Ministry of Industry and Trade of the Czech Republic as part of its Research project STRATECH No. FC-M2/05/00 and partly by the Ministry of Education, Youth and Sports of the Czech Republic as part of its research project No. MSM 0021627501.

References

1. Dlott DD (2005) Multi-phonon up-pumping in energetic materials. In: Shaw RW, Brill TB, Thompson DL (eds) Overviews of recent research on energetic materials. Adv Ser Phys Chem, vol 16. World Sci Publ Co Ptc Ltd., Singapore, p 303
2. Politzer P, Murray J (eds) (2003) Theor Comp Chem, vol 13. Energetic materials, Part 2, Detonation, combustion. Elsevier, Amsterdam
3. Fried LE, Manaa MR (2005) Modeling the reactions of energetic materials in the condensed phase. In: Shaw RW, Brill TB, Thompson DL (eds) Overviews of recent research on energetic materials. Adv Ser Phys Chem, vol 16. World Sci Publ Co Ptc Ltd., Singapore, p 275
4. Rice BM (2005) Application of theoretical chemistry in assesing energetic materials for performance and sensitivity. In: Shaw RW, Brill TB, Thompson DL (eds) Overviews of recent research on energetic materials. Adv Ser Phys Chem, vol 16. World Sci Publ Co Ptc Ltd., Singapore, p 335
5. Tarver CM, Manaa RM (2004) Chemistry of detonation waves in condensed phase explosives. In: Manaa RM (ed) Chemistry at extreme conditions. Elsevier, Amsterdam, p 495
6. Zeman S (2003) A study of chemical micro-mechanisms of initiation of organic polynitro compounds. In: Politzer P, Murray J (eds) Theor Comp Chemistry, vol 13, Energetic materials, Part 2, Detonation, combustion. Elsevier, Amsterdam, p 25
7. Zeman S (2006) J Hazard Mater A132:155
8. Zeman S, Friedl Z, Kočí J, Pelikán V, Majzlík J (2006) Cent Eur J Energ Mater 3(3):27
9. Zeman S (2002) Thermochim Acta 384:137
10. Zeman S (1999) Thermochim Acta 333:121
11. Manelis GB, Nazin GM, Rubtsov YI, Strunin VA (1996) Thermal decomposition and combustion of explosives and powders. Izdat Nauka, Moscow
12. Zeman S, Huczala R, Friedl Z (2002) J Energet Mater 20:53
13. Maksimov YY, Kogut EN (1979) Tr Mosk, Khim-Tekhnol Inst Mendeleeva 104:30
14. Shu Y, Korssounskii BL, Nazin GM (2004) Russ Chem Rev 73(3):293
15. Kuklja MM (2001) J Phys Chem B105:10159
16. Zeman S (2000) Propellants Explos Pyrotech 25:66
17. Chakraborty D, Muller RP, Dasgupta S, Goddard WA III (2000) J Phys Chem A104:2261
18. Zhang S, Nguyen HN, Truong TN (2003) J Phys Chem A107:2981
19. Lur'e BA, Svetlov BS (1967) Tr Mosk, Khim-Tekhnol Inst Mendeleeva 53:40
20. Nazin GM, Prokudin VG, Manelis GB (2000) Russ Chem Bull 49(2):234
21. Park J, Chakraborty D, Jamindra S, Xia WS, Lin MC, Bedford C (2002) Thermochim Acta 384:101
22. Sorescu DC, Alavi S, Thompson DL (2005) Theoretical and computational studies of energetic salts. In: Manaa MR (ed) Chemistry at extreme conditions. Elsevier, Amsterdam, p 431
23. Oxley JC, Smith JL, Chen H (2002) Thermochim Acta 384:91
24. Loebbecke S, Schuppler H, Schweikert W (2003) J Thermal Anal Calorim 72:453
25. Gong XD, Xiao HM, Dong HS (1998) Chin J Chem 16(4):311
26. Gong XD, Xiao HM, Dong HS (1999) Chem Res Chin Univ 15(2):152
27. Gu ZM, Fan JF, Xiao HM, Dong HS (2000) Chem Res Chin Univ 16:21
28. Xiao H, Fan J, Gu Z, Dong H (1998) Chem Phys 226:15
29. Fan J, Gu Z, Xiao H, Dong H (1998) J Phys Org Chem 11:177

30. Dubnikova F, Kosloff R, Almong J, Zeiri Y, Boese R, Itzhaky H, Alf A, Keiman E (2005) J Am Chem Soc 127:1146
31. Adri CT, van Duin ACT, Zeiri Y, Dubnikova F, Kosloff R, Goddard W (2005) J Am Chem Soc 127:11053
32. Okovytyy S, Kholod Y, Qasim M, Fredrickson H, Leszczynski J (2005) J Phys Chem A109:2964
33. Xu XJ, Xiao HM, Ju XH, Gong XD (2005) Youji Huaxue 25(1):536. Quoted in: Chem Abstr (2006) 144:173738
34. Manaa MR, Fried LF, Melius CF, Elstner M, Frauenheim T (2002) J Phys Chem A106:9024
35. Manaa MR, Fried LE, Reed EJ (2003) J Comput Aided Mater Des 10:75
36. Reed EJ, Fried LE, Manaa MR, Joannopoulos JD (2005) A multi-scale approach to molecular dynamic simulations of shock waves. In: Manaa RM (ed) Chemistry at extreme conditions. Elsevier, Amsterdam, p 297
37. Čapková P, Pospíšil M, Vávra P, Zeman S (2003) Characterization of explosive materials using molecular dynamics simulations. In: Politzer P, Murray J (eds) Theoretical and computational chemistry, vol 13, Energetic Materials, Part 1, Decomposition, crystal and molecular properties. Elsevier, Amsterdam, p 49
38. Pospíšil M, Vávra P (2005) In: Vágenknecht J (ed) Proc 8th Seminar New Trends in Research of Energetic Materials. Univ. of Pardubice, p 302
39. Sorescu DC, Rice BM, Thompson DL (2003) Molecular dynamic simulations of energetic materials. In: Politzer P, Murray J (eds) Theoretical and computational chemistry, vol 13, Energetic Materials, Part 1, Decomposition, crystal and molecular properties. Elsevier, Amsterdam, p 125
40. Tarzhanov VI (2003) Fiz Goreniya Vzryva 39(2):3
41. Bourne NK (2000) Proc R Soc London A457:1401
42. Iliushin MA, Tselinskii IV (2000) Zh Prikl Khim 73(3):1233
43. Aluker ED, Aduev BP, Belokurov GM, Tupitsin EV (2005) Fiz Goreniya Vryva 41(2):116
44. Aduev BP, Aluker ED, Belokurov GM, Zakharov YA, Krechetov AG (1999) J Experim Theor Phys 89(1):906
45. Kuklja MM, Aduev BP, Aluker ED, Krasheninin VI, Krechetov AG, Mitrofanov AY (2001) J Appl Phys 89(7):4156
46. Zakharov YA, Aluker ED, Aduev BP, Belokurova GM, Krechetov AG (2002) Predvryvnye yavleniya v azidakh tyazholykh metallov (Pre-explosion effects in the heavy metal azides), Tsentr ekonom. issled. Khimmash, Moscow
47. Aduev BP, Aluker ED, Krechetov AG, Mitrofanov AY (2003) Fiz Goreniya Vzryva 39(1):105
48. Aluker ED, Krechetov AG, Mitrofanov AY, Pashpekin AS (2004) Techn Phys Letters 30(4):772
49. Aluker ED, Aduev BP, Krechetov AG, Mitrofanov AY, Zakharov YA (2006) Early stages of explosive decomposition of energetic materials. In: Jiang SZ (ed) Decomposition of energetic materials, Focus on Combustion research. Nova Science Publ., Inc., New York
50. Lisitsyn VM, Oleshko VI, Tsipilev VP (2005) Rus Phys J 48(2):109
51. Zhilin AY, Ilyushin MA, Tselinskii IV, Kozlov AS, Lisker IS (2003) Rus J Appl Chem 76(4):572
52. Korepanov VI, Lisitsyn VM, Oleshko VI, Tsipilev VP (2004) Fiz Goreniya Vryva 40(1):126

53. Nagayama K, Inou K, Nakahara M (2001) In: Chiba A, Tanimura S, Hokamoto K (eds) Impact Engineering and Application. Elsevier, Amsterdam, p 515
54. Kubota S, Nagayama K, Shimada H, Matsui K (2001) In: Chiba A, Tanimura S, Hokamoto K (eds) Impact Engineering and Application. Elsevier, New York, p 521
55. Capellos C (1998) In: Proc 11th Detonation Symp, Snowmass Village, Colorado, p 3
56. Yu H, Hambir SA, Dlott DD (2006) Ultrafast dynamics of nanotechnology energetic materials. In: Thadhani NN, Armstrong RW, Gash AE, Wilson WH (eds) Multifunctional energetic materials. Mat Res Soc Symp Proc 896:71
57. Skinner D, Olson D, Block-Bolten A (1997) Propellants Explos Pyrotech 23:34
58. Sorescu DC, Rice BM, Thompson DL (1999) J Phys Chem B103:6783
59. Kuklja MM, Kunz AB (2000) J Appl Phys 87(1):2215
60. Manaa MR (2003) Initiation and decomposition mechanism of energetic materials. In: Politzer P, Murray J (eds) Theoretical and computational chemistry, Vol. 13, Energetic materials, Part 2, Detonation, combustion. Elsevier, Amsterdam, p 71
61. Goto N, Yamawaki H, Tonokura K, Wakabayashi K, Yoshida M, Koshi M (2004) Mat Sci Forum, Vols. 465–466. Trans Tech Publ, Switzerland, p 189
62. Rice BM, Hare J, Pai SV, Mattson W, Krasko G, Trevino SF, Sorescu DC, Thompson DL (2001) Khim Fiz 20(5):9
63. Armstrong RW, Elban WL (2004) Dislocation in energetic materials. In: Nabarro FRN, Hirth JP (eds) Dislocation in solids. Elsevier, Amsterdam, p 404
64. Armstrong RW, Elban WL (2006) Mater Sci Technol 22:381
65. Kunz AB, Kuklja MM, Botcher TR, Russell TP (2002) Thermochim Acta 384:279
66. Kuklja MM, Kunz B (2001) J Appl Phys 89:4962
67. Kuklja MM, Rashkeev SN, Zerilli FJ (2006) Appl Phys Lett 89:071904
68. Coffey CS (2003) Initiation due to plastic deformation from shock or impact. In: Politzer P, Murray J (eds) Theoretical and computational chemistry, vol 13, Energetic materials, Part 2, Detonation, combustion. Elsevier, Amsterdam, p 101
69. Coffey CS (1998) Khim Fiz 17(1):4
70. Kuklja MM, Kunz AB (2000) J Phys Chem 61:35
71. Kuklja MM, Stefanovich EV, Kunz AB (2000) J Chem Phys 112(7):3417
72. Kuklja MM, Kunz AB (1999) J Phys Chem B103:8424
73. Holmes W, Francis RS, Fayer MD (1999) J Chem Phys 110(7):3576
74. Dlott DD (2003) Fast molecular processes in energetic materials. In: Politzer P, Murray J (eds) Theoretical and computational chemistry, vol 13, Energetic materials, Part 2, Detonation, combustion. Elsevier, Amsterdam, p 125
75. McNesby KL, Coffey CS (1997) J Phys Chem B101:3097
76. Ye S, Tonokura K, Koshi M (2003) Combust Flame 132:240
77. Alper HE, Abu-Awwad F, Politzer P (1999) J Phys Chem B103:9738
78. Politzer P, Murray JS, Seminario JM, Lane P, Grice ME, Concha MC (2001) J Mol Struct (Theochem) 573:1
79. Politzer P, Boyd S (2002) Struct Chem 13(2):105
80. Murray JS, Lane P, Politzer P (1998) Mol Phys 93(2):187
81. Politzer P, Murray JS (2003) Sensitivity correlations. In: Politzer P, Murray J (eds) Theoretical and computational chemistry, vol 13, Energetic materials, Part 2, Detonation, combustion. Elsevier, Amsterdam, p 5
82. Politzer P, Lane P, Concha MC (2005) Computtional determination of the energetics of boron and aluminium combustion reaction. In: Manaa RM (ed) Chemistry at extreme conditions. Elsevier, Amsterdam, p 473
83. Rice BM, Sahu S, Owens FJ (2002) J Mol Struct (Theochem) 583:69

84. Fried LE, Manaa MR, Pagoria PF, Simpson RL (2001) Annu Rev Mater Res 31:291
85. Rice BM, Hare JJ (2002) J Phys Chem A106:1770
86. Edwars J, Eybl C, Johnson B (2004) Int J Quant Chem 100:713
87. Chen ZX, Xiao H, Yang S (1999) Chem Phys 250:243
88. Türker L (2005) J Mol Struct (Theochem) 725:85
89. Xiao MM, Fan JF, Gong XD (1997) Propellants Explos Pyrotech 22:360
90. Zhang C, Shu Y, Huang Y, Zhao X, Dong H (2005) J Phys Chem B109:8978
91. Zhang C, Shu Y, Wang X (2005) Energ Mater 23:107
92. Xiao HM, Li JS, Dong HS (2000) Acta Chim Sin 58(3):297
93. Loginov NP, Surkova SN (2006) Fiz Goreniya Vzryva 42(1):100
94. Vaullerin M, Espagnacq A (1998) Propellants Explos Pyrotech 23:237
95. Cho SG, No KT, Goh EM, Kim JK, Shin JH, Joo YD, Seong S (2005) Bull Korean Chem Soc 26(3):399
96. Keshavarsz MH, Jaafari M (2006) Propellants Explos Pyrotech 31:216
97. Wolker FE (1998) Khim Fiz 17(1):25
98. Zeman S (1999) J Energ Mater 17:305
99. Klimenko VY, Yakoventsev MA, Dremin AN (1993) Khim Fizika 12:671
100. Klimenko VY (1998) Khim Fizika 17:11
101. Dremin AN (1999) Toward detonation theory. Springer, Berlin Heidelberg New York
102. Hong X, Hill JR, Dlot DD (1995) Vibrational Energy Transfer in High Explosives: Nitromethane. In: Brill TB, Russell TP, Tao VC, Wardl RB (eds) Mat Res Soc Symp Proc 418:357
103. Tarver CM (1997) J Phys Chem A101:4845
104. Tarver CM, Utriew PA, Forbes JW (2001) Khim Fiz 20(3):38
105. Koshi M, Ye S, Widijaja J, Tonokura K (2001) Estimation of shock sensitivity based on molecular properties. In: Chiba A, Tanimura S, Hokamoto K (eds) Impact engineering and application. Elsevier, Amsterdam, p 175
106. Kuklja MM (2001) Electronic excitations in initiation of chemistry in molecular solids. In: Bulatov V, Cleri F, Colombo L, Lewi L, Mousseau N (eds) Advances in materials theory and modeling, bridging over ultiple-length and time scales. Mat Res Soc Symp Proc 677:AA2.4.1
107. Mathieu D, Martin P, La Hargue J-P (2005) Phys Scripta T118:171
108. Méreau R, Mathieu D, Elstner M, Frauenheim T (2004) Phys Rev B69:104101
109. White CT, Swanson DR, Robertson DH (2001) Molecular dynamics simulations of detonations. In: Dressler RA (ed) Chemical dynamics in extreme environments. World Sci., Singapore, p 547
110. van der Hejden AEDM, Bouma RHB, van der Steen AC (2004) Propellants Explos Pyrotech 29:304
111. Jindal VK, Dlott DD (1998) J Appl Phys 83(5):5203
112. Yoo CS, Holmes NC, Souers PC, Wu CJ, Ree FH, Dick JJ (2000) J Appl Phys 88(01):70
113. Rice BM, Mattson W, Trevino SF (1998) Phys Rev E57(1):5106
114. Engelke R, Blais NC, Sheffield SA, Sander RK (2001) J Phys Chem A105:6955
115. Andreev SG (1998) Fiz Khim 17(1):55
116. Wu CJ, Ree FH, Yoo CS (2004) Propellants Explos Pyrotech 29:296
117. Dreger ZA, Gruzdkov YA, Gupta YM (2002) J Phys Chem B106:247
118. Raikova VM, Likholatov EA (2005) Propellants Explos Pyrotech 30:250
119. Hu WF, He TJ, Chen DM, Liu FC (2002) J Phys Chem A106(32):7294
120. Nguen MT, Le HT, Hajgato B, Veszpremi T, Lin MC (2003) J Phys Chem A107:4286
121. Arenas JF, Otero JC, Pealez D, Soto J (2003) J Chem Phys 119(15):7814

122. Reed EJ, Joannopoulos JD, Laurence E (2000) Phys Rev B: Condens Matter Mater Phys 62:16500
123. Margetis D, Kaxiras E, Elstner M, Frauenheim T, Manaa MR (2002) J Chem Phys 117:788
124. Zhang XX, Bauer SH (1997) J Phys Chem B101:8717
125. Glaborg P, Bendsen AB, Miller JA (1999) Int J Chem Kinet 31:591
126. Winey JM, Gupta YM (1997) J Phys Chem B101:10733
127. Blais NC, Engelke R, Sheffield SA (1997) J Phys Chem A101:8285
128. Bouyer V, Darbord I, Herve P, Baudin G, Le Gallic C, Clément F, Chavent G (2006) Combust Flame 144:139
129. Leiber CO (2003) Assesment of safety and risk with a microscopic model of detonation. Elsevier, Amsterdam
130. Bharatam PV, Lammertsma K (2003) Nitro-*aci*-nitro tautomerizm in high-energetic compounds. In: Politzer P, Murray J (eds) Theoretical and computational chemistry, vol 13, Energetic materials, Part 1, Detonation, combustion. Elsevier, Amsterdam, p 61
131. Ferm EN (1999) Propellants Explos Pyrotech 24:1
132. Gruzdkov YA, Gupta YM (1998) J Phys Chem A102:2322
133. Woods E III, Dessiaterik Y, Miller RE, Baer T (2001) J Phys Chem A105:8273
134. Engelke R, Schiferel D, Storm CB, Earl WI (1988) J Phys Chem 92:6815
135. Nabatov SS, Zakushev VV, Dremin AN (1975) Fiz Goreniya Vryva 11:300
136. Engelke R, Sheffield SA, Stacy HL, Quintana JP (2005) Phys Fluids 17:096102
137. Cudzilo S, Trzcinski WA (1999) Pol J Tech Phys 40(2):223
138. Andreev SG, Babkin AV, Baum FN, Imkhovik NA, Kobylkin IF, Kolpakov VI, Ladov SV, Odintsov VA, Orlenko LP, Okhitin VN, Selivanov VV, Solovev VS, Stanyukovich KP, Chelyshev VP, Shekhte BI (2002) Fizika vzryva, Tom 1 (Physics of explosion, vol 1). Fizmatlit, Moscow
139. Licht H-H (2000) Propellants Explos Pyrotech 25:126
140. Yang S, Yue S (2003) HanNeg CaiLiao 11(4):231
141. Agrawal JP (1998) Progr Energy Combust Sci 24:1
142. Pagoria PF, Lee GS, Mitchell AR, Schmidt RD (2002) Thermochim Acta 384:187
143. Chapman RD, Wilson WS, Fronabarger JW, Merwin LH, Ostrom GS (2002) Thermochim Acta 384:229
144. Talawar MB, Sivabalan R, Asthana SN, Singh H (2005) Fiz Goreniya Vzryva 41(3):29
145. Zeman S (1980) Thermochim Acta 41:199
146. Zeman S (1998) In: Proc 29th Int Annual Conf on ICT, Karlsruhe, p 141/1
147. Zeman S, Krupka M (2003) Propellants Explos Pyrotech 28:301
148. Zeman S, Krupka M (2003) Propellants Explos Pyrotech 28:249
149. Clawson JS, Strohmeier M, Stueber D, Orend AM, Baric DH, Asay BA, Hiskey MA, Pugmire RJ, Grant DM (2002) J Phys Chem A106:6352
150. Zeman S, Pešková M (2005) Cent Eur J Energ Mater 2(3):71
151. Zeman S (1997) Thermochim Acta 290:199
152. Zeman S, Friedl Z (2004) J Thermal Anal Calorim 77:217
153. Zeman S, Kohlíček P, Maranda M (2003) Thermochim Acta 398:185
154. Kissinger HE (1957) Anal Chem 29:1702
155. Zeman S (2003) Propellants Explos Pyrotech 28:308
156. Kamlet MJ, Jacobs SJ (1968) J Chem Phys 48:23
157. Pepekin VI, Makhov NM, Lebedev YA (1977) Dokl Akad Nauk SSSR 230:852
158. TITAN (2001) v.1.0.8. Wavefunction. Schrödinger, USA

159. Zeman S, Zemanová E (2004) J Energ Mater 22:171
160. Zeman S, Friedl Z, Roháč M (2007) Thermochim Acta 451:105
161. Zeman S, Friedl Z (2004) Centr Eur J Energet Mater 1:1
162. Galwey AK, Brown ME (1999) Thermal decomposition of ionic solids. Elsevier, Amsterdam, p 22
163. Kohno Y, Ueda K, Imamura A (1996) J Phys Chem 100:4701
164. Bondi A (1968) Physical properties of molecular crystals, liquids, and glasses. Wiley, New York
165. Zeman S, Shu Y, Wang X (2005) Cent Eur J Energ Mater 2(4):47
166. Belmas R, Bry A, David C, Gautier L, Keromnés A, Thevenot G, Le Gallic C, Chenault J, Guillaument G (2004) Propellants Explos Pyrotech 29:282
167. Koroban VA, Maksimov YY (1968) Khim Khim Tekhnol 11:1032
168. Semenov NN (1958) Some problems of chemical kinetics and of reaction capability. USSR Acad Sci, Moscow, p 41, p 101
169. Zeman S, Varga R (2005) Cent Eur J Energ Mater 2(4):77
170. Zeman S, Roháč M (2006) HanNeng CaiLiao 14(1):361
171. Zeman S (1993) Thermochim Acta 216:157
172. Zeman S (2006) Proc 34th NATAS Annual Conf Thermal Anal Appl, Bowling Green, August 8th, p 074.1.05.208/1
173. Ou Y, Wang C, Pan Z, Chen B (1999) HanNeng CaiLiao 7:100
174. Zeman S, Valenta P, Zema V, Jakubko J, Kamenský Z (1998) HanNeng CaiLiao 6:118
175. Zeman V, Kočí J, Zeman S (1999) HanNeng CaiLiao 7:127
176. Zeman V, Kočí J, Zeman S (1999) HanNeng CaiLiao 7:172
177. Zeman S, Kočí J (2000) HanNeng CaiLiao 8:18
178. Kočí J, Zeman V, Zeman S (2001) HanNeg CaiLiao 9:60
179. Wang G-X, Xiao HM, Xu X-J, Ju XH (2006) Propellants Explos Pyrotech 31(2):102
180. Zeman S, Kočí J, Pelikán V, Majzlík J (2006) Cent Eur J Energ Mater 3(3):45
181. Auzanneau M, Roux M (1995) Propellants Explos Pyrotech 20:96
182. Varga R, Zeman S (2006) J Hazard Mater A132:165
183. Varga R, Zeman S, Kouba M (2006) J Hazard Mater A137:1345
184. Chen Z-X, Xiao H (2000) Int J Quantum Chem 79:350
185. Bulusu S, Autera JR (1983) J Energ Mater 1:133

Author Index Volumes 101–125

Author Index Vols. 1–100 see Vol. 100

The volume numbers are printed in italics

Subject Index

Printing: Krips bv, Meppel
Binding: Stürtz, Würzburg